The Principles and Practice of Antiaging Medicine for the Clinical Physician

Research and Business Chronicles: Biotechnology

Series Editor

Alain Vertès
Sloan Fellow
London Business School

Research & Business Chronicles: Biotechnology present accounts of the latest scientific advances in selected technology areas with the aim to facilitate the translation of innovation into commercial realities. Combining a deep and focused exploration of areas of basic and applied science with their fundamental business issues, the series highlights societal benefits, technical and business hurdles, and economic potentials of emerging and new technologies. In combination, the volumes relevant to a particular focus topic cluster analyses of key aspects of each of the elements of the corresponding value chain.

Aiming primarily at providing detailed snapshots of critical issues in biotechnology that are reaching a tipping point in financial investment or industrial deployment, the scope of the series encompasses various specialty areas including pharmaceutical sciences and healthcare, industrial biotechnology, and biomaterials. Areas of primary interest comprise immunology, virology, microbiology, molecular biology, stem cells, biologics, polymer science, formulation and drug delivery, renewable chemicals, manufacturing, and biorefineries. Each volume presents comprehensive review and opinion articles covering all fundamental aspect of the focus topic. The editors of each volume are experts in their respective fields and articles are carefully refereed.

For a list of other books in this series, please visit www.riverpublishers.com

The Principles and Practice of Antiaging Medicine for the Clinical Physician

Vincent C. Giampapa, M.D., F.A.C.S.

University of Medicine and
Dentistry of New Jersey
Newark
New Jersey

Aalborg

Published, sold and distributed by:
River Publishers
PO box 1657
Algade 42
9000 Aalborg
Denmark
Tel.: +4536953197

www.riverpublishers.com

ISBN: 978-87-92329-43-1

Dedication

For the last decade I have had the pleasure and honor to share new ideas and concepts about how we age with an extraordinary group of human beings.

This book is dedicated to these special physicians and pioneers who had the courage and vision to conceive of a new medical society and a new medical paradigm. They have changed how we view the aging process and unshackled our minds from the constraints of medical dogma that is hundreds of years old.

This special group of men has changed the quality and quantity of life for our present generation as well as for future generations to come. Because they have refused to accept the limits of contemporary medical minds as the limits of reality, we will all benefit.

I express my deepest thanks to

Eric Braverman
Ward Dean
Bob Goldman
Thierry Hertoge
Dharma Khalsa
Ron Klatz
Steve Novell
Chong Park
Steven Sinatra

for their tireless efforts in establishing the American Academy of Anti-Aging Medicine.

I would also like to extend very special thanks for the vision and inspiration of a great teacher and fellow plastic surgeon, Dr. James Smith, who encouraged me to continue on the path where no other plastic surgeon has gone.

This book is dedicated to all of these people.

Vincent C. Giampapa, M.D., F.A.C.S.

Acknowledgments

I would like to acknowledge the tireless effort and exceptional organizational qualities of Dolores June Reilly for heading this book project. Her limitless energy and enthusiasm are responsible for the final completion of this project.

I would also like to thank Frank Righter, General Manager, and Lois Lombardo, Project Manager, both of Nesbitt Graphics, Inc. for their assistance in the editing and fine tuning of my manuscript into a book. Their professionalism and dedication to achieving the highest quality of medical publishing standards are greatly appreciated.

I would also like to acknowledge Jose Maldonado for his graphic skills and contributing efforts in completing many of the illustrations and diagrams included in the text.

Finally, I would like to express my thanks to the members of the American Board of Plastic and Reconstructive Surgery as well as to the members of the American Academy of Anti-Aging Medicine for encouraging me to complete this project despite a long road filled with political obstacles and out-dated medical prejudices.

Vincent C. Giampapa, M.D., F.A.C.S.

About the Author

Vincent C. Giampapa is a member of the American Board of Plastic and Reconstructive Surgery as well as the American Academy of Anti-Aging Medicine. He is an assistant clinical professor at the University of Medicine and Dentistry of New Jersey, and a fellow of the American College of Surgeons. His research continues in the area of Stem Cell reprogramming and cellular aging, as well as applications of Adult Stem Cell use in reconstructive and aesthetic surgery. He has published many articles in peer reviewed medical journals and has filed over 20 patents with the US and international patent offices.

Dr. Giampapa is a recipient of the prestigious Science and Technology Award from the American Academy of Anti-Aging Medicine for his work in designing the first computerized evaluation of biomarkers of aging, and has served as consultant to NASA on the Space Shuttle. He has been the chairman for five meetings of The Annual International Conference on Anti-Aging Medicine, the director of the Giampapa institute for Age Management Therapies, and the Founder and Chief Medical Officer of Life Science Institute. Dr Giampapa was one of the founding members of the American Academy of Anti-Aging Medicine and the first President of the American Board of Antiaging Medicine. where he was responsible for creating the examination process to certify physicians in the clinical practice of anti aging medicine on an international basis. He has published five books on anti aging medicine for both the general public and medical doctors. He practices and resides in Montclair, New Jersey, USA.

Contents

Preface

Seneca the Younger stated: "Having good health is very different from not being sick"; John Ray stated: "Health is better than wealth." The subject of health and wealth was summed up by Charles C. Colton: "There is this difference between the two temporal blessings — health and money: Money is the most envied, but the least enjoyed; health is the most enjoyed, but the least envied ... the poorest man would not part with health for money, but the richest would gladly part with all his money for health."

Almost every middle-aged baby boomer would like to add 20 to 40 years of good health to the rest of his or her life. These people, now in their fifties, flee to gyms and health food stores and occupy the consultation seats in plastic surgeons' offices. They are intent on looking and feeling good and are determined in their search for the true fountain of youth.

Vincent C. Giampapa, M.D., F.A.C.S., has been a visionary plastic surgeon in the field of anti-aging medicine, as well as an original contributor to the field of cosmetic surgery. In the early 1990s, in order to bring science and validity to the nascent field of anti-aging medicine, he became one of the founding members of the American Academy of Anti-Aging Medicine and subsequently the first president of the American Board of Anti-Aging Medicine.

The longing to look and feel healthy and young is commonly heard by plastic surgeons. Is cosmetic surgery related to the field of anti-aging? Of course! Plastic surgeons were the first to perform kidney transplantation. They prolong the lives and help the well-being of burned, traumatized, and severely deformed patients. There has always been a direct link between the body image and the mind. It is no surprise that plastic surgeons were the first to obtain and isolate stem cells and, subsequently, to remove fat through liposuction techniques. Although critics of anti-aging science and age management complain that people should age naturally, most baby boomers certainly believe the opposite. Because everyone wants to look and feel younger, plastic surgeons can use the knowledge and guidance in this landmark text not only for their practice but also for themselves.

Because most physicians have little formal training in anti-aging medicine, this text is a very welcome addition to the medical, surgical and research libraries. Dr. Giampapa has clearly described the basic concepts and coming advances in this rapidly evolving field. Within the maze of anti-aging medicine, information and techniques, a guide is essential for directing the neophyte. Dr. Giampapa admirably fills this need and has even given us a new theory with which to conceptualize the aging process in an organized scientific manner. This extremely difficult task has been completed successfully.

In this readable and well-organized text, Dr. Giampapa provides 13 Chapters filled with clinically useful information and medical references for practicing physicians and plastic surgeons alike.

Chapters I–IV help answer the ever-present question, "What is aging and how can we modify it to improve the quality and quantity of our patients' lives?" Dr. Giampapa presents a new paradigm of aging supported by the latest research on DNA.

Chapters V–XI address specific treatment recommendations. Dr. Giampapa presents valid scientific data to support anti-aging medicine, the integrity of which was questioned throughout the 1990s. The relationship between longevity and caloric restriction, mitochondrial oxidation, chronic degenerative inflammation and immunopathology — along with low-glycemic diets and hormonal restoration, growth and rejuvenation factors, viral vector delivery systems and much more — is a well-proven reality today. How can these advances not help patients age optimally?

Chapter XII advises practitioners on the introduction of this exciting subject into their practices.

In Chapters XIII and XIV, we can all share in the excitement about what the future holds with regard to potential stem cell treatments; regenerative cellular medicine; implantation of futuristic artificial organs to treat diabetes and heart, liver and kidney diseases; and the regeneration of a damaged brain and body with computer interfaces.

We must put preexisting prejudices aside and look scientifically at this new specialty, which has the potential to add years of health and longevity to all our lives.

Dr. Giampapa must be congratulated on providing a catalytic text, a road map to a possible future in which cosmetic surgeons may provide this new specialty to their patients, in combination with the latest surgical procedures that have already been mastered. New and exciting research has shown that, in general, women who undergo facelifts live up to nine years longer than those who do not.*

---

*American Society of Plastic and Reconstructive Surgeons (ASPRS). 2002; Chicago, IL.

Without question, we must redefine what it means to be a cosmetic surgeon of the future.

Steven M. Hoefflin, M.D., F.A.C.S.
Associate Clinical Professor, Division of Plastic and
Reconstructive Surgery, UCLA School of Medicine

Contributing Authors

Stanley Burzynski, M.D., Ph.D., Burzynski Research Institute, Houston, Texas

Steven J. Greco, Ph.D., UMDNJ-New Jersey Medical School, Newark, New Jersey

Ronald Pero, Ph.D., Chief, Department of Ecogenetics, Professor of Cell and Molecular Biology, University of Lund, Sweden

Nicholas V. Perricone, M.D., Assistant Clinical Professor Dermatology, Yale University, New Haven, Connecticut

Pranela Rameshwar, Professor, UMDNJ-New Jersey Medical School, Newark, New Jersey

Oscar M. Ramirez, M.D., F.A.C.S., Assistant Clinical Professor, Johns Hopkins University, Baltimore, Maryland

Aristo Vojdani, Ph.D., M.T. Founder and Director Immunosciences Lab, Inc., Beverly Hills, California

Preface by Steven M. Hoefflin, M.D., F.A.C.S., Assistant Clinical Professor, Division of Plastic and Reconstructive Surgery, UCLA School of Medicine, Los Angeles, California

Introduction

Anti-Aging Medicine and Cosmetic Surgery

"The Connection"

Vincent C. Giampapa, M.D., F.A.C.S.

> Keep your mind open to change all the time. Welcome it. Court it. It is only by examining and reexamining your opinions and ideas that you can progress.
>
> Dale Carnegie

Aesthetic surgeons and physicians are used to observing the effects of aging on the face and body. As a new century dawns, they now have an opportunity to understand and interact with the causes of aging on the most fundamental level.

Since the early 1990s, many physicians have asked me, "How does a plastic surgeon become interested and involved in anti-aging medicine?" Aside from an innate interest in general, it became obvious to me, as I examined my patients for potential cosmetic surgical procedures, that I was witnessing the effects of what was occurring at a more intimate, deeper level: within the cellular structures of their bodies.[1] It was this sudden insight that has fueled my desire to understand what caused these effects. More important, I wanted to know what I could do to influence the origins of aging, rather than simply operate on the effects of aging.

Since the early 1990s, I have been intimately involved in the organization and the establishment of a medical society to help disseminate this information to all medical specialties. I and four other physicians met in Chicago and set the groundwork that would establish the American Academy of Anti-Aging Medicine. We defined anti-aging medicine as "the specialty of medicine that seeks to slow age-related disease process to improve both the quality and quantity of life." At that time, the concept of anti-aging medicine was considered far-fetched and beyond the space of respected

The Principles and Practice of Antiaging Medicine for the Clinical Physician, 1–12.

science. Despite the cool reception from the medical community, it was an idea whose time had come.

Since then, I have watched this organization grow from its original five members to its present state of over 10,000 physician members in 27 countries.

I have been actively involved in establishing the criteria and information base required by these physicians to help them observe, measure and clinically manipulate the effects of aging on their patients.

Many of these physicians both followed and helped design treatment protocols so that they have become much more efficient and, now, scientifically credible. It has been a time of trial and error, as well as an exercise in learning to think creatively and unconventionally. The advent of technology and improvements in laboratory diagnostics and testing have been the main reasons that anti-aging medicine has finally gained a credible foothold among the other medical specialties.

I have organized and held six international seminars on anti-aging medicine to present the latest breakthroughs in this new specialty. This information has been received with enthusiasm and excitement all over the world.

As a plastic surgeon, I would like to share with you this information and the basic concepts of anti-aging medicine and age management, as they are used at the beginning of the 21st century. They are sure to evolve, but we need no longer wait to have a marked impact on our patients' lives, as well as on our own. This book is merely an attempt to present the basic concepts. By no means does it cover every detail or complete science of anti-aging medicine.

Nonetheless, I believe that the information contained within this book will be more than enough to convince you of the validity and scientific nature of this new medical specialty and, more important, will provide you with an entirely new viewpoint and approach toward your cosmetic patients.

These are exciting times in science and in aging research. We have taken the giant step of mapping the human genome, the blueprint to create, repair and restore a human being (Diagram 1). Although our level of success can currently be seen as only the outline of this genetic map, there are daily additions to this information, and the understanding of the contents continues to deepen. The growth of anti-aging information is logarithmic, doubling every 2 years.

Since 1990, we have accumulated more information on aging than was obtained over the past 2,000 years. What was once viewed as "alchemy" has become molecular biology, and what was once "magic" is now seen as quantum physics and genetics.

We have also just begun to understand the important impact of our environment on health and longevity and, more specifically, our genes (Diagram 2).[1] Why do most people in the world have a life span of 60 to 70 years and a very few make it to 100 years of age or older with relatively good health? I believe we are in a

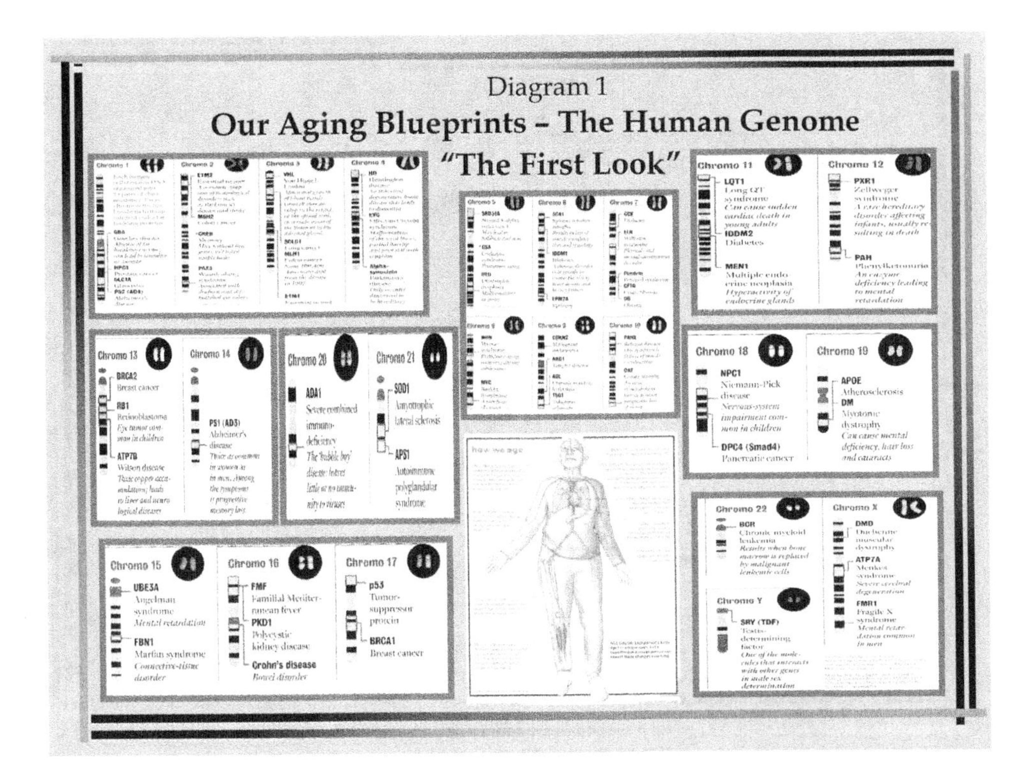
Diagram 1
Our Aging Blueprints – The Human Genome
"The First Look"

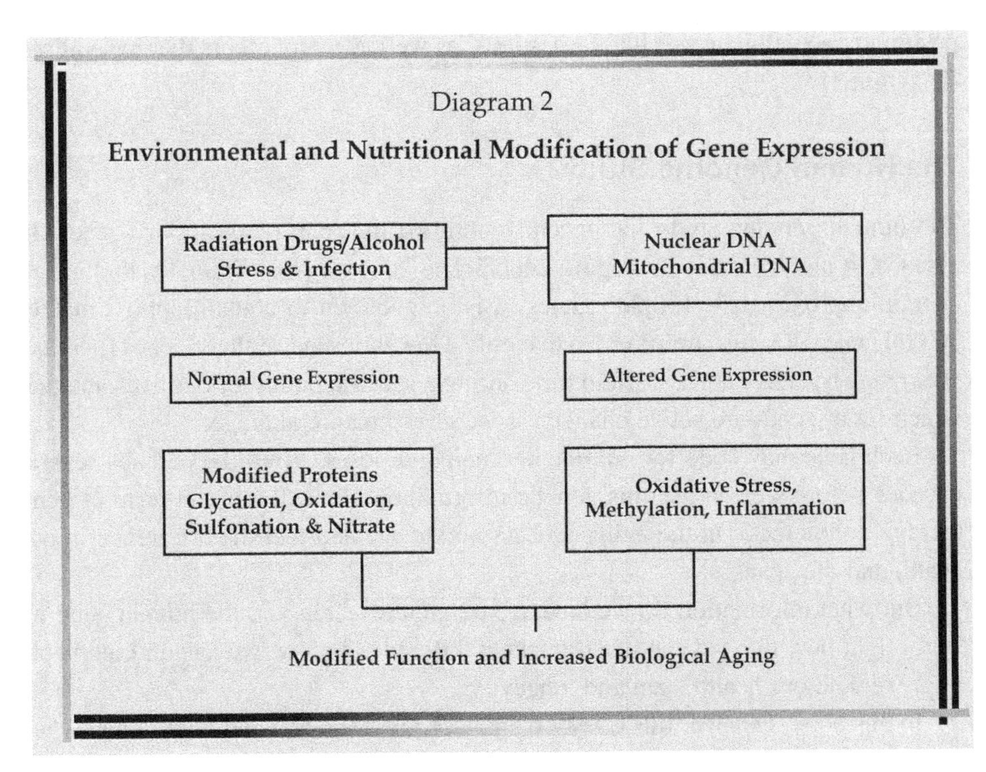
Diagram 2
Environmental and Nutritional Modification of Gene Expression
Radiation Drugs/Alcohol
Stress & Infection
Nuclear DNA
Mitochondrial DNA
Normal Gene Expression
Altered Gene Expression
Modified Proteins
Glycation, Oxidation,
Sulfonation & Nitrate
Oxidative Stress,
Methylation, Inflammation
Modified Function and Increased Biological Aging

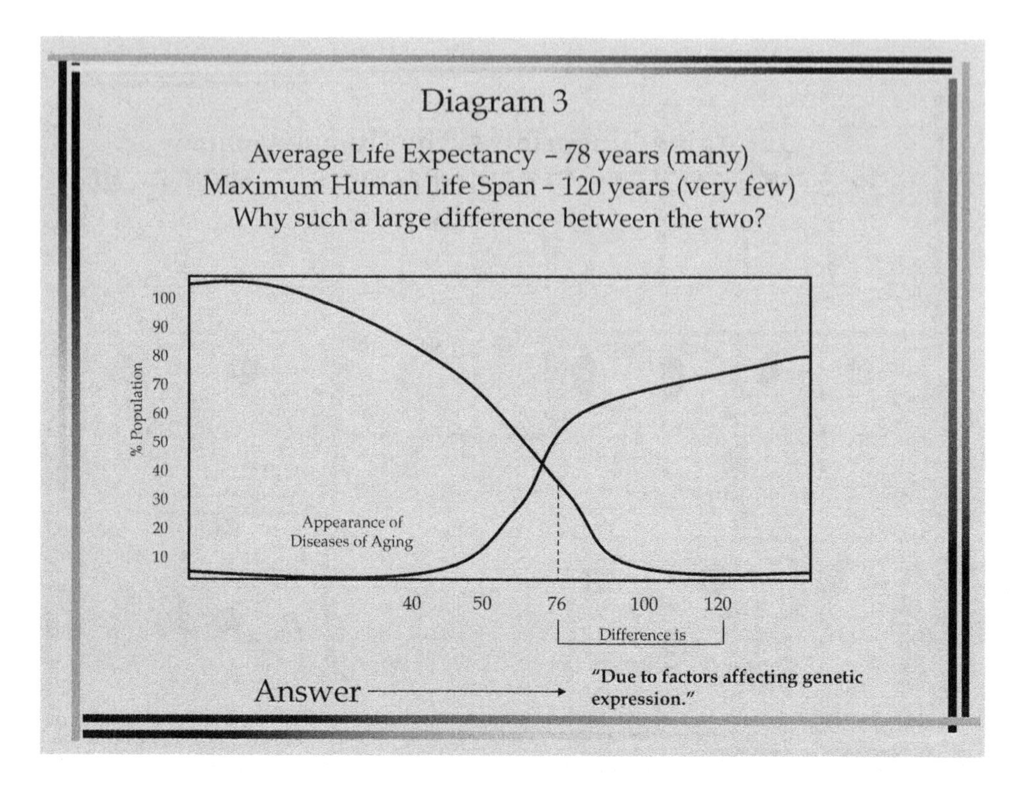

position to actually answer these questions, as well as many others that have arisen (Diagram 3).[2]

The Human Genome Study

The human genome study has recently outlined the map of these key genes. In essence, it also contains an "aging code." The genome was originally thought to contain approximately 100,000 genes; it is now known to contain approximately 30,000 genes. It is also now known that only a few thousand of these genes (perhaps 10%) may be active at any certain time during a specific period of our lives and that which set of genes are active changes as we grow, mature and age.

Each gene may code for not one but many purposes. It will be perhaps several decades before we can use this genetic information clinically, in the form of gene therapy, to intercede in the aging process and to see and feel the difference in our health and life spans.

But what information do we have from aging research and the human genome study right now that we can use to make a difference in how we age and, perhaps, improve both our health span and longevity?

In this book, I answer this question and also present a new paradigm on how we age. This is a model that can easily be incorporated into your practice or into your own personal lifestyle as a basis for improving how we age. I also present to you the radical concept that "*we are built for self-repair and not programmed to die*," in contrast to how we currently think. Information gathered during the 1990s seems to indicate that the number of cell divisions within our organs need not be finite.[1] This may mean that we have a potential to live longer, healthier lives under the right circumstances and in the right environment.

These first pieces of information from the human genome study have already revealed important information about what we may call the **aging code**.

Telomeres, located on the end of each chromosome, are now thought to be one of the central clocks that control the cellular aging process. They have been shown to still be present in humans alive at 90 years of age and older. These studies on "ancient cells" indicate telomere positions that certainly should be able to undergo many more cell divisions. These "control clocks" are not located at the end of the chromosome as in "old age" but are positioned in such a way as to indicate the potential of many more cell divisions.[1] Obviously, there occur cellular events that have not allowed us to reach our full potential as far as cell division repair and aging are concerned.

At the center of this new model of aging is the thought that if we can repair damage to DNA, as well as limit the damage in the first place, we would be able to allow genetic potential to reach its maximum. This should allow for a significantly longer period of health and a longer life span than is generally experienced at present.

The second most important piece of information that I focus on in this book has to do with the fact that embedded within our 46 chromosomes lies a "genetic code" that forms what can be viewed as an **aging equation** (Diagrams 4a and b). This equation seems to be responsible for controlling the cellular rate of aging, at least on a clinical observation level.

This code controls four main processes within the cell, which make up the basis of the aging equation. A review of the latest information[3] revealed that the genes of these four main processes, to date, are located on chromosomes 1, 2, 8, 14, 17, 19 and 22.

The equation and these genetic codes can be manipulated right now to successfully treat age-related changes that we observe in the body. They are also responsible for controlling the age-related disease processes we all experience as we grow older.[4] (For more detailed effects of these processes, see Diagram 5.)

Clinical anti-aging medicine now is, and for the next few decades will be, concerned with improving and controlling how efficiently the inherited genes that make up our individual aging codes actually function. To explain it in a more technical manner, what level of optimal genetic expression we are able to maintain, or induce, with age-management programs and anti-aging therapy will determine how we age, the quality of health of our lives and how long we live. From the viewpoint

Diagram 4a

The Aging Equation

Inadequate balance of antioxidants to inhibit free radical damage

GLYCATION

Changes in cell membranes and the intracellular environment
Compromised gene expression and DNA integrity

(i.e)> DNA damage
<DNA repair
>DNA copying
Errors

This creates increases in
OXIDATIVE STRESS

Sub-Optimal Levels and Imbalance of hormones

Altered **METHYLATION**

Increased INFLAMMATORY PROCESS

*EACH PROCESS SETS IN PLACE A SERIES OF EVENTS THAT INCREASES THE OTHER RELATED REACTIONS UTIMATELY CAUSING A CHANGE IN GENE EXPRESSION

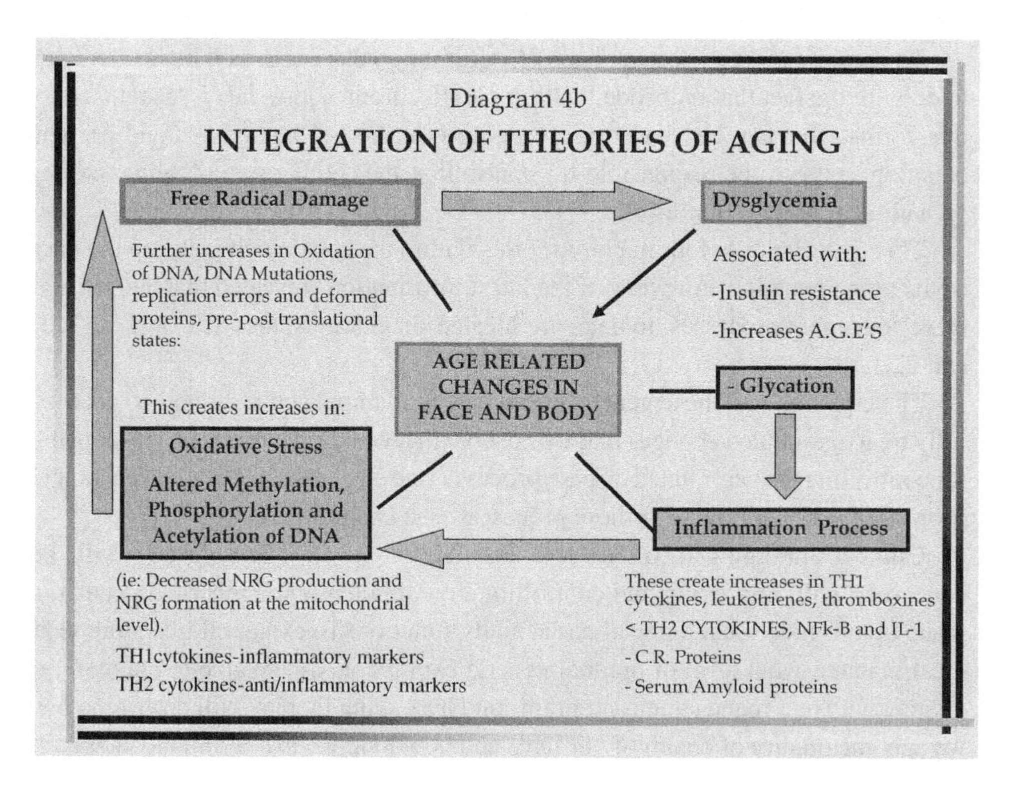

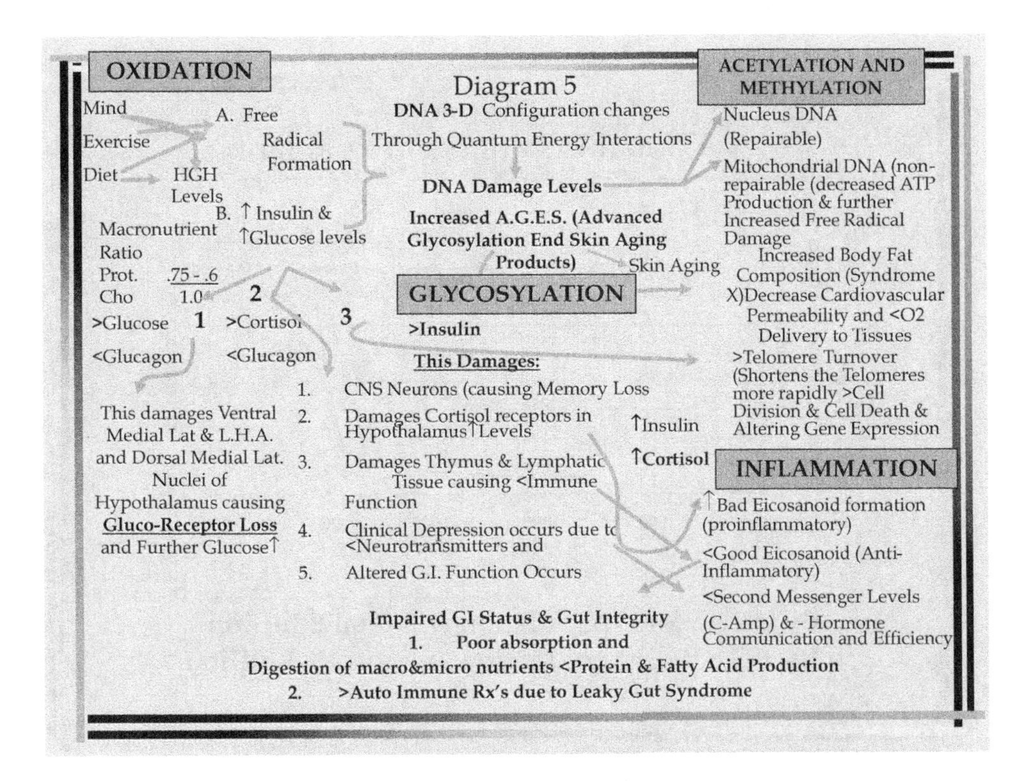

of cosmetic surgeons and physicians, it also determines how we physically look and how our appearance changes as time goes by.

Next, I would like to present a few facts that help emphasize why we need to focus now on anti-aging therapy and age-management programs. This is important, not just for our cosmetic patients, but in helping address a major world problem that is already presenting itself in every industrialized country throughout the globe (Diagram 6).

Fact: Every second, a baby boomer turns 50.

Fact: In 1995, the baby boomer generation saved $15 trillion for retirement. If the same disease patterns of aging continue, they will actually need $184 trillion to maintain their health as they age. There is not that much money on the planet.

Fact: Current medical anti-aging technology has the capability to provide at least 50% of the 76 million baby boomers with a healthy life expectancy of 90 to 100 years (Diagram 7a, 7b).[5]

Fact: A child born in 1997 in the U.S. can expect to live 76.5 years, which is 29 years longer than a child born in 1990.

Fact: Once we reach age 65, men can expect to live an additional 15.8 years, whereas women can expect to live an additional 17.6 years[4] (Diagram 8).

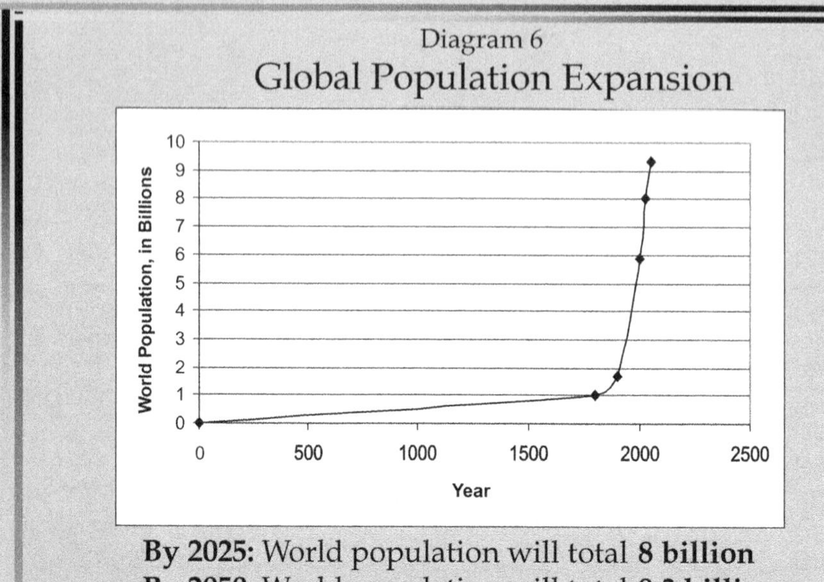

Diagram 7A

Anti-Aging Market

There are 76 million Baby Boomers in the U.S., and they:

- have accepted the concept of preventive health maintenance to maximize *"health span"*
- spend over $12 billion on non-prescription vitamins and supplements annually
- want to know which neutraceuticals they really need and how to measure the effect on their health and longevity.

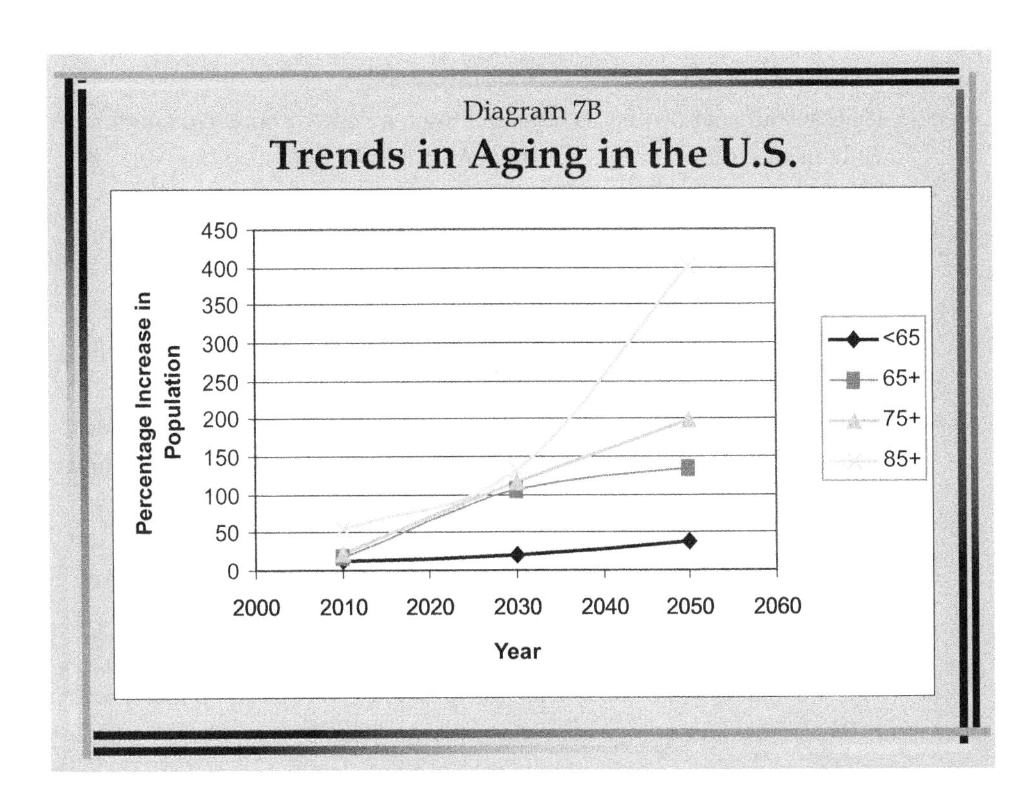
Diagram 7B
Trends in Aging in the U.S.
Percentage Increase in Population
450
400
350
300
250
200
150
100
50
0
2000
2010
2020
2030
2040
2050
2060
Year
<65
65+
75+
85+

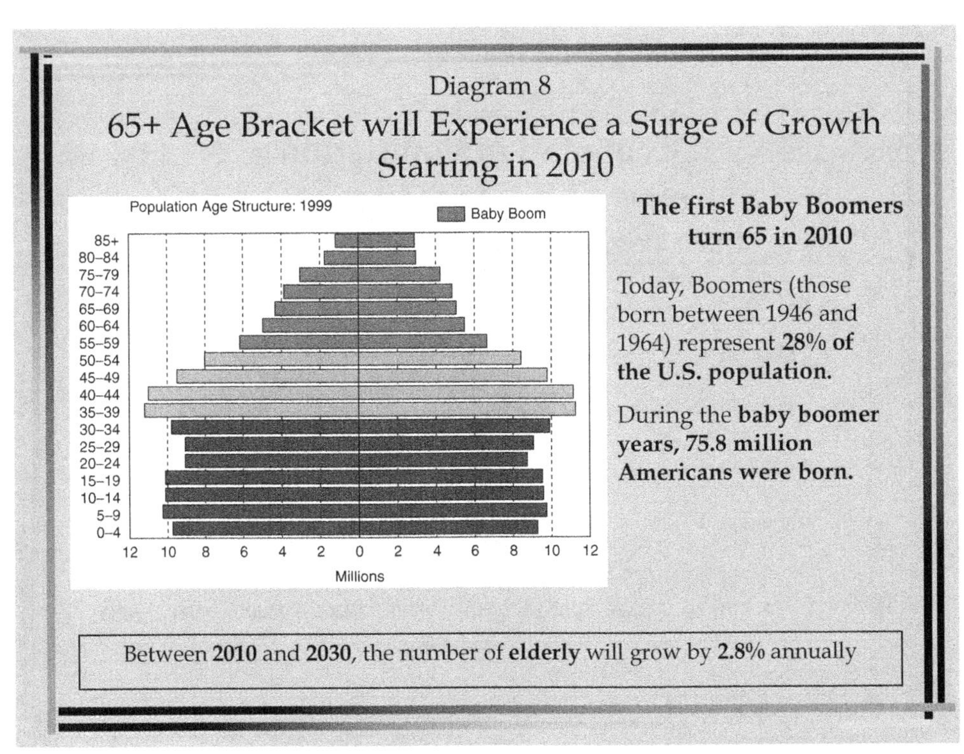
Diagram 8
65+ Age Bracket will Experience a Surge of Growth Starting in 2010
Population Age Structure: 1999
Baby Boom
85+
80-84
75-79
70-74
65-69
60-64
55-59
50-54
45-49
40-44
35-39
30-34
25-29
20-24
15-19
10-14
5-9
0-4
12 10 8 6 4 2 0 2 4 6 8 10 12
Millions
The first Baby Boomers turn 65 in 2010
Today, Boomers (those born between 1946 and 1964) represent 28% of the U.S. population.
During the baby boomer years, 75.8 million Americans were born.
Between 2010 and 2030, the number of elderly will grow by 2.8% annually

Fact: By the year 2003, 14% of the budget of the United States will be spent on geriatric medicine.[6]

Fact: If the health span — that is, maintaining a healthy productive condition without disease — of every American could be extended one year, the U.S. economy would save $1–3 trillion.

Fact: By the year 2025 in the U.S., there will be two 65-year-olds for every young teenager.[4]

Fact: By 2030, the elderly population in the United States will grow to between 59 and 78 million, or one fifth of the total U.S. population (United States Census Bureau).[4]

Fact: By the year 2050, there will be as many as 2.2 million Americans over the age of 100 (Diagram 9). The global population will total 9.3 billion (Diagram 10).

Fact: If the same disease patterns continue in the next 20 years, the health care system of the United States will become bankrupt! (Laurence Kotlifoff, Economist, Brown University.)

Cosmetic surgeons and physicians see the middle-aged population increase year after year. We are in a unique position not only to offer this population cosmetic procedures to slow and reverse the effects of aging but also to introduce to them the

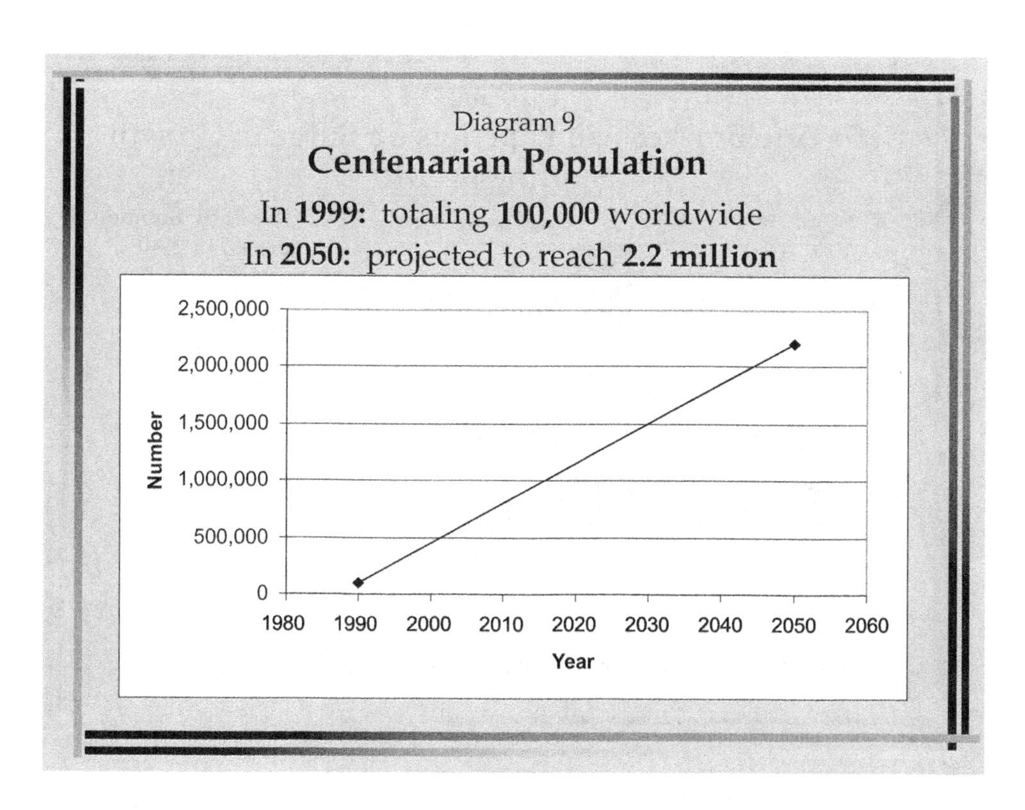

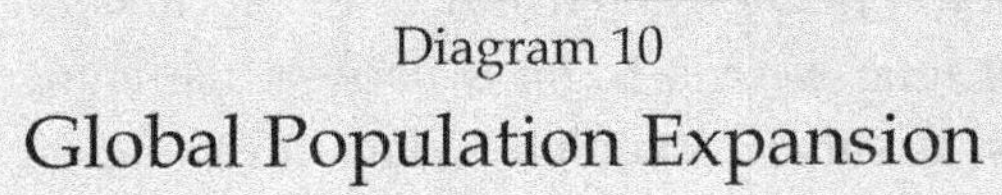

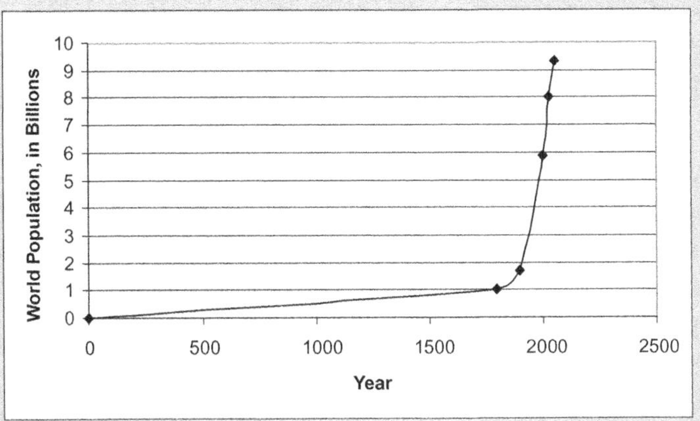

By 2025: World population will total **8 billion**
By 2050: World population will total **9.3 billion**

Diagram 11

Age Management Therapy Goals

1) **Control Environmental Effects** - Environment defined as = diet, exercise, mind state, toxic elements, pollution, local radiation.

2) **Improve the Function of the Aging Equation** – Control Glycation, Inflammation, Oxidation/Free Radical Levels and Methylation at the cellular level.

3) **Improve DNA Replication and Gene Expression** – Improve the ratio and DNA Repair over DNA damage therefore resulting in less cell mutations and more accurate cell copies during cell replication. *This preserves adult stem cell pools.*

ongoing science and programs to slow the causes of aging. We will become the source of age-management information and define its therapeutic goals (Diagram 11).

This is the first time we can begin to look at the aging process with a new perspective, not just as surgeons and physicians reversing the effects of aging, but as a key medical specialty responsible for treating the aging process not only from the outside but also from within at its most intimate level. In other words, with this new perspective, I believe that not only do we have the opportunity to ameliorate the signs of aging and improve the physical appearance of our patients by using our new array of mechanical and surgical technology, but also we may begin to markedly improve and inhibit the actual causes of these changes at the cellular level.

This perspective, along with the extensive array of surgical procedures that were developed over the last decades of the 20th century, will markedly improve our ability to have a positive impact on the aging process at the mental and emotional levels as well. This combined approach will most surely result in longer lasting and better surgical outcomes and in an improvement in our patients' quality of life and time.

The next decade will redefine what it means to be a cosmetic surgeon and physician.

References

[1] www.Weizman.CARD.5/2002.

[2] Giampapa VC. *Third International Symposium on Anti-Aging Medicine.* March 31–April 1, 2000, Newark, NJ.

[3] Mondello C, Petropoulis C, Monti D, Gonos ES, Franceschi C, Nuzzo F. Telo-mere length in fibroblasts and blood cells from healthy centenarians. *Exp Cell Res.* 1999; 248(1): 234–242.

[4] *Sixty-Five Plus in the United States.* [United States Census Bureau. Statistical Brief]. Washington, DC: Economics and Statistics Administration, U.S. Department of Commerce, May 1995.

[5] Data from statistics released by the Administration on Aging at http://www.aoa.dhhs.gov/aoa/stats/aging21.

[6] *Popular Science*, October 1999.

1

The Seven Basic Clinical Concepts of Anti-Aging Medicine and the Aging Equation

Vincent C. Giampapa, M.D., F.A.C.S.

Man's mind stretched to a new idea, never goes back to its original dimension.

Oliver Wendell Holmes

Optimizing, or more efficiently activating the "genetic code" is based on the ability to use the seven basic clinical anti-aging concepts.

The fundamental idea to grasp is that most age-related changes are caused by seven main processes as we age. These processes are as follows:

1. Glycation, the cross-linking of proteins (collagen, hemoglobin, and albumin), caused by elevated and poorly controlled blood glucose levels (Diagram I-1).[1–27]
2. Increased inflammatory processes, which result from abnormal balances of intracellular and extracellular compounds. These compounds include good and bad eicosanoids (prostaglandins), leukotrienes, cytokines, and thromboxanes. These are categories of age-accelerating compounds, which appear mainly as a result of the actions of free radicals. Poor fatty acid levels and ratios in the cell membranes are also responsible for increasing the inflammatory process and these compounds (Diagram I-2).[28–45]
3. Inappropriate intake and balance of extrinsic antioxidants to inhibit the action of free radicals,[7] as well as decreasing intrinsic antioxidant supplies[4] (e.g., superoxide dismutase, catalase, glutathione peroxidase) (Diagram I-3).

The Principles and Practice of Antiaging Medicine for the Clinical Physician, 13–20.

Diagram I-1

AGING AND GLYCATION

Glycation

- The cross-linking of proteins at the cellular and genetic levels is caused by unregulated glucose levels, insulin surges and insulin receptor insensitivity.
- This directly effects gene expression and protein synthesis.
- Glycosylation of the IgG molecule results in modified function of immunoglobulins which then have altered function and may contribute to "autoimmune reactions".

Diagram I-2

AGING AND INFLAMMATION

- **Inflammation** ⟶ initiated by acute phase proteins, creates an increase in key cytokines (TH1, CRP, and serum amyloid, interleukines). It initiates aging changes in the vascular tissues, painful joints, G.I. tract as well as modifies gene expression and post-translational protein formation.

Diagram I-3

AGING AND OXIDATIVE STRESS

- Oxidative Stress ⟶ The amount of free radical damage produced at the intra and extracellular level. This directly affects genetic structure and function, cell membrane and organelle function.

4. Improper methylation, acetylation and phosphorylation of DNA. These processes determine which genes are activated or inhibited and affect DNA masking and, therefore, gene expression[18,46] (Diagrams I-4 and I-5).
5. Changes in the cell membranes and the intracellular environment (pH levels, cell hydration and accumulation of cellular waste products), resulting in suboptimal protein turnover[18,19] caused by insufficient supply of repair building blocks (plasma amino acids, glycosaminoglycans, omega-3 and omega-6 fatty acids) and diminished protein synthesis in general. This leads to accumulation of damaged protein compounds in both the intracellular and extracellular compartments, our "**cellular soup**."
6. Abnormal ranges, as well as relative imbalances, of hormones (e.g., "increased insulin, increased cortisol, decreased thyroid hormone, decreased sex hormones, and decreased melatonin and growth hormone levels), resulting in poor cell signaling (signal transduction) and poor cell turnover and regeneration.
7. Compromised DNA structural integrity, resulting from the combination of increased DNA damage with decreased DNA repair. This results in the accumulation of DNA errors during cell replication to replace damaged

Diagram I-4

AGING AND METHYLATION OF DNA

- Methylation, the addition of a CH_3 group, controls the masking portion of specific regions of DNA and the unmasking of others. This alters how our genetic switch turns on to off or off to on, thereby modifying *GENETIC EXPRESSION.*
- The evidence emerging from recent research strongly suggests that brain aging and possibly cancer and cardiac aging are, in part, consequences of altered methylation patterns.

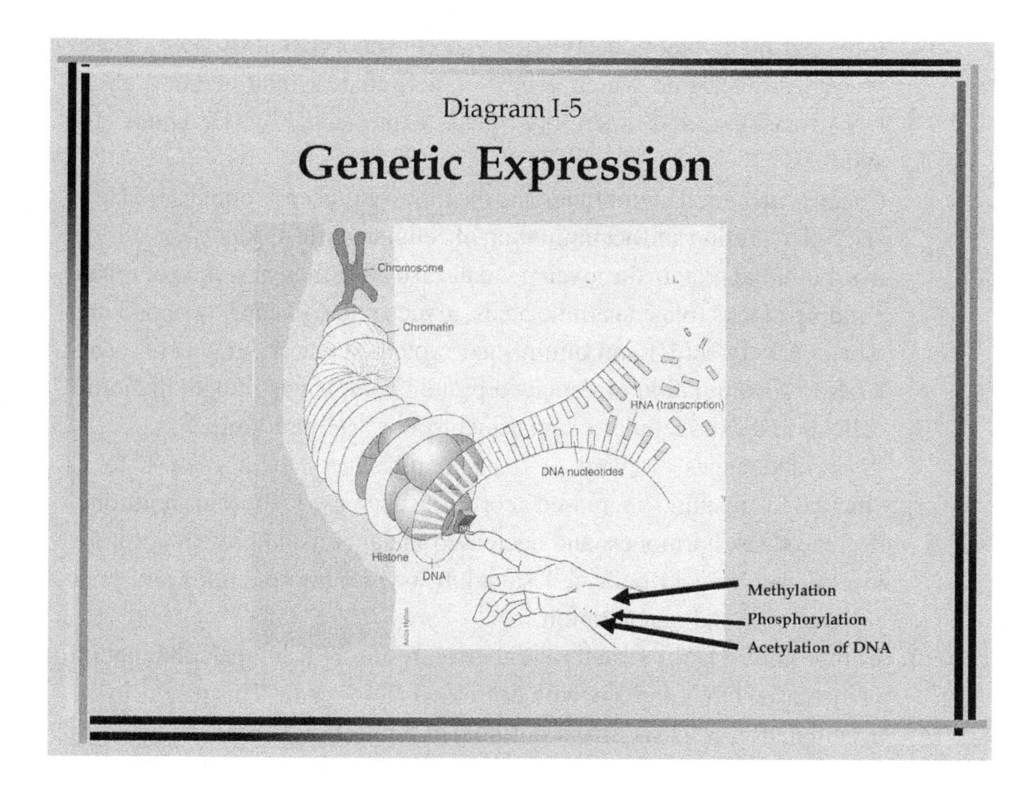

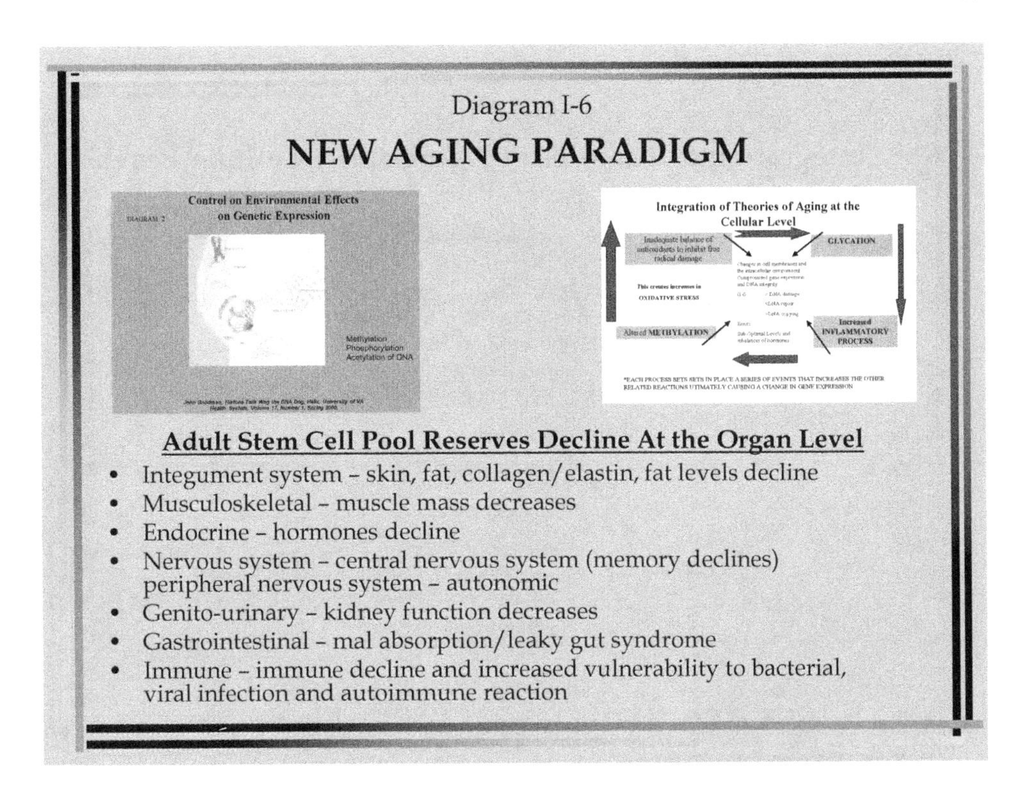

and aging tissue. This also results in faulty protein and enzyme production, which impairs the cellular machinery within each of the 100 trillion cells that make up the human body. It also results in deficiencies and mutations in stem cell reserves within all organ systems. **Stem cell reserves** are essential for maintaining optimal functional organ reserve as people grow older.

These seven fundamental processes affect genetic expression.[46] They form the bases of the "new aging paradigm" (Diagrams I-6 and I-7) and can be viewed as being interrelated, one having a direct impact on the other. **Genetic expression** is the process that regulates which genes are deactivated ("turned off") and which genes are activated ("turned on") (see Diagram I-5).[46] These processes can now be applied to an overall treatment concept that can be viewed as progressing from the intimate level of DNA within the cell and then outward to encompass total body homeostasis and integration (Diagram I-8). Before we begin to attempt to alter these fundamental processes, a review of past and present aging theories will allow for a deeper understanding of age management and anti-aging therapies.

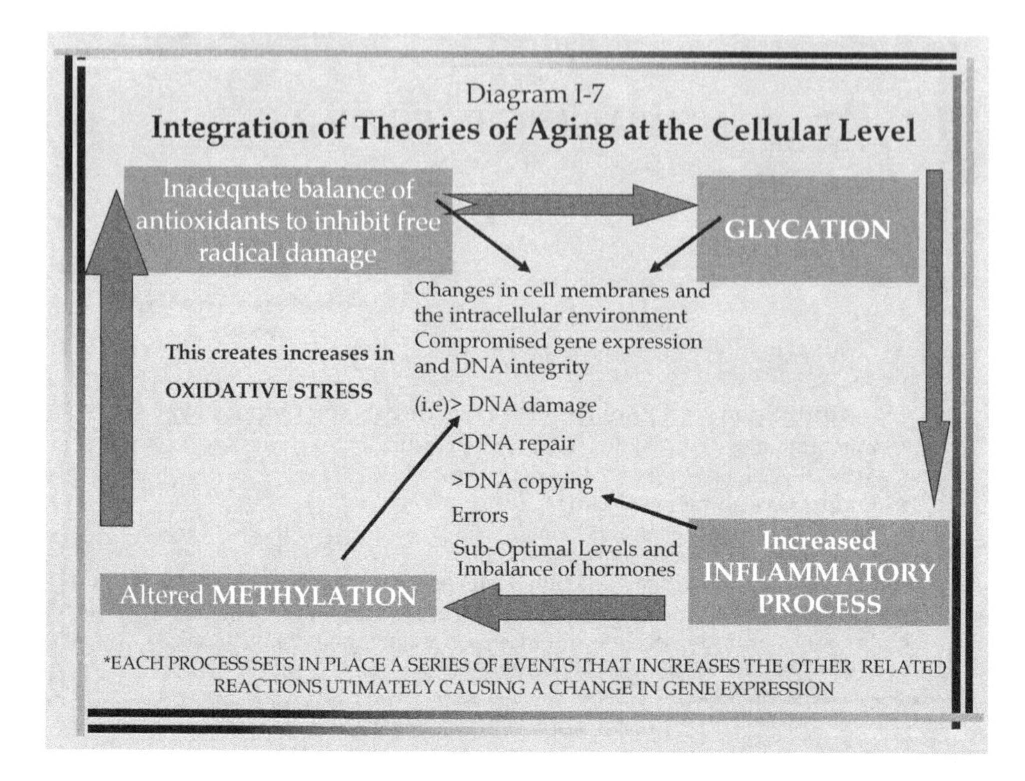
Diagram I-7
Integration of Theories of Aging at the Cellular Level
Inadequate balance of antioxidants to inhibit free radical damage
GLYCATION
Changes in cell membranes and the intracellular environment
Compromised gene expression and DNA integrity
This creates increases in OXIDATIVE STRESS
(i.e)> DNA damage
<DNA repair
>DNA copying
Errors
Sub-Optimal Levels and Imbalance of hormones
Increased INFLAMMATORY PROCESS
Altered METHYLATION
*EACH PROCESS SETS IN PLACE A SERIES OF EVENTS THAT INCREASES THE OTHER RELATED REACTIONS UTIMATELY CAUSING A CHANGE IN GENE EXPRESSION

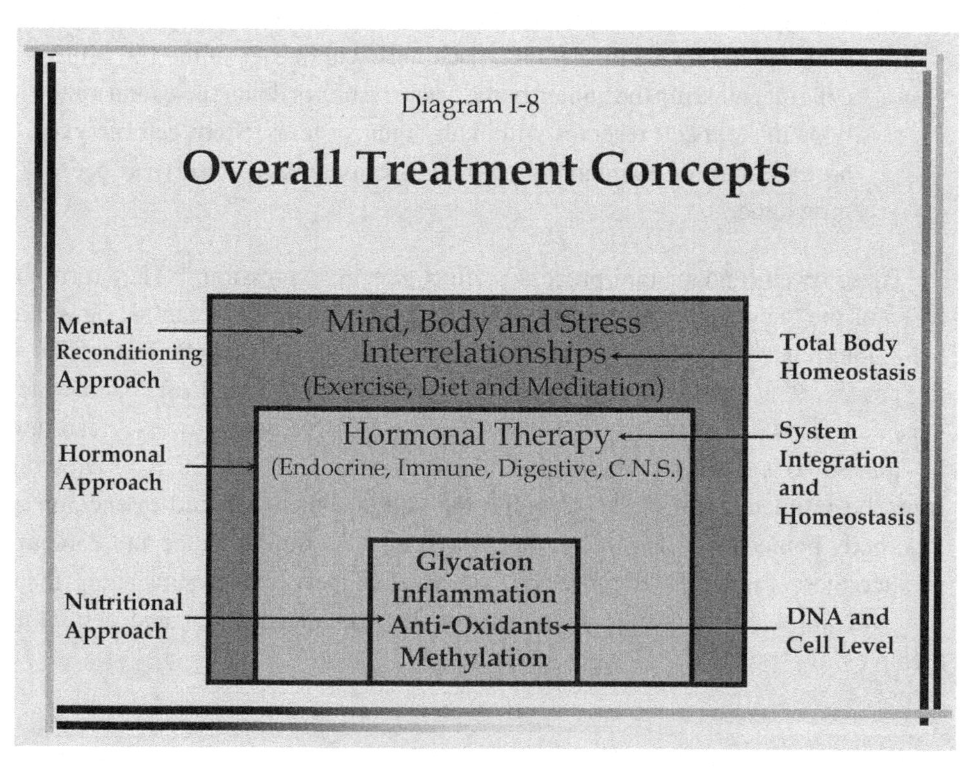
Diagram I-8
Overall Treatment Concepts
Mental Reconditioning Approach
Mind, Body and Stress Interrelationships
(Exercise, Diet and Meditation)
Total Body Homeostasis
Hormonal Approach
Hormonal Therapy
(Endocrine, Immune, Digestive, C.N.S.)
System Integration and Homeostasis
Nutritional Approach
Glycation
Inflammation
Anti-Oxidants
Methylation
DNA and Cell Level

References

[1] Roush W. Worm longevity gene cloned. *Science.* 1997; 277(5328): 897–898.
[2] Kimura KD, Tissenbaum HA, Liu Y, Ruvkun G. daf-2, An insulin receptor like gene that regulates longevity and diapause in *Caenorhabditis elegans. Science.* 1997; 277(5328): 942–946.
[3] Fleming JE, Quattrocki E, Latter G, Miouel J, Marcuson R, Zuckerkandl E, "Bensch KG. Age-dependent changes in proteins of *Drosophila melanogaster. Science.* 1986; 231: 1157–1159.
[4] Orr WC, Sohal RS. Extension of life span by overexpression of superoxide dismutase and catalase in *Drosophila melanogaster. Science.* 1994; 263: 1128–1130.
[5] King GL, Brownlee M. The cellular and molecular mechanisms of diabetic complications. *Endocrinol Metab Clin North Am.* 1996; 25(2): 255–270.
[6] Sternberg M, Urios P, Grigorova-Borsos AM. [Effects of glycation process on the macromolecular structure of the glomerular basement membranes and on the glomerular functions in aging and diabetes mellitus.] *C R Seances Soc Biol Fil.* 1995; 189(6): 967–985.
[7] Wolff SP, Jiang ZY, Hunt JV. Protein glycation and oxidative stress in diabetes mellitus and ageing. *Free Rad Biol Med.* 1991; 10: 339–352.
[8] Khan S, Rupp J. The effect of exercise conditioning, diet and drug therapy on glycosylated hemoglobin levels in type 2 (NIDDM) diabetics. *J Sports Med Phys Fitness.* 1995; 35: 281–288.
[9] Yamanouchi T, Akanuma Y, Toyota T, Kuzuya T, Kawai T, Kawazu S, Yoshioka S, Kanazawa Y, Ohta M, Baba S, et al. Comparison of 1,5-anhydroglucitol, HbA1c, and fructosamine for detection of diabetes mellitus. *Diabetes.* 1991; 40: 52–57.
[10] Meloni T, Pacifico A, Forteleoni G, Meloni GF. HbA1c levels in diabetic Sardinian patients with or without G6PD deficiency. *Diabetes Res Clin Pract.* 1994; 23(1): 59–61.
[11] Tahara Y, Shima K. Kinetics of HbA1c, glycated albumin, and fructosamine and analysis of their weight functions against preceding plasma glucose level. *Diabetes Care.* 1995; 18(4): 440–447.
[12] Broussolle C, Tricot F, Garcia I, Orgiazzi J, Revol A. Evaluation of the fructosamine test in obesity: Consequences for the assessment of past glycemic control in diabetes. *Clin Biochem.* 1991; 24: 203–209.
[13] Guillasseau PJ, Charles MA, Godard V, Timsit J, Chanson P, Paolaggi F, Peynet J, Eschwege E, Rousselet F, Lubetzki J. Comparison of fructosamine with glycated "hemoglobin as an index of glycemic control in diabetic patients. *Diabetes Res.* 1990; 13: 127–131.
[14] Knecht KJ, Dunn JA, McFarland KF, et al. Effect of diabetes and aging on carboxymethyllysine levels in human urine. *Diabetes.* 1991; 40: 190–196.
[15] Giugliano D, Ceriello A, Paolisso G. Diabetes mellitus, hypertension, and car diovascular disease: Which role for oxidative stress? *Metabolism.* 1995; 44(3): 363–368.
[16] Dills WL. Protein fructosylation: Fructose and the Maillard reaction. *Am J Clin Nutr.* 1993; 58(Suppl): 779S–787S.
[17] Yaylayan VA. Classification of the Maillard reaction: A conceptual approach. *Trends Food Sci Technol.* 1997; 8(1): 13–18.
[18] Rattan SI. Synthesis, modifications, and turnover of proteins during aging. *Exp Gerontol.* 1996; 31(1/2): 33–47.
[19] Rattan SI, Derventzi A, Clark BF. Protein synthesis, post-translational modifications, and aging. *Ann N Y Acad Sci.* 1997; 663: 48–62.
[20] Kimura T, Ikeda K, Takamatsu J, Miyata T, Sobue G, Miyakawa T, Horiuchi S. Identification of advanced glycation end products of the Maillard reaction in Pick's disease. *Neurosci Lett.* 1996; 219: 95–98.
[21] Coufturier M, Amman H, Des Rosiers C, Comtois R. Variable glycation of serum proteins in patients with diabetes mellitus. *Clin Invest Med.* 1997; 20(2): 103–109.
[22] Coussons PJ, Jacoby J, McKay A, Kelly SM, Price NC, Hunt JV. Glucose modification of human serum albumin: A structural study. *Free Rad Biol Med.* 1997; 22(7): 1217–1227.
[23] Suarez G, Maturana J, Oronsky AL, Raventos-Suarez C. Fructose-induced fluorescence generation of reductively methylated glycated bovine serum albumin: Evidence for nonenzymatic glycation of Amadori adducts. *Biochim Biophys Acta.* 1991; 1075(1): 12–19.

[24] Wu JT, Tu MC, Zhung P. Advanced glycation end product (AGE): Characterization of the products from the reaction between D-glucose and serum albumin. *J Clin Lab Analysis.* 1996; 10(1): 21–34.
[25] Miyata T, Hori O, Zhang JH, Yan SD, Ferran L, Iida Y, Schmidt AM. The receptor for advanced glycation end products (RAGE) is a central mediator of the interaction of AGE-$\beta_2$microglobulin with human mononuclear phagocytes via an oxidant-sensitive pathway. *J Clin Invest.* 1996; 98(5): 1088–1094.
[26] Wolff SP, Bascal ZA, Hunt JV. "Autooxidative glycosylation": Free radicals and glycation theory. *Prog Clin Bio Res.* 1989; 304: 259–275.
[27] Lapolla A, Poli T, Valerio A, Fedele D. Glycosylated serum proteins in diabetic patients and their relation to metabolic parameters. *Diabete Metabolisme.* 1985; 11: 238–242.
[28] Ballou SP, Lozanski GB, Hodder S, Rzewnicki DL, Mion LC, Sipe JD, Ford AB, Kushner I. Quantitative and qualitative alterations of acute-phase proteins in healthy elderly persons. *Age Aging.* 1996; 25: 224–230.
[29] Cooper GJ, Tse CA. Amylin, amyloid and age-related disease. *Drugs Aging.* 1996; 9(3): 202–212.
[30] Hilliquin P. Biological markers in inflammatory rheumatic diseases. *Cell Mol Biol.* 1995; 41(8): 993–1006.
[31] Rook GA, Zumia A. Gulf War syndrome: Is it due to a systemic shift in cytokine balance towards a Th2 profile? *Lancet.* 1997; 349: 1831–1833.
[32] Chandra RK. Nutrition and the immune system: An introduction. *Am J Clin Nutr.* 1997; 66(2): 460S–463S.
[33] Sprietsma JE. Zinc-controlled Th1/Th2 switch significantly determines development of diseases. *Med Hypotheses.* 1997; 49(1): 1–14.
[34] Mendall MA, Patel R, Ballam L, Strachan D, Northfield TC. C-reactive protein and its relation to cardiovascular risk factors: A population based cross sectional study. *BMJ.* 1996; 312(1038): 1061–1065.
[35] Haverkate F, Thompson SG, Pyke SD, Gallimore JR, Pepys MB. Production of C-reactive protein and risk of coronary events in stable and unstable angina. *Lancet.* 1997; 349(9050): 462–466.
[36] Cabana VG, Siegel JN, Sabesin SM. Effects of the acute phase response on the concentration and density distribution of plasma lipids and apolipoproteins. *J Lipid Res.* 1989; 30: 39–49.
[37] Hasdai D, Scheinowitz M, Leibovitz E, Sclarovsky S, Eldar M, Barak V. Increased serum concentrations of interleukin-1b in patients with coronary artery disease. *Heart.* 1996; 76(1): 24–28.
[38] Pelletier JP, Martel-Pelletier J. [Role of synovial inflammation, cytokines and IGF-1 in the physiopathology of osteoarthritis.] *Rev Rhum Ed Fr.* 1994; 61(9, pt 2): 103S–108S.
[39] Moulton PJ. Inflammatory joint disease: The role of cytokines, cyclooxygenases and reactive oxygen species. *Br J Biomed Sci.* 1996; 53(4): 317–324.
[40] Kaufman W. Niacinamide therapy for joint mobility: Therapeutic reversal of a common clinical manifestation of the normal aging process. *Conn State Med J.* 1953; 17: 584–589.
[41] Jonas WB, Rapoza CP, Blair WF. The effect of niacinamide on osteoarthritis: Pilot study. *Inflamm Res.* 1996; 45(1): 330–334.
[42] Miesel R, Kurpisz M, Kroger H. Modulation of inflammatory arthritis by inhibition of poly(ADP ribose) polymerase. *Inflammation.* 1995; 19(3): 379–387.
[43] Busse E, Zimmer G, Schopohl B, Kornhuber B. Influence of α-lipoic acid on intracellular glutathione *in vitro* and *in vivo. Arzneimittelforschung.* 1992; 42(6): 829–831.
[44] Firestein GS, Zvaifler NJ. Anticytokine therapy in rheumatoid arthritis. *N Engl J Med.* 1997; 337(3): 195–197.
[45] Clancy RM, Abramson SB. Nitric oxide: A novel mediator of inflammation. *Proc Soc Exp Biol Med.* 1995; 210(2): 93–101.
[46] Goodman J. Histone tails wag the DNA dog. *Helix.* [University of VA Health System]. Spring 2000; 17(1).

2

Theories of Aging

Old and New Concepts of the Aging Process

Vincent C. Giampapa, M.D., F.A.C.S.
Stanley Burzynski, M.D., Ph.D.
Ronald Pero, Ph.D.

Errors, like straws, upon the surface flow; He who would search for pearls must dive below.

John Dryden, *All for Love*

The New Aging Paradigm

Since the 1960s, a number of theories of aging have evolved. They are based on the degree of knowledge present at a given time. Today this degree of knowledge is proceeding exponentially! In reality, there is no single cause of aging. **Aging is a complex series of events that changes from year to year for each individual as he or she interacts with the environment.** It can best be conceptualized as a change in a different series of gene activity, as we move through childhood to "adolescence to adulthood and eventually old age. A review of the theories of aging popular in the past gives a general idea of some of the fundamental events involved.[1] Familiarity with these general theories should result in an understanding of the basic concepts behind the clinical treatment of aging and the aging codes (Diagram II-1).

Since the mid-1990s, some startling new information has led me to the formation of a new theory of aging that is based and focused more on the ultimate source of aging itself: DNA. This theory, along with the information on the importance of stem cells and their contribution to the aging process, constitutes this new paradigm of aging. It will be discussed in more detail at the end of this chapter.

The Principles and Practice of Antiaging Medicine for the Clinical Physician, 21–32.

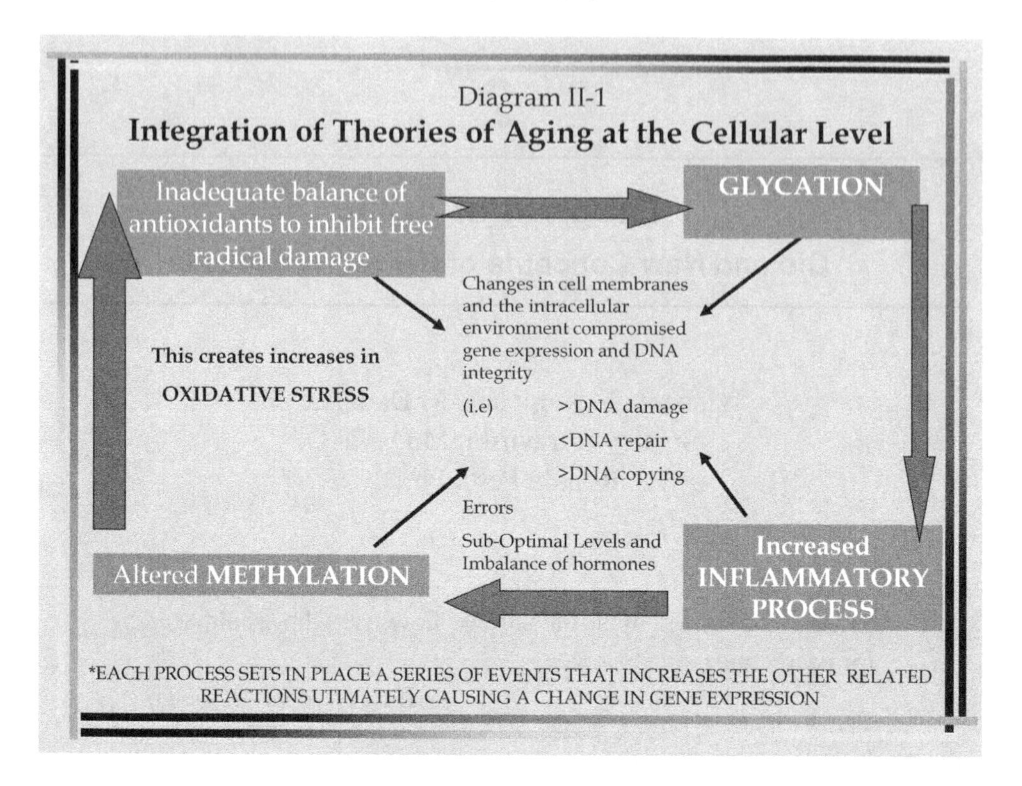

Past Theories

Wear-and-Tear Theory The wear-and-tear theory of aging was originally presented by Dr. August Weismann, a German biologist. This theory focused on the fact that the body and its cells are damaged by overuse and abuse; that is, the specific organs are worn down by toxins in diet and in the environment. Most specifically, Dr. Weismann's theory focused on the detrimental consumption of fat, sugar, caffeine and alcohol, as well as the harmful effects of the sun's ultraviolet rays. The wear-and-tear process is not confined to our organs; it also takes place at the cellular level.

Neuroendocrine Theory of Aging Developed by Vladimir Dilman, Ph.D., a Russian research scientist, this theory elaborated on the wear-and-tear theory by focusing more specifically on the neuroendocrine system. Dr. Dilman's concept of aging focused on the complicated network of biochemicals that controls the release of hormones and, therefore, regulates other vital body functions. Dr. Dilman stated that during youth, "our hormones work together to regulate many bodily functions, including our responses to stress from the environment, as well as changes in our diet and mental state." Dr. Dilman further stated that "the hypothalamus governs the release of key hormones that regulate virtually all organs in the body through a complex series of chain reactions." According to this theory, the hormones are vital for repairing and regulating body functions, and when aging causes a drop in these key hormone productions, it alters the body's ability to repair and regenerate itself as well. Most important, Dr. Dilman stressed how hormone production is highly interactive: that is, the drop in the production of any one of the hormones will probably have a feedback effect on the whole mechanism, signaling other organs to alter their release of hormones as well. In essence, the neuroendocrine system is like a symphony, irreparably ruined by discord of any one instrument. Because of Dr. Dilman's theory, the concept of hormone replacement remains a key component of any anti-aging treatment.

Genetic Control Theory of Anti-Aging According to this theory, obsolescence is programmed into our genes and therefore encoded in our DNA. The theory is that everyone is born with a unique genetic code and a fixed predisposition to certain types of age-related diseases. It stresses that genetic inheritance has the preponderance of effects on how quickly a person ages and how long a person lives. It is a deterministic theory, comparing humans to a machine that is preprogrammed to self-destruct. As a result of this theory, anti-aging medicine concepts began to stress the importance of supplying the building blocks of DNA within each cell. Although many aspects of this theory have been supported by research on DNA, the one of complete determinism by a preprogrammed set of genetic sequences has been shown not to be completely true. What is true, is that we inherit genetic *tendencies*, not genetic *certainties*.

Free Radical Theory of Anti-Aging Originally introduced by Dr. Gershchman in 1954, this theory was further expanded by Dr. Denham Harman of the University of Nebraska College of Medicine. A **free radical** is any molecule that differs from conventional molecules in that it possesses a free electron. This property makes it react with other molecules in a highly volatile and destructive way. In a conventional molecule, the electrical charges are balanced. Electrons normally reside in pairs so that electrical charges can neutralize each other. Atoms that are missing electrons combine with atoms that have extra electrons, therefore creating a stable molecule with evenly paired electrons, and the result is a neutral electrical charge.

A free radical, however, has an extra electron, creating a negative (−) charge. Because of this unbalanced electrical energy, the free radical tends to attach itself to other molecules in order to obtain a matching electron and attain electrical equilibrium. Some scientists speak of the free radicals as "thieves," which break up the normal pairing of electrons in neighboring molecules in order to steal a new electron for themselves. This leads to the creation of more free radicals and extensive body damage mainly to the cell membranes, as well as to genetic material. Free radicals, however, are important as a normal aspect of healthy body status: without free radical activity, the body would not be able to produce energy, maintain immunity, use hormones or even contract muscles. This electrical production within the body enables the body to perform these key functions. This key quantity of electricity comes from the unbalanced electron activity of free radicals.

The most significant targets of free radical damage are cell membranes, DNA and RNA. Damage to nuclear DNA can result in faulty copies during cell reproduction and damage to mitochondrial DNA; this in turn results in a decrease in energy production, which is essential for running the nuclear genetic machinery necessary to make proteins, enzymes and even hormones.

Free radical damage begins at birth and continues until death. Early in life, its effects are relatively minor because the body has extensive repair and replacement mechanisms in place that allow the cells and organs to work optimally. Over time, however, the accumulative effects of free radical damage begin to take their toll.

Cosmetic surgeons and physicians are used to seeing the effects of free radical damage on collagen and elastin, the substance that keeps skin smooth, moist, flexible and elastic. The most obvious effects of free radical damage occur in the skin in the forms of deep wrinkles, loss of elastic properties of muscle and a decrease in the content of facial fat.

One of the key components of anti-aging therapy is to prescribe and create a host of natural and manufactured antioxidants to help combat the effects of aging. The body possesses an inherent ability to produce intrinsic antioxidants, the most common ones being superoxide dismutase, catalase and glutathione peroxidase. These are produced and stored mainly at the intracellular level of the mitochondria, which is another key focus of anti-aging therapy.

Waste Accumulation Theory According to this theory, cells produce more waste than they can properly dispose of. This waste may include many different toxins, which accumulate in the cell and therefore interfere with normal cell function, ultimately killing the cells. Results of a great deal of new research support this theory, especially with regard to the production of a waste product called **lipofuscin**. This compound is frequently found in the nerve and heart cells. Lipofuscin is formed by a complex reaction that binds fat and proteins within the cells.

Membrane Hypothesis of Aging[2,3] A recent elaboration on the waste accumulation theory is this hypothesis (Diagram II-2), which was popularized by Dr. Imre Zs-Nagy.[2] In this theory, free radicals damage the cell membrane, and it becomes unable to let nutrients in and waste products out. Waste products focused on in this theory include lipofuscin and cellular salts in the form of potassium, which build up within the cell, causing the cell to become dehydrated. This interferes with the normal flow of electrolytes in and out of the cell, allowing the toxins and the cell's metabolic waste products to accumulate and eventually cause the death of the cell. This theory also accounts for the effects seen in aging resulting from a loss of hydration in both the facial skin and body skin. This is discussed further in Chapter VII, on skin and aging.

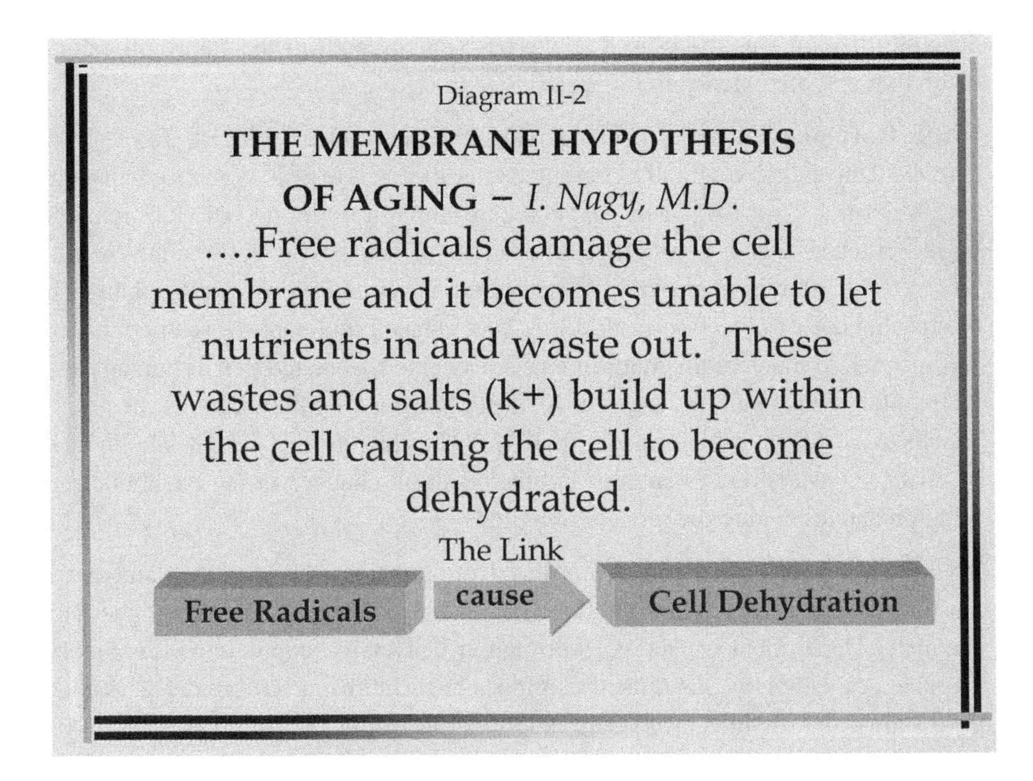

"Limited Number of Cell Division" Theory According to this theory, the number of cell divisions is directly affected by the accumulation of the cells' waste products. The more waste that is accumulated within our cells, the faster these cells degenerate.

This theory was originally proposed by Dr. Alexis Carroll, a French surgeon who was able to keep two chicken hearts alive in saline solution for over 28 years. He believed he had achieved this by disposing of the waste products within the cell. The theory was discounted by Dr. Leonard Hayflick when it was found that fresh cells had inadvertently been added to the cultures, making the chicken heart cells seem immortal.

Hayflick Limit Theory In 1961, two cell biologists, Dr. Morehead and Dr. Leonard Hayflick, made one of the most important contributions to aging theories. Dr. Hayflick[1] theorized that the aging process was controlled by a biological clock contained within each living cell. In his research, he studied human fibroblast cells from the lung, skin, muscle and heart and found them to have a limited life span: they divided approximately 50 times over a period of years and then suddenly stopped.

Nutrition seemed to have an effect on the number of cell divisions. Whereas overfed cells made up to 50 divisions in a year, underfed cells took up to three times as long as normal (or overfed) cells to make the same number of divisions.

This information was to be used as the basis of research in the future on caloric restriction and life extension, in general.

Death Hormone Theory Dr. Donner Denckle, an endocrinologist, formerly of Harvard University, originally presented this theory. He was convinced that the "death hormone," or decreasing oxygen consumption hormone (DECO), released by the pituitary gland, contributed to the loss of neurons in the brain. When he removed the pituitary glands of rats, their immune systems were revitalized, the rate of crosslinking or glycation in the cells was reduced and cardiovascular function was restored to much more youthful levels. Denckle speculated that as humans age, the pituitary cells begin to release this "death hormone," which inhibits the ability of cells to use the thyroid hormone (thyroxin). This affects the basic metabolic rates of cells to convert food to energy, and the resulting changes in the metabolic rate bring on and accelerate the process of aging.

Thymic Stimulating Theory According to this theory, the thymus gland is the master gland of the immune function. Dr. Alan Goldstein, Chairman of the Biochemistry Department of George Washington University, popularized this theory. The size of the thymus gland itself continues to shrink from birth to death. Studies have shown that thymic factors are helpful in restoring the immune function, as well as in rejuvenating poorly functioning immune systems in both old and young patients. Thymic hormones probably play a role in stimulating and controlling the production of neurotransmitters and key hormones that affect brain function, which therefore affects central pacemakers of aging.

Mitochondria Theory of Aging This theory is similar to the free radical theory, in general, and stresses the fact that free radicals directly damage mitochondrial DNA. Mitochondria are the energy-producing organelles (subcellular units or micro organs) within the cells that are responsible for producing adenosine triphosphate (ATP), which is the primary source of cellular energy.[4] ATP fuels the nuclear and mitochondrial genetic machinery to allow cell reproduction to occur. At the same mitochondrial location, cells produce energy but also produce the vast amounts of potentially damaging free radicals. Mitochondria, because they have limited or very poor DNA repair mechanisms, are an extremely sensitive target for free radical damage.

Errors and Repair Theory In 1963, Dr. Leslie Orgel, of the Salk Institute, suggested that because the machinery for making proteins in cells is so essential to life, an error in this genetic machinery could be catastrophic. He focused on the production and reproduction of DNA and stated, "Over time this is not carried out with accuracy." Because the body's protective mechanisms to restore DNA are poor from birth on and decrease in number and efficacy with age, this repair system is incapable of making perfect repairs on damaged DNA, and this results directly in disease and other age-related changes. This theory seems to have gained credibility, specifically in reference to the more recent research and literature documenting the importance of DNA repair.[5–8]

Identical DNA Theory This theory suggests that there are genetically identical "sequences of DNA within our genetic codes that take over after the initial sequences are worn out or damaged. Dr. Medvedev, at the National Institutes of Medical Research in London, proposed that in different species, life span may be a function of the degree of these repeated genetic sequences in the nucleus.

Crosslinkage Theory of Aging This theory was initially proposed in 1942 by "Johann Bjorksten. The crosslinking phenomenon, he stated, "occurs in neighboring molecules, and as we age, the number of crosslinks increases, causing, for instance, skin to shrink and collagen to become less soft and pliable." He proposed that these crosslinks begin to obstruct the passage of nutrients and waste products between cells. He also proposed that the immune system is incapable of cleaning out this excess crosslinked material, which is found in a specific form of glucose molecules. These glucose molecules react with proteins, causing further crosslinking and the subsequent formation of destructive free radicals. His research has caused more recent anti-aging scientists to focus on the importance of glycation as a key component of the aging process.

Autoimmune Theory According to this theory, as the body ages, the body's ability to produce necessary antibodies to fight key diseases declines. In another sense, the immune system also becomes self-destructive and reacts against its own proteins. More recent literature has lent credibility to this theory, especially in view of the effects of the aging gastrointestinal tract and leaky gut syndrome (LGS). LGS has been shown to be directly related to the autoimmune phenomenon as people age.

Caloric Restriction Theory This theory was originally proposed by renowned gerontologist Dr. Roy Walford, of the University of California, Los Angeles, Medical School, who stated, "Undernutrition without malnutrition can dramatically retard the functional as well as the chronological aging process." Although Dr. Walford could not explain the underlying biological processes, he stated that "cutting caloric restriction had a direct effect on obtaining maximum health and life span." He stressed the importance of not only caloric restriction but also moderate vitamin and mineral supplement intake, coupled with regular exercise. The discovery of the *Sir2P* gene has verified the importance of such supplements and exercise.

Gene Mutation Theory In the 1940s, scientists investigated the role of mutations in aging. Mutations are changes that occur in the genetic structure of DNA. Genes are critical for every aspect of cellular life. In experiments with radiation, it was observed that radiation not only increased animals' genetic mutation but also accelerated the aging process.

Rate of Living Theory This theory was originally proposed by a German physiologist, Max Rudner, who discovered the relationship among metabolic rate, body size and longevity. He first introduced the theory in 1908 and stated simply that "We are each born with a limited amount of 'energy.' When we use this 'energy' slowly, our rate of aging is slowed. If the energy is used quickly, aging is hastened."

Telomerase Theory of Aging According to this theory, the one most recently proposed, aging is based on a specific sequence of base pairs located at the end of each chromosome. The end portion of a chromosome is called a **telomere**. This information was first discovered by a group of scientists at Geron Corporation of Menlo Park, California. It was observed that telomeres were sequences of base pairs extending from the ends of the chromosomes. They maintain the integrity of chromosomes during cell division. Every time a cell divides, telomeres are shortened and more distal to the end of the chromosome. This occurs until the cell stops dividing and, eventually, dies. Scientists discovered that the key element in rebuilding disappearing telomeres is the immortalizing enzyme called **telomerase**. This is an enzyme found only in germ cells, cancer cells and another cell group now known as stem cells. Telomerase appears to repair and replace telo-meres, restoring them to a less distal position on the chromosome and thereby manipulating the clock mechanism that controls cellular aging.

The Methylation Control of Gene Activation and Silencing Theory

According to Dr. Stanley Burzynski, adult cells in the body have an established methylation pattern in their DNA that is essential to the aging program. But Dr. Burzynski also states that in the very first days of life this methylation pattern is erased. Most genes in our DNA are active during the initial embryonal development. Then they begin to become blocked through methylation as their expression is no longer needed. Many genes are silenced in this manner after birth, including the one for hemoglobin F (Fetal Hemoglobin). This trend accelerates especially after age 25 with increasing numbers of genes silenced as we grow older. If for any reason the silencing affects tumor suppressor genes, the aging person may develop cancer. Continuous silencing of genes in the aging bodies is a major factor leading to progressive aging and ultimately death. Dr. Burzynski has isolated a number of substances that can reactivate these tumor suppressor genes in cancer patients as well as reactivate other genes that are involved in symptomatic aspects of aging. He has termed these molecular switches **antineoplastons**. While activating tumor "suppressor genes to help alleviate the potential cancer, these antineoplastons "also "turn on" other genes, which can provide aging patients with a number of benefits. Therefore, Dr. Burzynski believes it is realistic to contemplate that we may be able to stop and reverse the pattern of methylation that occurs in normal aging by using properly designed molecules to act as "switches" that deactivate certain genes and activate more youthful genes.

Unified DNA Damage Theory of Aging (UDDTA)

After a review of all these theories of aging, it becomes obvious that there are many different concepts, which can affect aging and many overlaps among these theories.

The new theory which I have conceptualized over the last decade and which I now describe focuses on a combination of these theories. This theory stresses that the basis of control for the aging process in cells lies in the accurate reproduction of DNA and its resulting essential products — stem cells. DNA is essential for life, because it is directly responsible for every molecular change occurring within the body. Damage is inflicted on DNA every day by the environment, diet and both physical and emotional stress that humans endure. This damage is the primary cause of aging and diseased states. As bodies age, people become more susceptible to many serious diseases. The most common age-related diseases today in the United States that cause death are heart disease, stroke, cancer and Alzheimer's disease. But what causes these diseases and how do they start? In essence, they can all be traced back to the initial breakdown of, and damage to, DNA.

In this theory, **both the rate and quality of DNA damage most likely triggers key genetic control processes like methylation and therefore are responsible for changing gene profiles** as we advance from fetus through childhood, adolescence, adulthood and old age.

The importance of a healthy state of DNA cannot be overemphasized. Within each human body, there are an estimated 100 trillion cells. Within each of these cells is approximately 5 feet of DNA. That means there are billions of miles of DNA within each human body. Just improving a small percentage of this extraordinary and vast amount of genetic material can make a major difference in the quality of health, well-being and how a person ages.

In essence, DNA is "life's blueprint," or, in more focused terms of this discussion, "aging software." An individual code for health and life expectancy is genetically passed on to each person. It is **inherited** by each person from his or her parents, grandparents and so forth. New research has documented that life spans in both animals and humans are directly correlated with DNA repair rates.[6,9]

Throughout life, DNA reproduces and replaces itself continually. In optimal conditions, DNA copies itself over and over again, making perfect reproductions. This is very close to the state that people are in when they are young and healthy. As people age, however, their DNA is damaged continually through ongoing bombardment by excess free radicals, environmental effects and radiation. The DNA begins to reproduce poorly and ultimately stops reproducing completely, which results in cell death.

Think of it as making a photocopy. If a well-maintained machine is used to copy the original document, it produces an excellent copy. But if the machine is poorly maintained, it will ultimately produce poor copies. Furthermore, if the machine continues to make poor copies of poor copies, the degradation becomes worse and worse with each successive copying cycle until eventually the copies are illegible. In essence, the same thing happens with DNA. If the body cannot produce clean and accurate copies of its DNA, health and longevity are directly affected. Therefore, the key to optimal health is to keep DNA clean and healthy so that it produces ideal, clean copies of itself. Currently, this can be done by helping the body neutralize excess free radicals and, at the same time, strengthen and nourish cells and the building blocks for DNA.

New literature has documented that even in old people, the telomeres (the sequence of base pairs at the end of our chromosomes) that seem to control the number of cell divisions have still not reached their terminal positions, which should signal final cell death.[10] In essence, humans have not reached their full genetic potential, which is stored in each cell.

According to this new theory, **humans are not irreversibly programmed to age and die, as is currently thought; in contrast, humans are programmed for self-repair and longevity.** The *key* to optimal health, therefore, is the ability to keep DNA healthy in order to produce, ultimately, as clean a cell copy as is possible. This new theory is referred to as the **unified DNA damage theory of aging** (UDDTA).

The essential concepts in this new theory emphasize the **ability to improve the ratio of DNA repair to DNA damage.** This ratio of repair to damage is much higher early in life, when DNA repair can easily neutralize the damage sustained by DNA. In both nonprimate mammals and primates, it has been shown that the greater the DNA repair levels are, the longer the life span is.[6] As animals age, this ratio becomes inverse, and the damage rates overcome the DNA repair capabilities. This inversion of the ratio of DNA repair to damage is linked directly to a number of key processes that are encoded within genes (see Diagram II-1). These processes, as mentioned in Chapter I, are glycation, inflammation, oxidation and methylation. The origin of these chemical processes lies within the genetic blueprints, the DNA. During the aging process, they slowly become uncontrolled, and each reaction contributes to further damage produced by the others. In essence, the loss of balance in this aging equation is responsible for altering the ratio of DNA repair to DNA damage. The loss of the subsequent accurate copies of DNA and the damage to the enzymes responsible for the DNA repair are the causes of the age-related changes that are seen in the form of both the microscopic and macroscopic effects on the body.

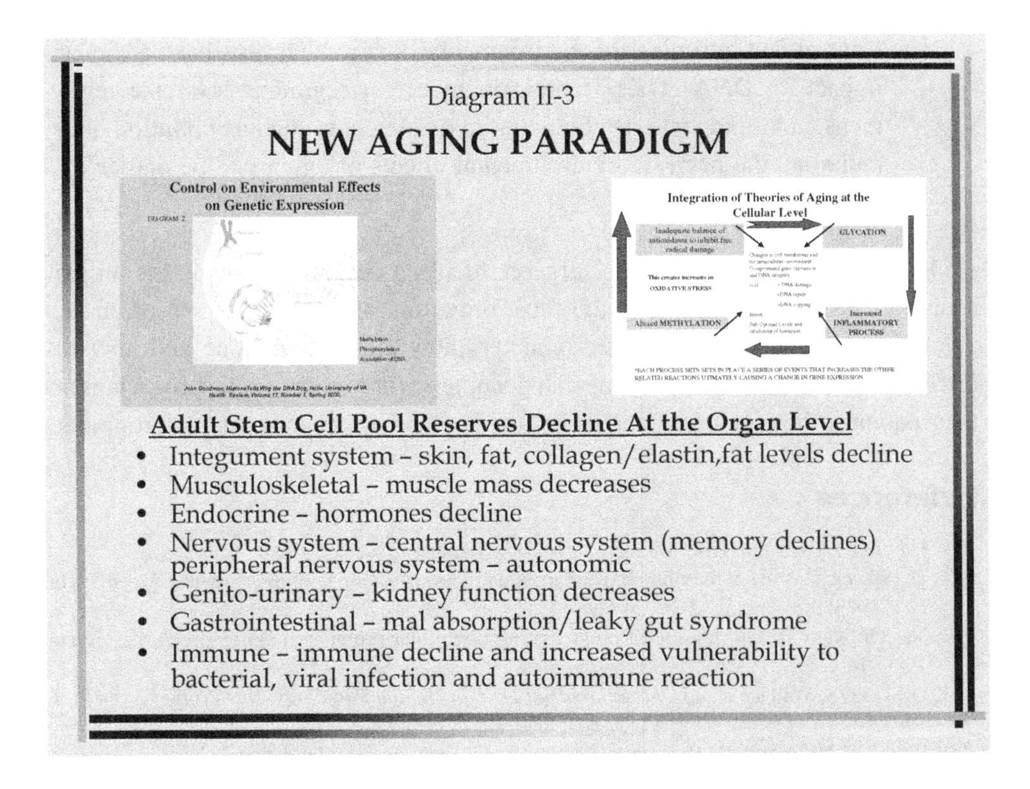

Furthermore, the damage to DNA has a direct impact on another key component that is responsible for aging: stem cell pool reserves. **Adult stem cells** have been shown to be present in many different body tissues, including adipose tissue, neuronal tissue, immune tissue and gastrointestinal tissue. Adult stem cells have the potential to redifferentiate and restore the aging and nonfunctioning tissues that are damaged by the many aspects of aging itself. With continued damage and poor repair of DNA, the genetic machinery in the DNA of stem cells is also injured, and its regenerative and restorative capacities are therefore lost over time[10] (Diagram II-3).

In summary, the key goal of anti-aging therapy should be to accomplish the following:

1. Improve the ratio of DNA repair over DNA damage. This will result in fewer cell mutations and more accurate cell copies during cell replication, will preserve the stem cell pool and, ultimately, will optimize the aging process within all the key organs.
2. Improve the function of what has been called the aging equation (see Diagrams II-1, II-3); that is, control the processes of glycation, inflammation, oxidation and methylation. All of these components directly affect DNA function, DNA repair and DNA damage.

3. Control and optimize the environmental factors that also have a direct impact on DNA. These factors include diet regimens, exercise regimens and mind state. By reducing toxic elements such as pollution and radiation, the negative environmental effects of aging can be markedly decreased as well.

In summary, for the cosmetic surgeon, focusing on and improving the ultimate source of aging itself, DNA, is the most important and effective goal of an anti-aging program. We clarify in later chapters how this can be done and how new breakthroughs in laboratory testing can document, for both cosmetic physicians and their patients, the efficacy of anti-aging therapy and age-management "programs.

References

[1] Hayflick L. *Why and How We Age.* New York: Ballantine Books; 1994.
[2] Zs-Nagy I. The role of membrane structure and function in cellular aging: A review. *Mech Ageing Dev.* 1979; 9(3–4): 237–246.
[3] Nagy I, Nagy K. On the role of cross-linking of cellular proteins in aging. *Mech Ageing Dev.* 1980; 14(1–2): 245–251.
[4] DiMauro S, Wallace D, eds. *Mitochondrial DNA in Human Pathology.* New York: Raven Press; 1993.
[5] Pero RW, Holmgren K, Perssen L. Gamma-radiation induced ADP-ribosyl transferase activity and mammalian longevity. *Mutation Res.* 1985; 142: 69–73.
[6] Grube K, Burke A. Poly(ADP-ribose) polymerase activity in mononuclear leukocytes of 13 mammalian species correlates with species-specific life span. *Proc Natl Acad Sci USA.* 1992; 89: 11759–11763.
[7] Sandoval-Chacón M, Thompson JH, Zhang XJ, Liu X, Mannick EE, Sadowicka H, Charbonet RM, Clark DA, Miller MJ. Antiinflammatory actions of cat's claw: The role of NF-kB. *Aliment Pharmacol Ther.* 1998; 12(12): 1279–1289.
[8] Sheng Y, Bryngelsson C, Pero RW. Enhanced DNA repair, immune function and reduced toxicity of C-MED 100, a novel aqueous extract from *Uncaria tomentosa. "J Ethnopharmacol.* 2000; 9: 115–116.
[9] Sheng Y, Li L, Holmgren K, Pero RW. DNA repair enhancement of aqueous extracts of *Uncaria tomentosa* in a human volunteer study. *Phytomedicine.* 2001; 8(4): 275–282.
[10] Mondello C, Petropoulis C, Monti D, Gonos ES, Franscheschi C, Nuzzo F. Telomere length in fibroblasts and blood cells from healthy centenarians. *Exp. Cell Res.* 1999; 248(1): 234–242.

3

Biomarkers of Aging

Understanding Gene Expression

Vincent C. Giampapa, M.D., F.A.C.S.
Aristo Vojdani, Ph.D.

All disease is genetic.
Paul Boerg, Co-inventor of Genetic Engineering

The first question anti-aging physicians ask is "How can we measure our rate of aging and the success of our treatments for aging?" They are measured by what are called **biomarkers**.

A biomarker is a measurable chemical substance, or physiological valve, that is known to change during aging (Diagram III-1). Some of the more important clinical biomarkers that have been documented to improve quality of life, as well as appearance, are the endocrine biomarkers, such as dihydroepiandrosterone (DHEA), growth hormone (HGH) and testosterone levels[1] (Diagram III-2). Both Dr. Vladimir Dilman and Dr. Ward Dean, in their book entitled *Neuroendocrine Theory of Aging*,[2] have supported the importance of biomarkers.

Another important concept is **gene expression**, the process that determines which genes are actively operating at a specific time because of factors present in their immediate environment. The final effects of genetic expression depend on a number of key factors that are unique for each individual (Diagram III-3).

Still another important concept is **gene plasticity**, the process that occurs when the combination of genetic potential with environmental factors results in the outcomes seen both in laboratory test results and in the physical body (Diagram III-4).

In addition to laboratory data, which document altered gene expression, subjective improvement is an integral part of an anti-aging, or age-management, evaluation. By recording the combination of both objective laboratory data and subjective data

The Principles and Practice of Antiaging Medicine for the Clinical Physician, 33–48.

Diagram III-1

Why and how do we measure our rate of aging and success of our treatments for aging?

- Biomarker ⟶ is a measurable chemical substance, or Value that we know chains as we grow from younger to older
  - DHEA Levels
  - Skin thickness and collagen amount
  - Lung Capacity
  - Muscle Mass/Fat ratio

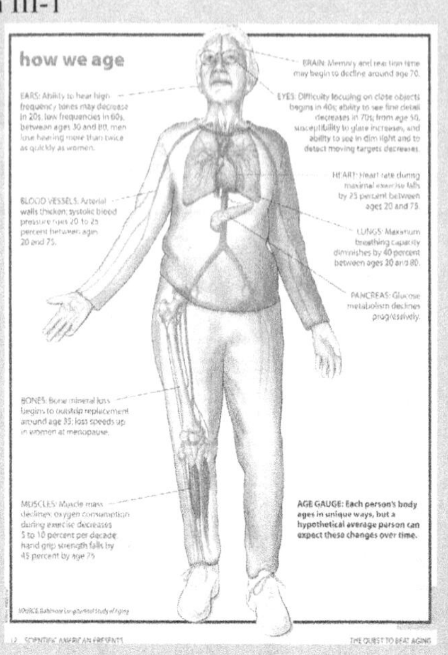

Diagram III-2

Endocrine Biomarkers

- Measure serum hormone levels in 56 exceptionally healthy males aged 20 to 84.
- Compare results with well established cognitive and physical markers of biological aging.
- Serum levels of free testosterone, DHEA & the ratios of IGF-1:HGH correlate best to age-related functional deficits.

Diagram III-3

Aging Evaluation Summary

Measure and Document

1 – Objective Biomarkers

2 – Subjective Biomarkers

3 – Modifiable Biomarkers

4 – Non-modifiable Biomarkers

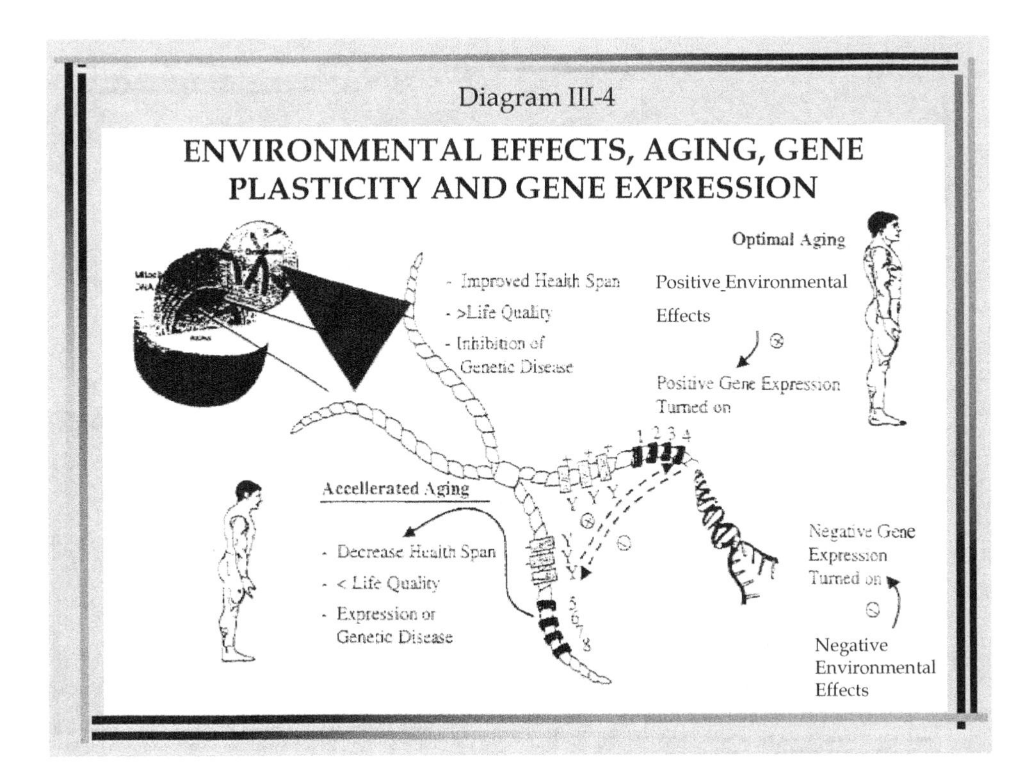

(the patient's recognition of change), the utility and effectiveness of an anti-aging program can be validated with a high degree of scientific credibility.

Organizing Biomarkers

Biomarkers can be organized into a number of different categories[3–8] (Diagram III-5). For the purposes of this book, they are viewed with relevance to the aging equation.

Objective biomarkers are biomarkers that are measurable by laboratory methods and standardized testing protocol. These include the following:

1. Hormonal panel.
2. Glycation, inflammation, oxidation and methylation panel.
3. Cardiovascular laboratory panel.
4. Immune panel.
5. Bone metabolism panel.
6. Brain metabolism panel.
7. Ultra Fast CT Heart.
8. Body Mass Index and pH level.
9. Body Composition Data.

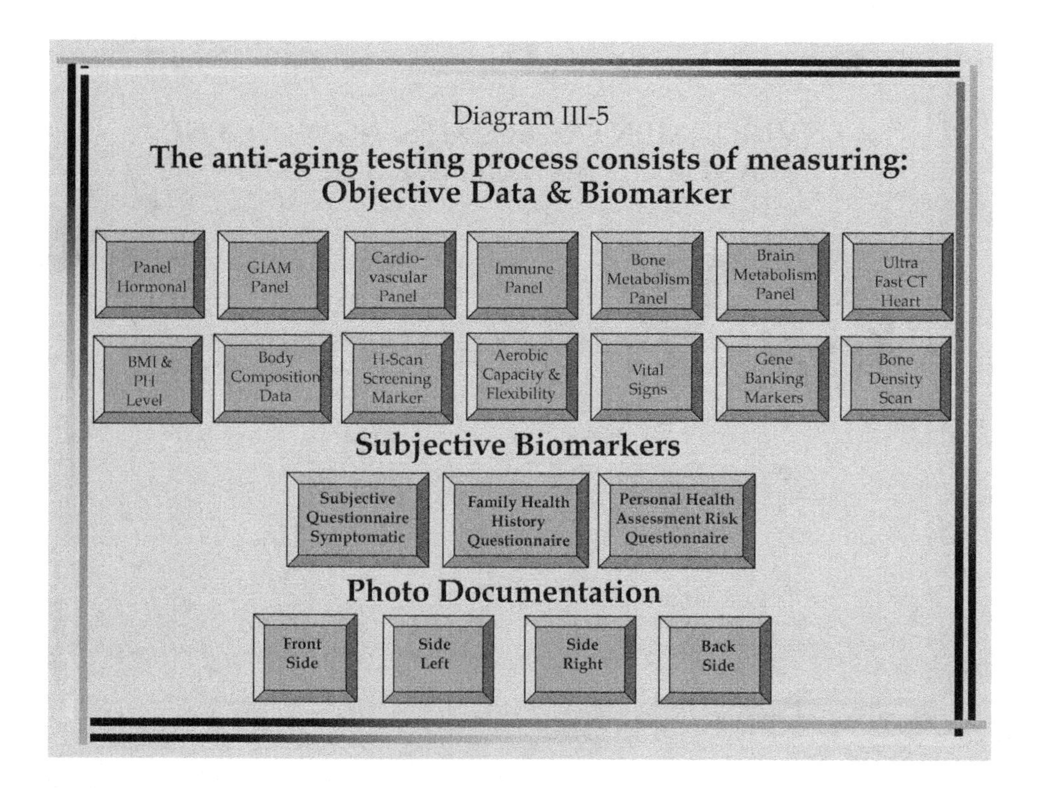

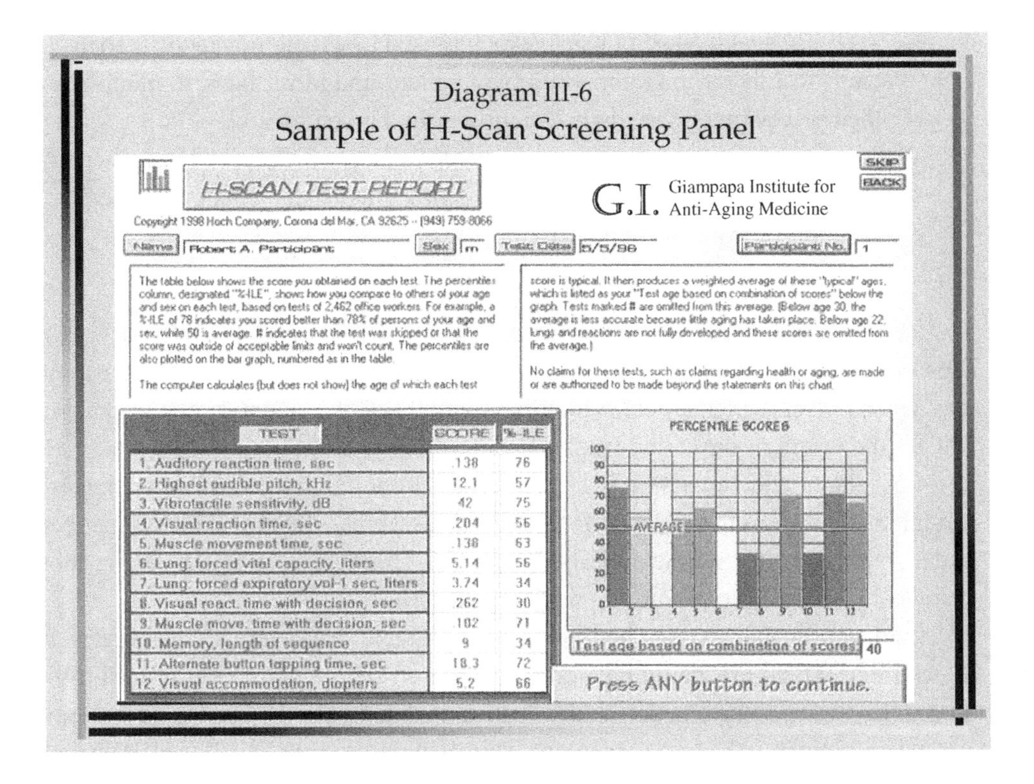

Diagram III-6
Sample of H-Scan Screening Panel

SKIP
BACK

H-SCAN TEST REPORT

G.I. Giampapa Institute for Anti-Aging Medicine

Copyright 1998 Hoch Company, Corona del Mar, CA 92625 -- (949) 759-8066

Name: Robert A. Participant | Sex: m | Test Date: 5/5/98 | Participant No. 1

The table below shows the score you obtained on each test. The percentiles column, designated "%-ILE", shows how you compare to others of your age and sex on each test, based on tests of 2,462 office workers. For example, a %-ILE of 78 indicates you scored better than 78% of persons of your age and sex, while 50 is average. # indicates that the test was skipped or that the score was outside of acceptable limits and won't count. The percentiles are also plotted on the bar graph, numbered as in the table.

The computer calculates (but does not show) the age of which each test score is typical. It then produces a weighted average of these "typical" ages, which is listed as your "Test age based on combination of scores" below the graph. Tests marked # are omitted from this average. (Below age 30, the average is less accurate because little aging has taken place. Below age 22, lungs and reactions are not fully developed and these scores are omitted from the average.)

No claims for these tests, such as claims regarding health or aging, are made or are authorized to be made beyond the statements on this chart.

TEST	SCORE	%-ILE
1. Auditory reaction time, sec	.138	76
2. Highest audible pitch, kHz	12.1	57
3. Vibrotactile sensitivity, dB	42	75
4. Visual reaction time, sec	.204	56
5. Muscle movement time, sec	.138	63
6. Lung: forced vital capacity, liters	5.14	56
7. Lung: forced expiratory vol-1 sec, liters	3.74	34
8. Visual react. time with decision, sec	.262	30
9. Muscle move. time with decision, sec	.102	71
10. Memory, length of sequence	9	34
11. Alternate button tapping time, sec	18.3	72
12. Visual accommodation, diopters	5.2	66

PERCENTILE SCORES

AVERAGE

Test age based on combination of scores: 40

Press ANY button to continue.

10. Computerized H-Scan Screening Marker* (Diagram III-6).
11. Aerobic capacity and flexibility.
12. Vital signs.
13. Gene banking markers.
14. Bone density scans.

Objective biomarkers are hard data, which are measured and documented with standardized and accepted laboratory parameters and technology. Objective data and biomarkers can be viewed as direct measurements of altered gene expression; that is, changes in the functions of genes and their subsequent protein production are responsible for creating the molecular changes that we see within our objective data measurements.

Subjective biomarkers are revealed by specific questionnaires describing age-related symptoms and body changes. They are indirect measurements of gene expression. The questionnaires include the following:

1. A general subjective questionnaire, which reveals key symptoms of age-related diseases.

*H-Scan from Hoch Company; see Resource section at the end of the book.

2. A family health questionnaire, which is used to screen for genetic tendencies most likely to be inherited by a patient and most likely to manifest themselves in a given environment within a given lifestyle.
3. A personal health assessment risk questionnaire, which is used to evaluate the patient's lifestyle, behavioral tendencies and "personal environment" within which his or her genetic potential is expressed.

Subjective biomarkers, in general, are a reflection of indirect secondary gene expression changes; that is, they are measures of the symptoms and effects a patient may sense or feel, which are related indirectly to the molecular and genetic changes that are the origin of these findings.

Photographic documentation is an extremely important tool in a comprehensive age-management evaluation. It is the final physical representation, which can be seen as the combined effect of the underlying secondary and primary genetic expression changes that have occurred during a patient's lifetime.

Another and more convenient way of looking at biomarkers of aging is in determining which are modifiable and which are nonmodifiable. **Nonmodifiable biomarkers** are genetic characteristics that cannot be modified by diet, pharmaceuticals, nutraceuticals or lifestyle at present. This includes body height, bone length, etc. **Modifiable biomarkers** are biomarkers that respond relatively quickly to changes in lifestyle and diet and to environmental changes. For instance, decreased muscle mass and aerobic capacity are strongly correlated with lower biological functional age. But they are also some of the key biomarkers that can be improved quickly with changes in exercise and with diet and nutrition programs.

Dr. Ward Dean described other biomarkers in his book *The Biological Aging Measurement — Clinical Applications*.[9] For instance, muscle strength, basal metabolic rate, body/fat ratio, glucose tolerance, cholesterol, high-density lipoprotein (HDL) levels, blood pressure and bone density are also modifiable biomarkers that respond well to changes in lifestyle interventions, including changes in diet, nutrient supplements and prescription medications.

In the future, many more modifiable biomarkers are sure to be identified; they include gene banking (Diagram III-7). This topic is discussed in the final chapter of this book, "Anti-Aging Technologies: Present and Future Trends."

For the cosmetic surgeon interested in presenting to a patient a simple but effective anti-aging program, or age-management system, it is important to choose the biomarkers that are most easily measurable and yet most informative.

In accordance with anti-aging clinical goals — that is, emphasizing maintenance of optimal DNA function by decreasing DNA damage, increasing DNA repair, augmenting the immune function, and optimizing gene expression — a core or basic anti-aging program can easily be established.

Diagram III-7

Gene Banking

- Preserves your DNA at the present, so it may be used more optimally with the technologies of the immediate future.
- Gene banking will most likely be the basis of personalized anti-aging therapies (e.g. therapeutic cloning) in the near future.

First, however, why are these clinical goals important? Maintaining optimal DNA function as a primary goal for an age-management program is the essential concept in any anti-aging program.

It must be remembered that the reason why people do not reach their full health and longevity potential is not because their DNA is faulty but because their DNA repair processes are. This results in poor genetic copies within each generation of cell populations, causing mutations in proteins and errors in enzyme production. These processes induce apoptotic pathways to cell programs to kill abnormal cells and also interfere with stem cell pool reserves. They continue until selected organ systems fail and people lose functional organ reserve (Diagram III-8) and, eventually, die.

It is also essential to keep in mind the effects of today's progressively hostile environment, which also causes more oxidative stress (more free radicals), as well as damage to both nuclear and mitochondrial DNA banks. This contributes to loss of both health and longevity, as well as the physical changes[10] that cosmetic surgeons have been trained to observe and operate on with cosmetic procedures (Diagram III-9).

The consequences of poor DNA repair become quite obvious in view of the sequence of events with both adequate and inadequate DNA repair mechanisms (Diagrams III-10 and III-11). If the repair of DNA damage is complete, a person basically maintains the optimal use of his or her inherited genetics, and little or no

Diagram III-8

AGING AND ORGAN RESERVE

Organ reserve - The quality of an organism to mobilize a margin of defense of functional ability where it is exposed to various stress factors which allows the organism to return its function back to normal physiology. This results in resilience and health in younger individuals.

As individuals age they lose organ reserve. Therefore, maintenance of organ reserve is a critical factor in the prevention of age – related disease and, is an essential feature in "anti-aging medicine."

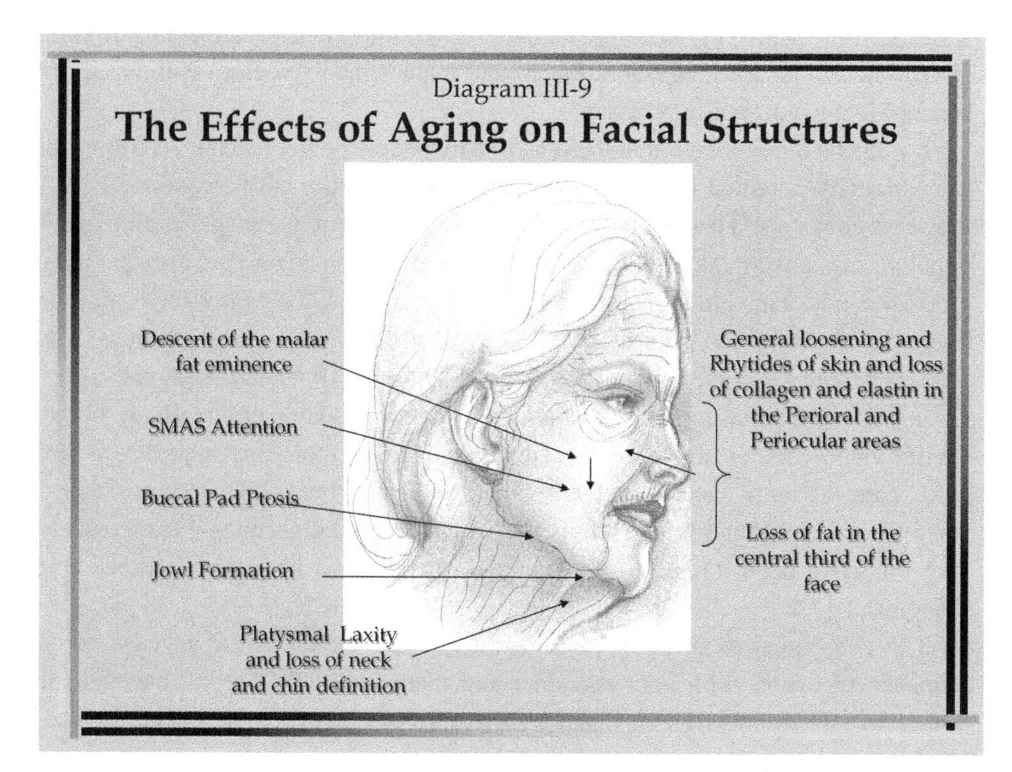

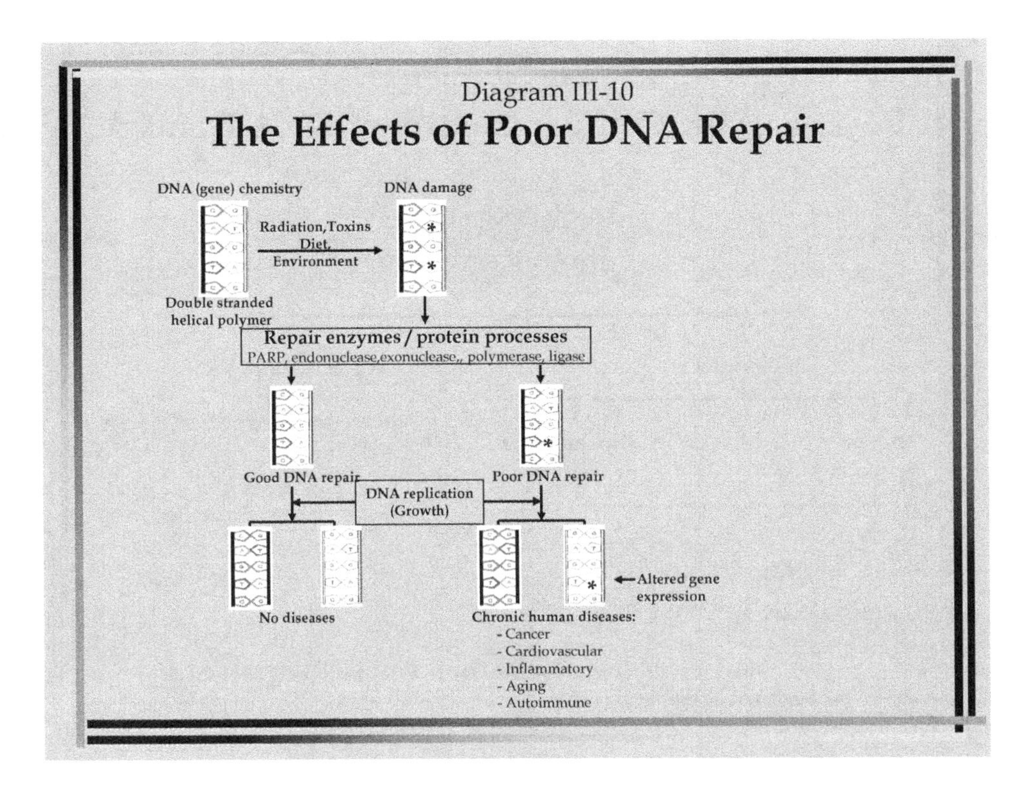
Diagram III-10
The Effects of Poor DNA Repair
DNA (gene) chemistry
DNA damage
Radiation,Toxins
Diet,
Environment
Double stranded
helical polymer
Repair enzymes / protein processes
PARP, endonuclease,exonuclease,, polymerase, ligase
Good DNA repair
Poor DNA repair
DNA replication
(Growth)
Altered gene
expression
No diseases
Chronic human diseases:
- Cancer
- Cardiovascular
- Inflammatory
- Aging
- Autoimmune

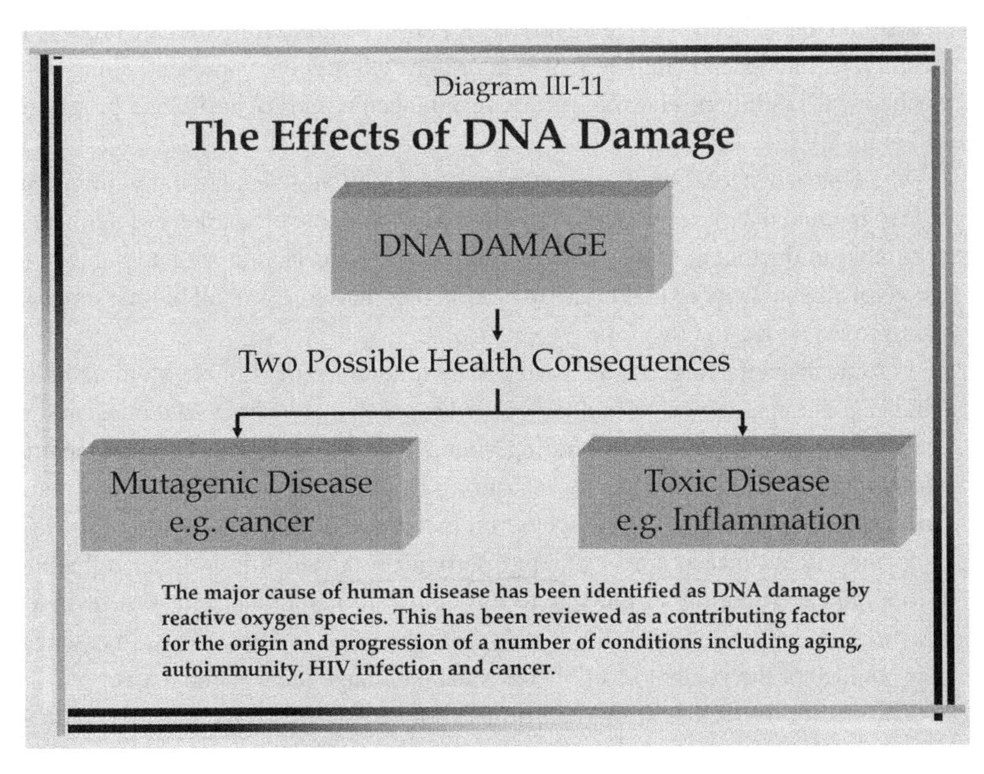
Diagram III-11
The Effects of DNA Damage
DNA DAMAGE
Two Possible Health Consequences
Mutagenic Disease
e.g. cancer
Toxic Disease
e.g. Inflammation
The major cause of human disease has been identified as DNA damage by reactive oxygen species. This has been reviewed as a contributing factor for the origin and progression of a number of conditions including aging, autoimmunity, HIV infection and cancer.

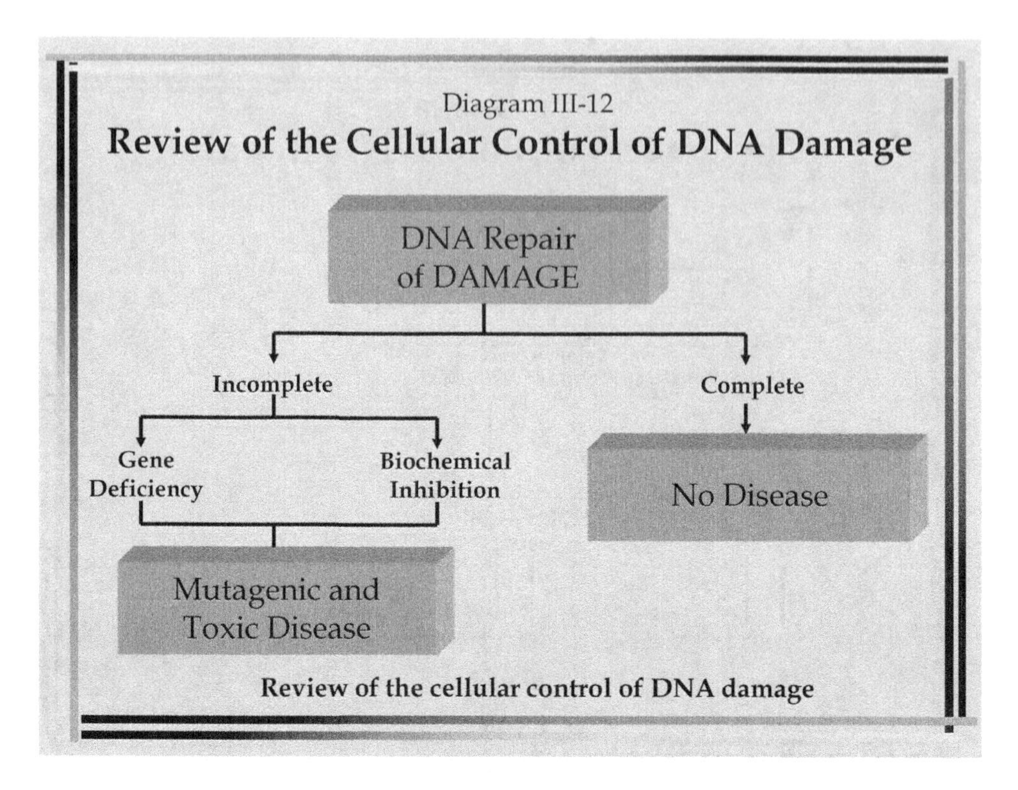

Review of the cellular control of DNA damage

disease manifests itself for a relatively long period of time. If this repair process is incomplete, then genetic deficiency, or poor gene expression, is present, along with biochemical inhibition, and this results in mutagenic and toxic disease processes (Diagram III-12).

The effects of a toxic environment can also be seen. It is, again, vitally important to keep in mind that the main goal is to improve modifiable biomarkers of aging and to be able to document these changes with appropriate laboratory and subjective data. Not all genetic potential is easily modifiable; therefore, not all biomarkers can be improved, at least at this time.

For the cosmetic surgeon, the easiest way to monitor modifiable biomarkers of aging is to use an office-based biomarker test kit which measures DNA damage rates. These rates are obtained by measuring changes in 8-hydroxy-2'-deoxyguanosine levels, which is the standard test for measuring DNA damage inside the cell nucleus. The second most effective and targeted biomarker, also available in the office test kit format, is the measurement of lipid peroxidation, which is a direct reflection of free radical oxidation, or damage to the lipid cell membrane. This test involves measuring 8-epi prostaglandin $F_{2\alpha}$ levels from the urine or blood. This, in essence, gives an idea of the general level of free radical damage present and active.

The ability to document marked drops in both DNA damage and free radical levels with these two tests forms the hallmark of a core anti-aging program for the cosmetic surgeon. The significance of these two biomarkers has been discussed in Chapter II on theories of aging. The ease of sampling through blood or urine tests, as well as the ease of monitoring the resulting two values, has been positively correlated with marked improvements in quality of health and sense of well-being.[11]

Other key modifiable biomarkers that can be measured are changes in degrees of the four key concepts mentioned previously in the aging equation: rates of glycation, methylation, oxidative stress and inflammation. Improvements in these laboratory data are also directly related to overall improvements in gene expression occurring at the cellular level and directly affect cell signaling.

In summary, for the cosmetic surgeon, an introductory anti-aging program, or age-management evaluation, needs to consist of nothing more than measurement of and a therapeutic improvement in the following biomarkers:

1. DNA damage.
2. Free radical levels.
3. Physiological changes, as documented by a symptom-related questionnaire (see Resource section).

This simple approach can have a tremendous impact on aging efficiency (Diagram III-13). With specialized physicians actively involved in a more comprehensive anti-aging program and evaluation, a much more comprehensive and

Diagram III-13

Age Management Therapy Goals

1)Control Environmental Effects -Environment defined a = diet, exercise, mind state, toxic elements, pollution, local radiation.

2) Improve the Function of the Aging Equation -Control Glycation, Inflammation, Oxidation/Free Radical Levels and Methylation at the cellular level.

3) Improve DNA Replication and Gene Expression -Improve the ratio and DNA Repair over DNA damage therefore resulting in less cell mutations and more accurate cell copies during cell replication. This preserves adult stem pools.

exhaustive list of subjective, objective and modifiable biomarkers should be compiled in an overall evaluation. This book does not present details of a comprehensive evaluation, because that requires a degree of knowledge and time commitment that most busy cosmetic surgeons and physicians do not have. Instead, this book presents the format for an essential core biomarker evaluation that can be used in a cost-efficient manner and can give an age-management program as much credibility and effective impact as possible for the busy cosmetic practice.

Along with the use of these key biomarkers, an overall age-management program should include essential key anti-aging supplements[12–30] to accomplish the key goals previously discussed. With the right balance of nutraceuticals, it is possible to markedly improve key modifiable biomarkers, as follows:

1. Decreasing DNA damage rates.
2. Increasing DNA repair rates.
3. Improving immune function.
4. Regulating the key concepts of the aging equation: the processes of glycation, methylation, oxidation and inflammation.

This approach also helps balance the biorhythmic cycle and pattern of the autonomic nervous system and aids in regulating hormonal release patterns.

Further improvement in the assimilation and use of basic food nutrients ingested with each meal can be accomplished as well by aiding the digestion with the right digestive enzymes and supplemental intestinal flora support.

Improvement in the pH levels of both the extracellular matrices ("cellular soup" — that is, the fluid around the cells) and the intracellular matrix (the fluid within the cells) markedly enhances the biochemical efficiency of the cell machinery and improves the aging process at this microscopic level.

Another main concept to keep in mind is the delivery of these supplements throughout the day: that is, a phased delivery system. The use of as many naturally occurring plant and enzymatic complexes as possible, rather than synthetic supplements, is also important. It will be made clearer later that the natural components in products are much more effective than synthetically manufactured or designed supplements.

In summary, the overall effects of this nutraceutical approach are as follows:

1. Improved gene expression[31] (Diagrams III-14 and III-15).
2. Improved quality of life.
3. Improved modifiable biomarkers of aging.
4. A noticeable effect and improvement in the overall physical signs of aging that plastic surgeons recognize.

Diagram III-14

Neutraceuticals → **Influence Gene Expression** →

(i.e. Vitamins, minerals, phytochemicals,enzyme complexes, amino acids, etc.)

Down regulating certain genes and upregulating other genes

Alter Cell Signaling or Signal Transduction

(Cytokines, eicosanoides other growth factors etc.)

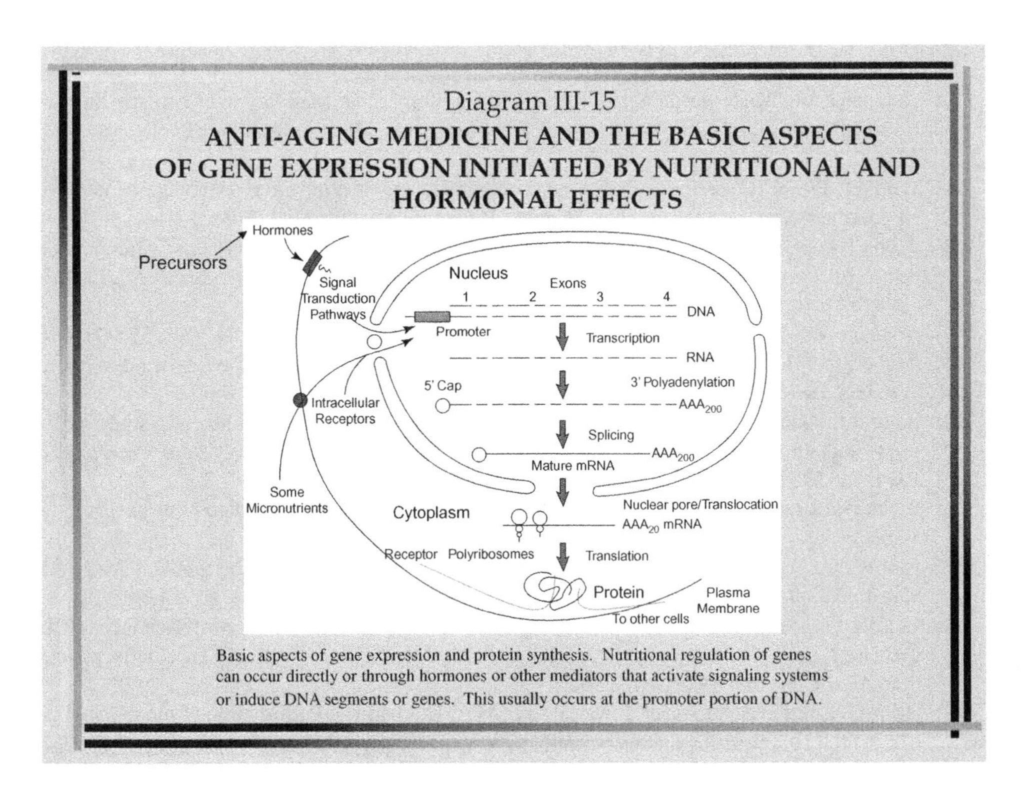

Basic aspects of gene expression and protein synthesis. Nutritional regulation of genes can occur directly or through hormones or other mediators that activate signaling systems or induce DNA segments or genes. This usually occurs at the promoter portion of DNA.

Since the early 1990s, this strategy has been shown to be very successful in all patients who have been treated with this approach.

It is, without question, time to realize that this information and these simple laboratory tests are at physicians' disposal to improve the well-being of patients' quality of life, as well as their overall physical appearances.

References

[1] Morley JE, Kaiser F, Raum WJ, Perry HM 3rd, Flood JF, Jensen J, Silver AJ, Roberts E. Potentially predictive and manipulable blood serum correlates of aging in the healthy human male. *Proc Natl Acad Sci USA.* 1997; 94: 7537–7542.

[2] Dilman V, Dean W. *Neuroendocrine Theory of Aging.* Pensacola, FL: Center for Bio-Gerontology; 1992.

[3] *The Duke Longitudinal Studies of Normal Aging 1955–1980: An Overview of History, Design, and Findings.* New York: Springer Publishing Co.; 1985.

[4] Evans W, Rosenberg IH. *Biomarkers.* New York: Simon & Schuster; 1991.

[5] Hayflick L. *How and Why We Age.* New York: Ballantine Books; 1994.

[6] *Older and Wiser: The Baltimore Longitudinal Study of Aging.* [NIH Publication No. 89-2797]. Washington, DC: U.S. Government Printing Office; 1989.

[7] Timiras PS, ed. *Physiological Basis of Aging and Geriatrics.* 2nd ed. Boca Raton, FL: CRC Press; 1994.

[8] Timiras PS, Quay WB, Vernakdakis A, eds. *Hormones and Aging.* Boca Raton, FL: CRC Press; 1995.

[9] Dean W. *The Biological Aging Measurement—Clinical Applications.* Pensacola, FL: Center for Bio-Gerontology; 1988.

[10] Giampapa VC, Klatz R., Goldman R. Anti-Aging Surgery: A Step Beyond Cosmetic Surgery. In: *Advances in Anti-Aging Medicine.* Vol. 1. New York: Mary Ann Liebert Publishers; 1996: 57–60.

[11] Wild CP, Pisani P. Carcinogen DNA and protein adducts as biomarkers of human exposure in environmental cancer epidemiology. *Cancer Detect Prev.* 1998; 22: 273–283.

[12] Alpha Tocopherol, Beta Carotene, Cancer Prevention Study Group. The effect of vitamin E and beta carotene on incidences of lung cancer and other cancers in male smokers. *N Engl J Med.* 1994; 330: 1029–1035.

[13] Arsenian MA. Magnesium and cardiovascular disease. *Prog Cardiovasc Dis.* 1993; 35: 271–310.

[14] Ascherio A, Hennekens CH, Willett WC. Trans-fatty acid intake and risk of myocardial infarction. *Circulation.* 1994; 89: 94–101.

[15] Baggio E, Gandini R, Plancher AC, Passeri M, Camosino G. Italian multicenter study on the safety and efficacy of coenzyme Q10 as adjunctive therapy in heart failure. *Molec Aspects Med.* 1994; 15: S287–S294.

[16] Block G, Patterson B, Safar A. Fruit, vegetables and cancer prevention. *Nutr Cancer.* 1992; 18: 1–29.

[17] Blot WJ, Li JY, Taylor PR, Gauo W, Damsey SM, Wang GQ, Yang CS, Zheng F, Gail M, Li GY. Nutritional intervention trials in Linxion, China. *J Natl Cancer Res.* 1993; 85: 1483–1492.

[18] Colditz GA, Branch LG, Lipnic RJ, Willett WC, Rosener B, Posner BM, Hennekens CH. Increased green and leafy vegetable intake and lowered cancer deaths in an elderly population. *Am J Clin Nutr.* 1985; 41: 32–36.

[19] Hill EG, Johnson SB, Lawson LD, Mahfouz MM, Holman RT. Perturbation of the metabolism of essential fatty acids by dietary partially hydrogenated vegetable oil. *Proc Natl Acad Sci USA.* 1982; 79: 953–957.

[20] Lindheim SR, Presser SC, Ditkoff EC, Vijod MA, Stranczyk FZ, Lobo RA. A possible bimodal effect of estrogen on insulin sensitivity in postmenopausal women and the attenuation effect of added progestin. *Fertil Steril.* 1993; 60: 664–667.
[21] Maurer K, Ihl R, Dierks T, Frolich L. Clinical efficacy of gingko biloba special extract EGb 761 in dementia of Alzheimer type. *J Psychiatr Res.* 1997; 31: 645–655.
[22] Mohr D, Bowry VW, Stocker R. Dietary supplementation with coenzyme Q10 results in increased levels of ubiquinol-10 within circulating lipoproteins and increased resistance of human low-density lipoproteins to the initiation of lipid peroxidation. *Biochim Biophys Acta.* 1992; 1126: 247–254.
[23] Murray MT. *Encyclopedia of Nutritional Supplements.* Rocklin, CA: Prima Publishing; 1996.
[24] Polyp Prevention Group. A clinical trial of antioxidant vitamins to prevent colorectal adenoma. *N Engl J Med.* 1994; 331: 141–147.
[25] Rimm EB, Stampfer MJ, Acherio A, Giovannucci E, Colditz GA, Willett WC. Vitamin E consumption and risk of coronary heart disease in men. *N Engl J Med.* 1993; 328: 1450–1456.
[26] Roberts HJ. *Aspartame: Is It Safe?* Philadelphia: Charles Press; 1990.
[27] Stampfer MJ, Hennekens CH, Mason JE, Colditz GA, Rosner B, Willett WC. Vitamin E consumption and risk of coronary disease in women. *N Engl J Med.* 1993; 32: 1444–1449.
[28] Shekelle RB, Lepper M, Liu S. Dietary vitamin A and risk of cancer in the Western Electric Study. *Lancet.* 1981; 2: 1185–1190.
[29] Steinmetz KA, Potter JC. Vegetables, fruit and cancer. I. Epidemiology. *Cancer Causes Control.* 1991; 2(5): 325–357.
[30] Willett WC, Stampfer MJ, Manson JE, Colditz GA, Speizer FE, Rosner BA, Sampson LA, Hennekens CH. Intake of trans fatty acids and risk of coronary heart disease among women. *Lancet.* 1993; 341: 581–585.
[31] Shils ME, Olson JA. *Modern Nutrition in Health and Diseases.* 9th ed. Baltimore: Lippincott Williams & Wilkins; 1999: 573–603.

4

DNA: The Core of Our Aging Blueprint

Vincent C. Giampapa, M.D., F.A.C.S.
Ronald Pero, Ph.D.

When you teach your son, you teach your son's son.

Within each of the 100 trillion cells that make up the human body is a specific area termed the **nucleus**, or **control center**, of the cell. Within each nucleus, there are so many base pairs of the DNA that if they were laid end to end, there would be trillions of miles of DNA within each person (Diagram IV-1).

Primer of Basic Genetics for the Aesthetic Surgeon and Physician

Within the nucleus, there exists one set of genes passed on from the mother and one set from the father. Each complete pair includes approximately 30,000 to 80,000 genes. People inherit one copy of each set of these genes from each parent. The process of combining these sets from both the mother and father is referred to as **recombination**.

The **human genome** has been compared by a number of authors to a book. If the human genome is described in this manner, the description may be as follows:

1. The human genome is composed of 23 chapters called **chromosomes**.
2. Each chapter contains several thousand stories. These stories are called **genes**.
3. Each story is made up of paragraphs called **exons**, which are interrupted by advertisements, or **introns**.
4. Each paragraph also contains other words, called **codons**.

The Principles and Practice of Antiaging Medicine for the Clinical Physician, 49–61.

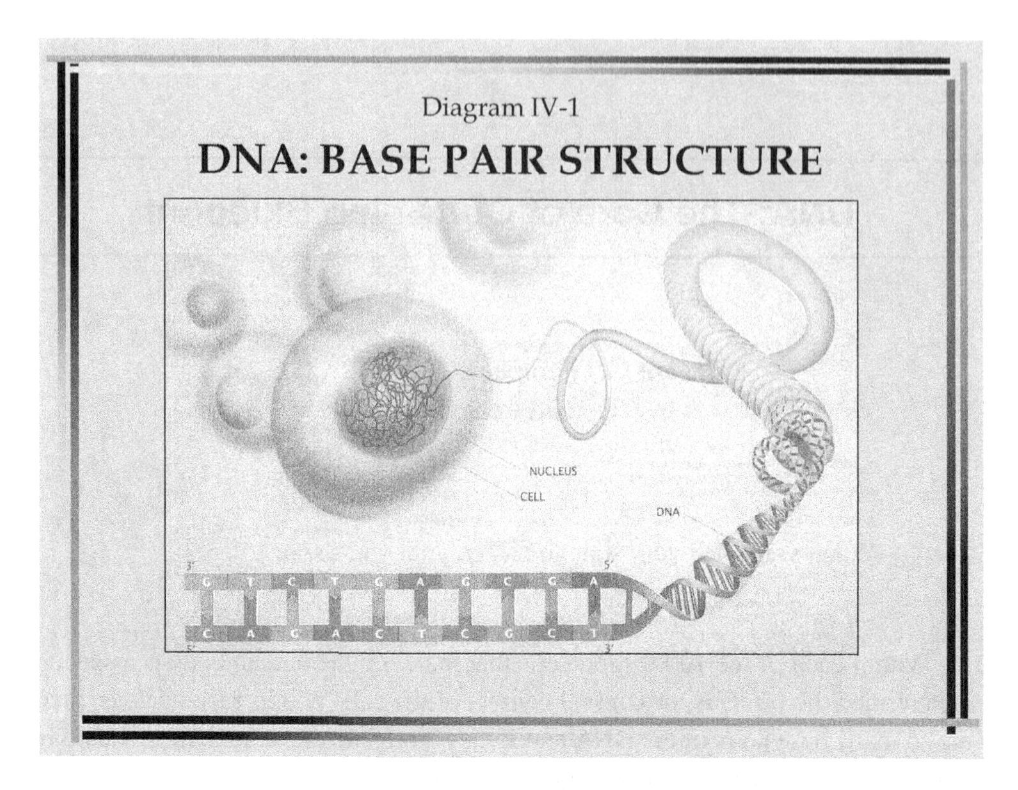

5. Each word is written in letters, called **bases**. There are four bases, which consist of these words: **adenine**, **cytosine**, **guanine** and **thymine**,
6. Each chromosome can be considered a pair of extremely long DNA molecules.
7. The human genome, under the correct conditions, can both read and "photocopy" itself. The photocopying process is known as **replication**, and the reading process is known as **translation**.
8. The base pairs bond to each other in a complementary manner[1]: Adenine (A) always binds with thymine (T), and guanine (G) always binds with cytosine (C). In order for a single strand of DNA to copy itself, it constructs a complementary strand with all the Ts opposite all the As and all the Gs opposite all the Cs. In this manner, replication is accomplished.
9. **Transcription**, on the other hand, is a process by which the information of a gene is transcribed into a copy by the same **base pairing** combination. The process of transcription makes **RNA** and not another molecule of DNA. This molecule is very similar in chemical structure to DNA; the only difference is that RNA contains **uracil** (U) in place of thymine.

The RNA copy form is called **messenger RNA**. At this point, removal of all the introns, and the splicing together of all the exons, modifies it. Messenger RNA is then transported out of the nucleus and into the "cellular soup," or extracellular fluid. A molecular machine referred to as a **ribosome**, which is also made partly of RNA, moves along the messenger RNA, translating each three-letter **codon**, or sequence of base pairs. Messenger RNA functions with a different alphabet made of 23 amino acids. At this stage, messenger RNA interacts with **transfer RNA**. As each amino acid is produced by the messenger RNA and transfer RNA complex, it is attached to the other newly produced amino acids in order to form a chain in the same order as the codons were positioned on the original messenger RNA molecule. When the whole message from the messenger RNA has been translated, this long chain of amino acids folds three-dimensionally upon itself into a characteristic shape.

These folded amino acid sequences are now known as **proteins**. It is essential to understand the importance of proteins. The complete body is primarily made up of proteins or its processes which are governed by proteins. Also, in essence, every protein is a **translated gene**.

Proteins called **enzymes** catalyze the body's chemical reactions. Proteins are also responsible for switching genes on and off — stimulating or inhibiting their action — by physically attaching themselves to the promoters' or enhancers' regions near the start of the genes themselves. Different genes are switched on and off during different periods of a person's life and in different parts of the body.

When different parts of the gene are copied mistakenly or a certain sequence of these genetic letters is left out, the result is known as a **mutation**.

It is important to keep in mind that aside from the nucleus, which contains these 23 pairs of chromosomes from the mother and father, other genes are also present outside the nucleus within the cellular organelle called a **mitochondrion**. The mitochondrial genes are inherited only from the mother. Therefore, humans are slightly more like their mothers than like their fathers with regard to genetic makeup.

Not all genes appear to be made of DNA. Some viruses use RNA instead. Certain genes do not produce protein. Some genes are transcribed into RNA and not into protein. This type of RNA may form part of the ribosome or part of transfer RNA. Some reactions within the cell are catalyzed by RNA instead of by proteins. Some of the three-letter codons specify start or stop commands as well. This information is useful in the continuing discussion concerning **DNA, the core of our aging blueprint**. A detailed account of this information is available in *Molecular Cell Biology*, by Harvey Lodish.[2] First, there are a few important points to be emphasized.

Genes act as repositories of coded information for the synthesis of proteins, enzymes and, eventually, hormones. Some of these key proteins actually regulate the repair of DNA or determine which segments of genes are functioning, or active, at a specific time. This determination is referred to as **gene expression**.

Specific environmental agents, such as radiation, sun exposure and pollution, can modify how all of these genetic characteristics are actually expressed or activated.

Diagram IV-2

Human Genetic Structure

22% - devoted to RNA + protein synthesis
12% - to cellular division
12% - to cellular signaling and communication
12% - to immune defense
17% - to metabolism
8% - to structure
17% - unknown

Specific environmental agents can modify how all of these characteristics are expressed. Our goal is to manipulate gene expression related to healthy maintenance of structure and function throughout the aging process.

The fundamental goal of anti-aging therapy is to manipulate gene expression as it relates to the healthy maintenance of structure and function of an individual as he or she ages.

Laboratory analysis has documented the relative content of the human genetic structure (Diagram IV-2):

1. 22% of DNA is devoted to RNA and protein synthesis.
2. 12% is devoted to cellular division.
3. 12% controls cell signaling and communication.
4. 12% is directly related to immune function.
5. 17% affects cell metabolism.
6. 8% is responsible for cellular cytostructure.
7. 17% has no known function. This portion of DNA has been termed **junk DNA** by research scientists at present. It is not actually "junk"; its purpose is simply not yet understood. Theories of junk DNA have evolved and included the concept that this extra DNA may be simply leftover DNA from humans' evolutionary past, or perhaps it may be redundant repeat units, which function as extra codes[3–5a] (Diagrams IV-3 and IV-4). Other theorists state that this junk DNA may actually be affected by the earth's electromagnetic fields, human emotional states, or both.

Diagram IV-3

Junk DNA – Old Theories

The Past

- DNA not used at present due to previously needed proteins not required at present.
- Redundant repeat units, which function at extra codes.

Diagram IV-4

Junk DNA – New Theories

DNA is turned on by:

- Emotional states *the E.M.Field and Resonance Fields
- Key trace "elements" (platinum and gold groups) which we ingest from natural foods and manufactured sources.
- The message from DNA to RNA can only be transmitted with the appropriate solution of Bio water solvent.

E – Earth's field M – Human's Field

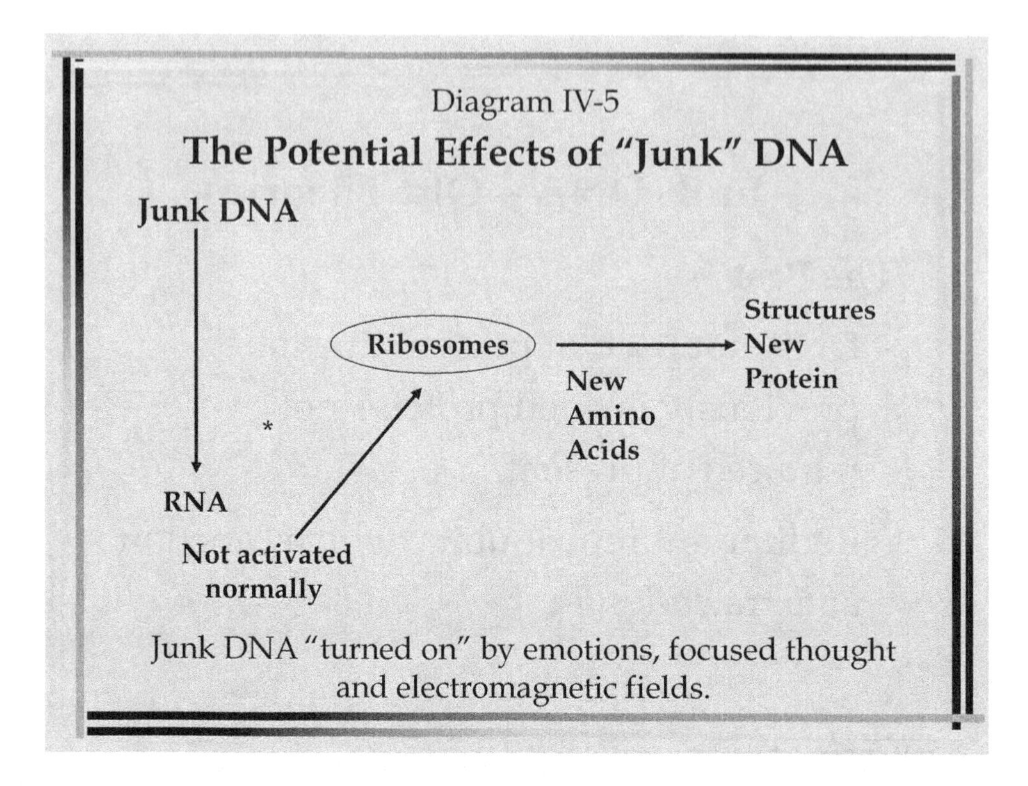

There is also mounting scientific evidence that this unknown "nonfunctional" segment of DNA may actually respond to "focused intentional thought" which is generated from the human electromagnetic field itself and may be the basis of the mind-body healing interaction seen in the form of "miraculous cures" or "spontaneous remissions" (Diagram IV-5; see Diagram IV-4.) At the moment, all of this is conjecture, but each of these potential theories is backed by scientific evidence.

Another key concept to keep in mind is that there are two basic types of DNA: nuclear DNA and mitochondrial DNA. Gerontologists such as Doctors Lee, Weindruch and Aiken[6] believe that "one of the *central features of biological aging* is the *alteration of mitochondrial function* that occurs as a consequence of *'free radical'* damage" (Diagram IV-6).

One of the key features of DNA is the ability of nuclear DNA to repair itself as it suffers damage from the environment and free radicals.

Nuclear DNA is also different from mitochondrial DNA in that it has a protective coat, or layer of proteins called **histones**, that absorb much of the free radical damage, which protects its essential genetic structure and codes. In fact, at the level of the nucleus, there are a number of DNA repair mechanisms that have evolutionarily

Diagram IV-6

Quenching Reactive Oxygen Species in the Mitochondrion

- About 10% of the air we breathe is converted into free radicals in the mitochondria of our cells.
- Hydrogen peroxide, hydroxyl radical, superoxide, lipid peroxide and singlet oxygen.
- Mitochondria
  Highest free radical production/levels of antioxidants
- Vitamin E, glutathione, co-enzyme Q10, lipoic acid, carnitine and taurine

evolved to prevent severe permanent genetic damage.[7] The amount of DNA repair activity correlates directly with life span across different species.[8a–8c]

Mitochondrial DNA is 2,000 times more susceptible to oxidative damage from free radicals than is nuclear DNA. It contains no (known) DNA repair systems and does not replicate itself. It also has no protective histone coat. Mitochondrial DNA is also unique in that it is a ring-shaped structure rather than a double helix, as is present in the nuclear DNA compartment.

Mitochondrial DNA is also much more susceptible to damage than nuclear DNA because mitochondria are at the site where most free radicals are formed. This occurs during the process of energy production for the cell. This energy process, which is the formation of adenosine triphosphate (ATP), is essential for all cellular functions, as well as cell replication (Diagram IV-7). Without ATP, cell repair slows down or stops (Diagram IV-8).

In the more recent analogies, mitochondria are compared to "semiconductors," or "chemical transducers," that convert the potential chemical energy in food to potential metabolic energy. Mitochondria accomplish this by stripping off the electrons found in the molecules of food and causing them to move through a complex compartment of cellular membranes, as well as through the genes themselves.

In essence, mitochondria may be viewed as **quantum energy devices** that remove energy from matter — that is, the food we eat — and transfer this energy

Diagram IV-7

Mitochondria

- A single ATP molecule must be recycled within a mitochondrion approximately 1,000 times per day for the body to maintain its energy supply.

Diagram IV-8

Mitochondria and Aging
The Link – ATP Production

Mitochondria are semiconductors or "*chemical transducers*" converting to potential chemical NRG in food to potential metabolic NRG by stripping off electrons and causing them to move through cellular membranes and the genes themselves. As the integrity of this process is lost and electrons can no longer move down the "wire" of the ETC of the mitochondria efficiently, the electrons jump off the wire causing damage to the mitochondria themselves, to other cellular structure and to both nuclear and mitochondrial DNA.

to different components of the cells, including the nucleus, in order to create new matter such as proteins and enzymes. The proteins then allow cells and organs to grow, reproduce and maintain health and youthful function.

Mechanisms of DNA Repair and DNA Damage: The Fundamental Concept of the New Aging Paradigm

According to this new aging paradigm, humans are not programmed to age and die; instead, they are programmed, genetically, for cell repair and longevity. In other words, humans are programmed to live longer than they actually do, if they can use their full potential. That potential is related directly to the ratio of DNA repair to DNA damage.[8a–8c,9] Within each of the 46 chromosomes that make up the human body, there is a series of proteins that the DNA is wound about. These proteins, called *histones*, help stabilize the three-dimensional structure of the double helix but also, more important, protect the genetic content of each chromosome from free radical damage and other sources of damage as well. The initial process of damage to DNA begins when there is a break in the DNA strand[10,11] (Diagram IV-9). This event occurs most commonly because free radicals penetrate the protective his-tone coating and cause a break in the double helix structure. Damage to the underlying DNA strand causes the release of adenosine diphosphate ribosyl transferase (ADPRT).

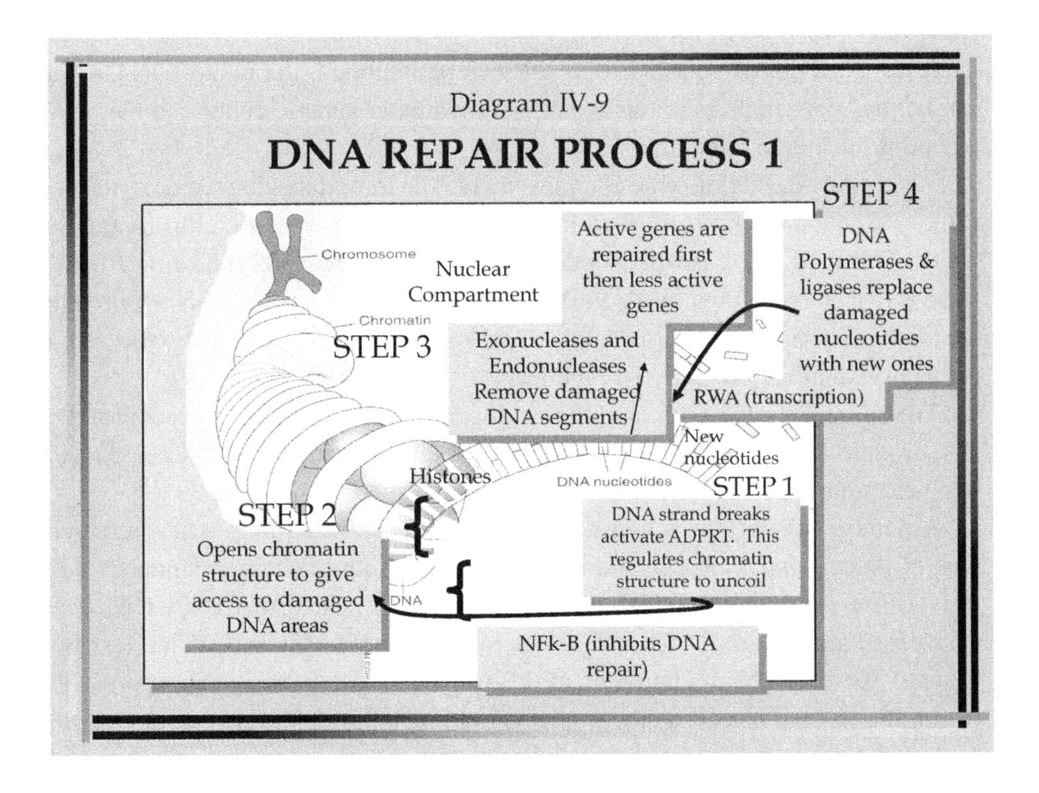

The break in the DNA strand caused by the free radicals causes this compound to be released suddenly. ADPRT then reacts with DNA and opens the protective coating of chromatin surrounding the chromosome. The chromatin layer resembles a Slinky (toy). When the coils open up, the genes are exposed to the fluid within the cell and, more specifically, within the nucleus. When ADPRT opens up this protective coating, a number of other events occur: the damaged segments of DNA are attacked by endonucleases and exonucleases that can enter into the damaged segment of the chromosome. The exonucleases cut out other damaged base pairs of the DNA. At this stage of DNA repair, DNA polymerase and DNA ligase replace the damaged base pairs with new ones from the cell fluid-like environment. Therefore, ADPRT is essential in order to open up the chromatin to allow the damaged base pairs of DNA to be exposed and then be fixed by the whole sequence of DNA stitching and repairing enzymes mentioned earlier, the endonucleases.

Also residing in the intracellular fluid is a very important compound called nuclear transcription factor kappa B. NF-κB which is activated by a series of events already described: glycation, inflammation, methylation and oxidative stress. NF-κB inhibits ADPRT from opening up the chromatin so that DNA repair can commence and be completed (see Diagram IV-9). Hence, in this specific situation it is important to understand that NF-κB controls DNA repair. It also results in a series of other molecular responses that interfere with DNA repair. The consequences of poor DNA repair and excessive DNA damage are summarized throughout the rest of this chapter (Diagrams IV-10, IV-11 and IV-12).

It has been discovered that if NF-κB can be inhibited, the rate of DNA repair can be markedly improved; this would help maintain genetic codes and the aging blueprint in optimal condition.[12–16]

Within the ADPRT enzyme complex, there is an important zinc region. Attached to the zinc region is a thiol group, which is a chemical structure with sulfur containing bonds. This thiol group is what bonds to the damaged DNA sites, allowing ADPRT to be activated (see Diagram IV-9). When a high level of free radicals are present, the thiol group cannot bond to damaged DNA segments and it becomes oxidized. It is therefore inhibited directly by high levels of free radicals.

The consequences of this are far-reaching. In essence, if the body is depleted of zinc or antioxidants and there are many free radicals, this key complex is directly inhibited from binding to DNA to initiate the whole repair sequence.

A number of substances will stimulate DNA repair according to new research. One of them is *niacinamide*, and another is *zinc*. Both of these compounds help the ADPRT complex link up to the damaged DNA sites so that DNA repair can commence and be completed at a faster rate and more efficiently. The next two chapters show that the inhibition of NF-kB is one of the key therapeutic approaches within the science of anti-aging medicine; this includes aging of both the interior body and the external skin.

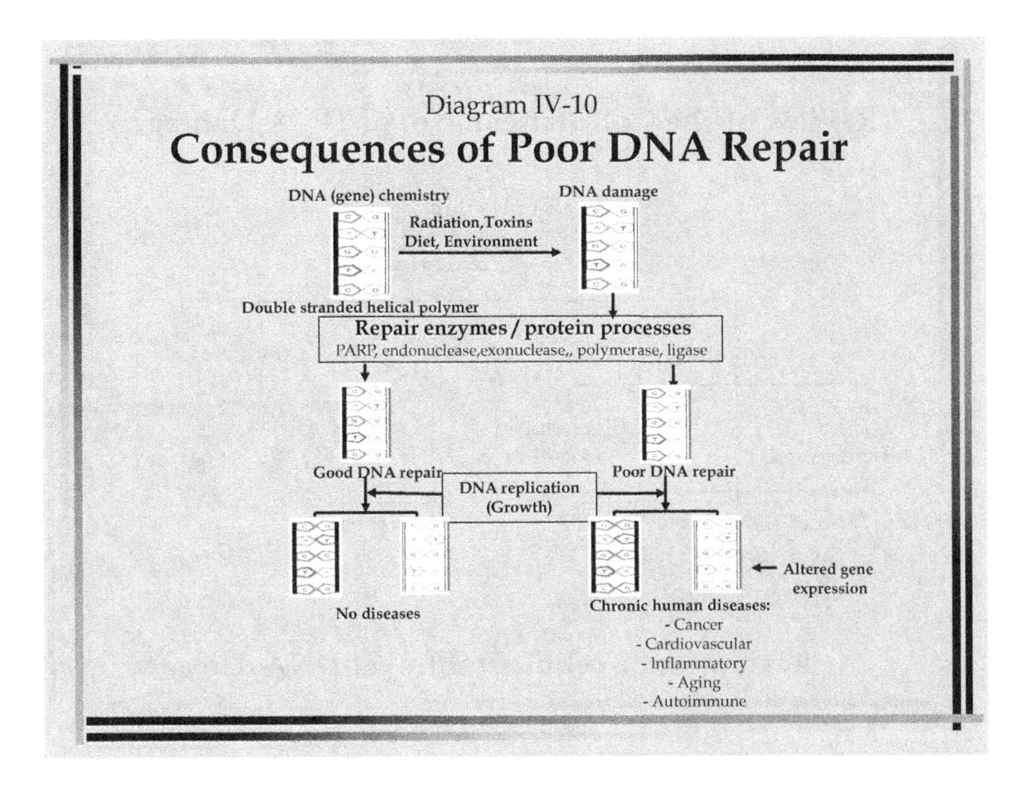
Diagram IV-10
Consequences of Poor DNA Repair
DNA (gene) chemistry
DNA damage
Radiation,Toxins
Diet, Environment
Double stranded helical polymer
Repair enzymes / protein processes
PARP, endonuclease,exonuclease,, polymerase, ligase
Good DNA repair
Poor DNA repair
DNA replication
(Growth)
Altered gene expression
No diseases
Chronic human diseases:
- Cancer
- Cardiovascular
- Inflammatory
- Aging
- Autoimmune

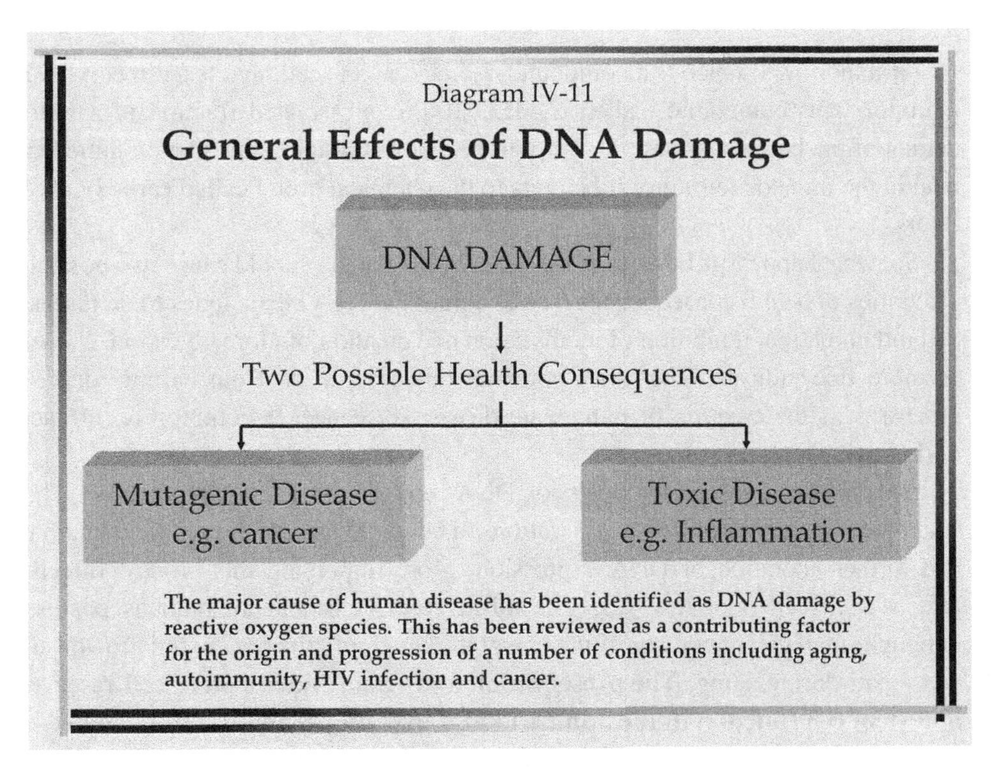
Diagram IV-11
General Effects of DNA Damage
DNA DAMAGE
Two Possible Health Consequences
Mutagenic Disease
e.g. cancer
Toxic Disease
e.g. Inflammation
The major cause of human disease has been identified as DNA damage by reactive oxygen species. This has been reviewed as a contributing factor for the origin and progression of a number of conditions including aging, autoimmunity, HIV infection and cancer.

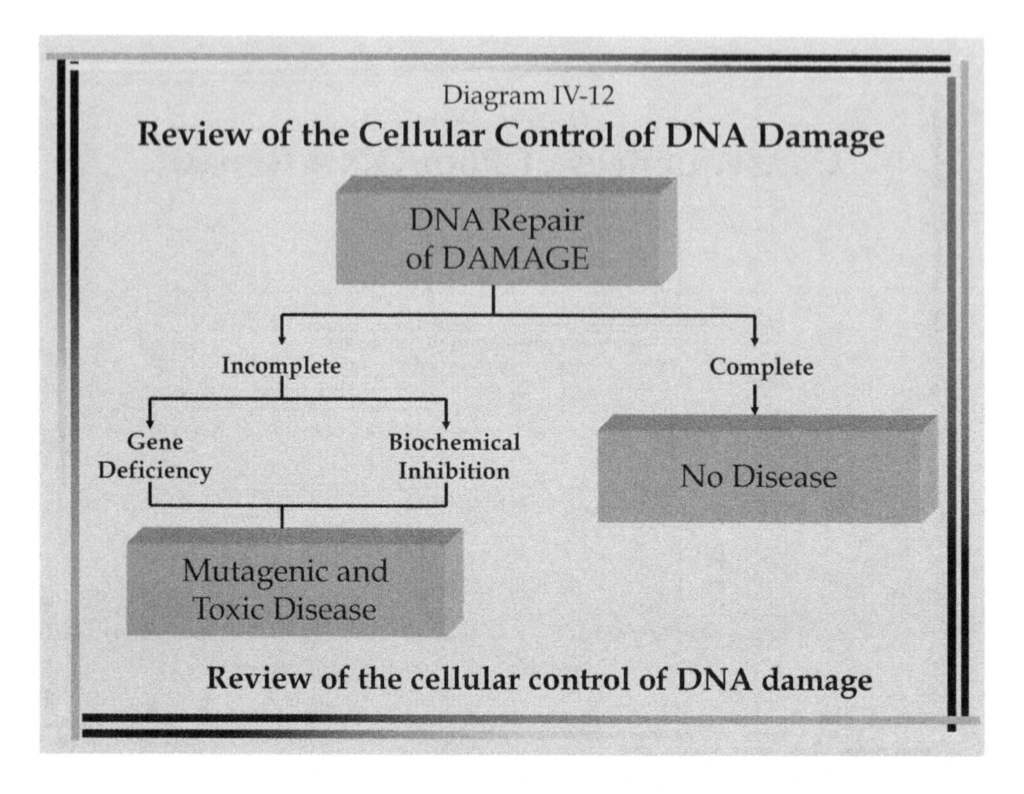

Review of the cellular control of DNA damage

Of major importance is a compound, isolated and tested, that is a direct NF-κB inhibitor. This compound, called C-MED-100, is an isolated fraction of a water extract of the herb cat's-claw.[14–16] It will be a key anti-aging therapeutic nutraceutical in the immediate future. It belongs to the chemical group called carboxy alkyl estors.

So, what happens if DNA repair is poor? DNA damage could cause two possible categories of health consequences (see Diagram IV-11). One includes toxic disease and inflammation, inhibition of methylation and creation of more glycation, as well as more free radicals. The other potential outcome, aside from chronic disease processes, is the creation of mutagenic disease, or cancer, a common occurrence during aging.

Another key consequence of poor DNA repair is poor DNA replication. This results in poor copies of future generations of cells and can result in more mutations and further alteration of DNA expression. More important, this process directly affects stem cell pool reserves. The **adult stem cells** that all humans possess, although small in number, constitute one of the key reservoirs of new cell growth and cell repair during aging. **The preservation and restoration of stem cell reserves and stem cell function in the adult will also take on paramount importance in the field of anti-aging medicine in the immediate future**.

The measure of DNA damage and free radical levels are the two most easily obtained biomarkers that are clinically useful. These measurements are easily obtained from a small urine or blood sample and supply the key information necessary to document the effectiveness of an age-management program. The next few chapters discuss how these measurements can be used. They should be considered the two core laboratory tests for aesthetic surgeons and physicians as they begin to incorporate anti-aging medicine into their practice.

References

[1] Mullis KB. The unusual origin of the polymerase chain reaction. *Sci Am.* [Revolution in Science issue]. 1990; 262(4): 56–61, 64–65.

[2] Lodish H. *Molecular Cell Biology.* New York: WH Freeman & Co; 2000.

[3] Gesteland RF, Atkins JF, eds. *The RNA World.* Cold Spring Harbor, NY: Cold Spring Harbor Laboratory Press; 1993.

[4] Woese C. The Universal Ancestor. *Proc Nat Acad Sci USA.* 1998; 95: 6854–6859.

[5] Poole AM, Jeffares DC, Penny D. The path from the RNA world. *J Mol Evol.* 1998; 46: 1–17.

[5a] Jeffares DC, Poole AM, Penny D. Relics from the RNA world. *J Mol Evol.* 1998; 46: 18–36.

[6] Lee CM, Weindruch R, Aiken JM. Age associated alterations of the mitochondrial genome. *Free Rad Biol Med* 1997; 22(7): 1259–1269.

[7] Chapter 12 in Lodish, H. *Molecular Cell Biology.* New York: WH Freeman and Co; 2000.

[8a] Grube K, Burkle A. Poly(ADP-ribose) polymerase activity in mononuclear leukocytes of 13 mammalian species correlates with species-specific life span. *Proc Natl Acad Sci USA.* 1992; 89: 11759–11763.

[8b] Pero RW, Holmgren K, Persson L. Gamma-radiation induced ADP-ribosyl transferase activity and mammalian longevity. *Mutation Res* 1985; 142: 69–73.

[8c] Pero RW, Hoppe C, Sheng Y. Serum thiols as a surrogate estimate of DNA repair to mammalian life span. *J Anti-Aging Medicine*; 2000: 3(3): 241–249.

[9] Mondello C, Petropoulis C, Monti D, Gonos ES, Franceschi C, Nuzzo F. Telo-mere length in fibroblasts and blood cells from healthy centenarians. *Exp Cell Res.* 1999; 248(1): 234–242.

[10] Pero RW, Roush GC, Markowitz MM, Miller DG. Oxidative stress, DNA repair, and cancer susceptibility. *Cancer Detect Prev.* 1990; 14: 555–561.

[11] Pero RW, Anderson MW, Doyle GA, Anna CH, Romagna F, Markowitz B, Bryngelsson C. Oxidative stress induces DNA damage and inhibits the repair of DNA lesions induced by *N*-acetoxy-2-acetylaminofluorene in human peripheral mononuclear leukocytes. *Cancer Res.* 1990; 50: 4619–4625.

[12] Sheng Y, Pero RW. DNA repair enhancement by a combined supplement of carotenoids, nicotinamide, and zinc. *Cancer Detect Prev.* 1998; 22(4): 284–292.

[13] Pero RW, Olsson A, Sheng Y, Hua J, Moller C, Kjellen E, Killander D, Marmor M. Progress in identifying clinical relevance of inhibition, stimulation and measurements of poly ADP-ribosylation. *Biochimie* 1995; 77: 385–393.

[14] Sheng Y, Bryngelsson C, Pero RW. Enhanced DNA repair, immune function and reduced toxicity of C-MED-100, a novel aqueous extract from *Uncaria tomentosa. J Ethnopharmacology.* 2000; 69: 115–126.

[15] Lamm S, Sheng Y, Pero RW. Persistent response to pneumococcal vaccine in individuals supplemented with a novel water soluble extract of *Uncaria tomentosa*, C-MED-100. *Phytomedicine.* 2001; 8(4): 267–274.

[16] Sheng Y, Li L, Holmgren K, Pero RW. DNA repair enhancement of aqueous extracts of *Uncaria tomentosa* in a human volunteer study. *Phytomedicine.* 2001; 8(4): 275–282.

5

Diet and Aging

Vincent C. Giampapa, M.D., F.A.C.S.

> Every man takes the limits of his own field of vision for the limits of the world.
>
> Schopenhauer

The Basic Facts

The primary purpose of altering our diet as we age is essentially an attempt to activate more of our genetic potential by *positively* influencing our gene expression with the food we eat (Diagram V-1). This means that unhealthy eating habits result in poor gene activation and "*turn on*" the genes responsible for *accelerated* aging and disease. At present, the only scientifically documented means of slowing aging is through dietary changes and more specifically caloric restriction. The essential component of this approach is decreasing glucose utilization at the cellular level.[1–12] The good news is that changing our diet can be as easy as upgrading our computer software.

The second important concept to keep in mind is that an *anti-aging diet* is significantly different than a *weight loss diet*. The key efforts of an anti-aging diet focus on not losing weight but, more importantly, on altering body composition to a more youthful level. This altered body composition focuses mainly on *decreasing* body fat and *increasing* muscle mass. When we accomplish this type of alteration in body composition, a number of key events take place:

1. There is an improvement in our ability to burn calories more efficiently because, with increased muscle mass and less body fat, caloric burn rates are much higher, as in younger people.

The Principles and Practice of Antiaging Medicine for the Clinical Physician, 63–92.

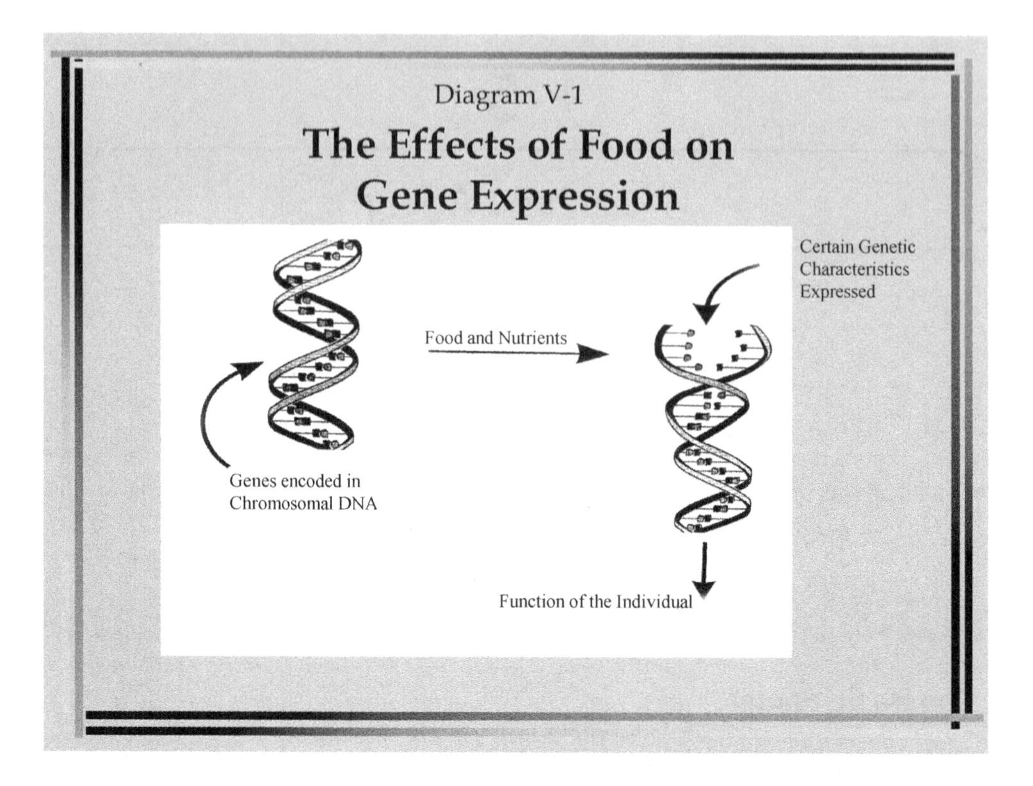

2. We attain a more optimal **body mass index** (BMI) and body composition, resulting in a decreased number of age-related diseases (Diagrams V-2a to V-2c).
3. We experience an improved level of functionality, or quality of life as we age.
4. We maintain the optimal production and levels of key anti-aging hormones, which have a direct impact on the rate and quality of aging. These key hormones, basically, are **growth hormone, insulin-like growth factor (IGF)–1, insulin, glucagon**, and **cortisol**.

Another essential goal of an anti-aging dietary program is to improve genetic expression by protecting deoxyribonucleic acid (DNA) from damage by excessive free radical (oxidative stress) production.

An often underemphasized and essential component of this special anti-aging diet is maintaining ideal pH levels at both the intracellular and the extracellular levels. The importance of this point cannot be emphasized enough!

Acidification of cellular fluids results in premature aging at the molecular, genetic and tissue levels. An increased incidence of viral, bacterial and even cancerous growths occurs when the **biological terrain** becomes overactivated. Computerized tools to measure the biological terrain have been available for years but are

Diagram V-2a

Body Mass Index (BMI):

The Way to Identify At-Risk Patients

- BMI can be used to determine healthy weights for patients. This mathematical formula, which correlates highly with body fat, is expressed as, BMI = kg/m^2.
- Cutoff points for desirable BMI ranges can be identified based on the relationship between BMI and health outcome.
- Since a relative weight of 120% of ideal weight roughly corresponds to a BMI of $27kg/m^2$, a BMI greater than or equal to $27kg/m^2$ is considered to be overweight.

Diagram V-2b

Body Mass Index (BMI):

The Way to Identify At-Risk Patients

BMI and weight-related health risks

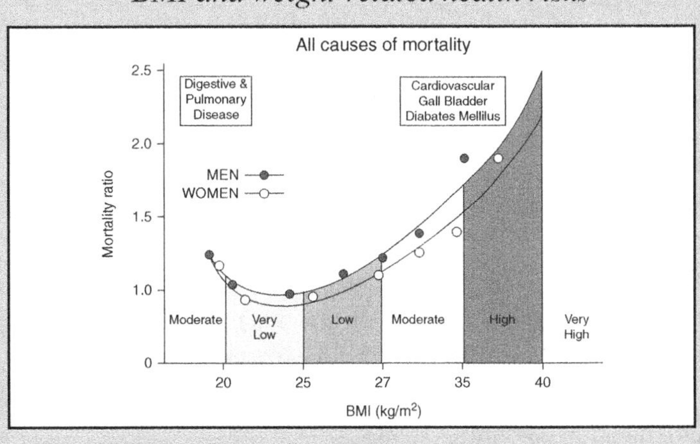

Diagram V-2c

Health Risks Associated with Obesity and B.M.I. greater than 27.

- Insulin resistance
- Non-insulin-dependent diabetes mellitus
- Hypertension
- Dyslipidemia
- Gallstones
- Cholecystitis
- Respiratory dysfunction
- Certain forms of cancer
- Cardiovascular disease

just now becoming accepted in mainstream medicine.[13] This is the most prevalent problem in America and in all industrialized countries in general.

The purpose of food, in general, is to *create* energy for the cellular processes that are required to maintain normal body function and maintenance as well as cell regeneration.

Utilizing these key concepts in an anti-aging diet supplies the information needed to fine-tune hormonal responses as well as our genetic potential. By taking a few essential steps, we can easily improve the efficiency of our aging metabolism as well as our quality of health.

For most people, decreasing body fat and maintaining a more youthful body composition with more muscle mass requires a few relatively simple steps:

1. Optimizing pH and anti-aging hormonal levels to increase our levels of energy.
2. Optimizing metabolism to stop storing fat and start burning it.
3. Helping the body get rid of toxins from a myriad of metabolic processes and environmental exposures that have accumulated over the years. This is accomplished by improving liver function (phase I and phase II liver detoxification enzyme systems).

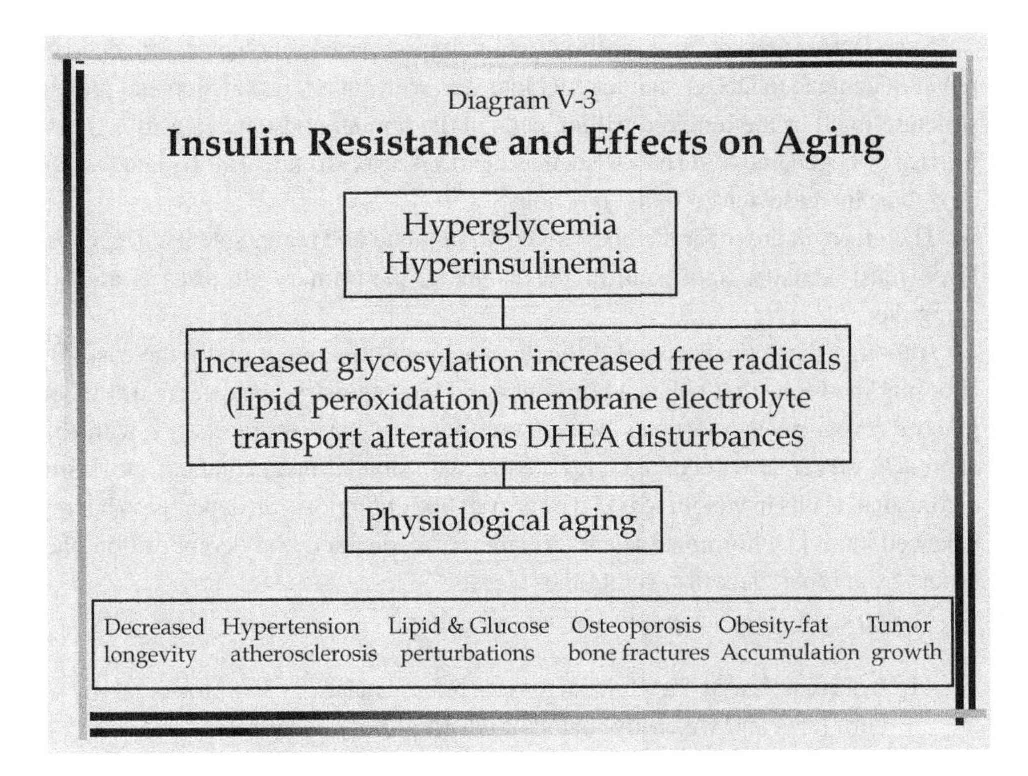

These three steps will upgrade the ability to utilize our *inherited* genetic potential more optimally. The final result is an improved sense of well-being and an improved quality of life.

The key metabolic goal with any anti-aging diet can be summarized in the following statements. Optimize insulin levels and glucose utilization (Diagram V-3). *Glucose* plays a *direct role* in controlling the action of many anti-aging hormones such as insulin, cortisol and dehydroepiandrosterone DHEA.[14] Further, many hormonal levels and the intensity of the signals that they send play an important role in maintaining a youthful body. These key hormones (insulin, glucagon, DHEA, growth hormone and IGF-1) do not function efficiently when too many glucose molecules are attached to them. This process is called **glycation** (see earlier). Glycation plays a role in determining the structure and function of both lipids and proteins that we need as well. Excess sugars can also crosslink molecules together. This process inactivates it. When protein or lipid function is compromised, DNA is directly affected. For genes to be copied or expressed accurately, fully functional membranes and proteins are required. When these membranes and proteins are partially inactivated by high sugar (glucose) levels, then the most fundamental processes of life cannot take place efficiently.

Even DNA repair requires optimally functioning protein molecules. Inadequate repair of damage to DNA can markedly interfere with energy production and protein structure in all of the hundred trillion cells in the human body. It can also *increase* the risk of degenerative diseases, such as heart disease, stroke, arthritis and cancer, as well as increase autoimmune reactions.

Therefore, in order for the body to repair, regulate and regenerate itself, glucose levels must be under tight control. This is one of the primary purposes of an anti-aging diet.[15]

Although these goals sound difficult to achieve, this is not really the case. By choosing the foods that balance hormones and biochemistry, we can control blood glucose levels much more easily. Not only can we have more energy with this approach, we can also become more satisfied with smaller meals. In fact, successful dieting that results in weight loss and improved body composition depends on having balanced these key hormonal levels. An anti-aging diet and body composition plan differs from other diets in several other ways.

An anti-aging diet accomplishes the following:

1. It decreases body fat while increasing muscle mass. This improves caloric burn rates and we easily maintain our ideal body weight.
2. It stresses changes in body composition over weight loss. The greater ratio of muscle to fat, the greater our metabolic rate.
3. It improves the functional capacity of key organ systems, such as the endocrine and digestive systems.
4. It optimizes pH in the biological fluids in the body in both the intra-cellular and extracellular fluid spaces (the "cellular soup") emphasizing alkalinity.
5. It supplies large quantities of plants and foods containing natural antioxidants and anti-cancer phytochemicals (plant chemicals). These compounds directly affect DNA function and have recently been shown to affect the promoter region of DNA itself.[16]
6. It balances insulin, glucagon and cortisol levels.
7. It increases growth hormone secretion.
8. It suppresses cortisol while maintaining DHEA.
9. It balances thyroid hormones, thus elevating triiodothyronine (T_3) levels.
10. It elevates "good" cholesterol (high-density lipoprotein, or HDL) and lowers "bad" cholesterol (low-density lipoprotein, or LDL).
11. It augments omega-3 fatty acid intake relative to omega-6 fatty acid intake and controls inflammation **(eicosanoid hormones)** at the subclinical and clinical levels.

How Diet Affects Key Anti-Aging Hormones

Insulin The essential hormone to remember is **insulin**. Elevated levels in an older person have been associated with accelerated aging and age-related diseases in many areas.[17–30] Insulin is known as the "fat storage hormone," and its control is the key to lowering blood sugar. Insulin's job is to convert glucose into **glycogen**. Glycogen is stored in the liver and muscles and is our chief source of cellular fuel. If the blood glucose levels rise too quickly, not all of the glucose can be stored as glycogen. In this case, glucose is converted to fat and stored in our adipose (fat) tissues. Unfortunately, fat is stored mainly in the waist, thighs, hips and buttocks. This is what we have come to recognize as **lipodystrophy** in our cosmetic surgery patients. This scenario results from overeating, excessive dietary carbohydrate, and a low protein intake. When blood glucose rises too quickly, *excessive* insulin production is the result.

The next event in the cascade is the production of the inflammatory hormones, or the "bad" eicosanoids. These eicosanoids create constriction in the blood vessels (**vasoconstriction**), making less oxygen available to the tissues to help burn fat. If the body *cannot* burn fat sufficiently, it must store it. This is the principal cause of weight gain (Diagram V-4).

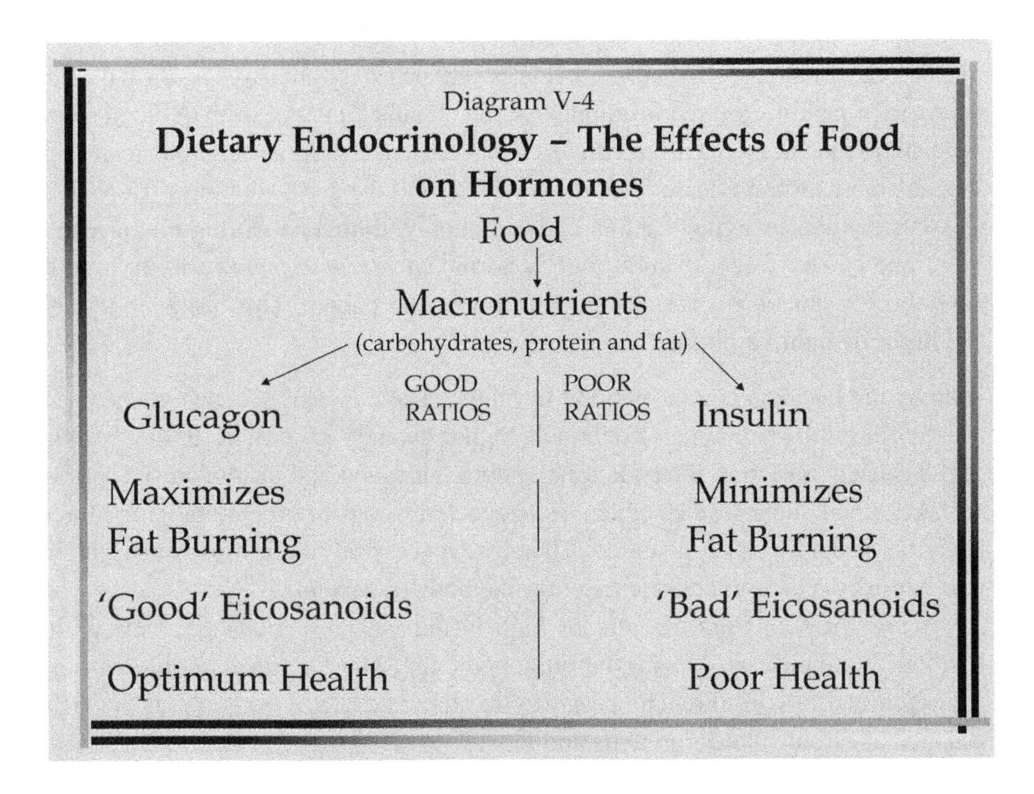

In summary, elevated insulin levels are caused by three main dietary factors:

1. Overeating.
2. Meals and snacks with excess carbohydrate as well as high-glycemic foods.
3. Meals and snacks with too little protein.

There are four main reasons why elevated insulin levels prevent us from burning body fat:

1. Stored carbohydrates are burned for energy instead of stored fat.
2. Excess blood glucose is converted into additional fat.
3. Insulin induces bad eicosanoids, which constrict blood vessels and make less oxygen available to burn fat efficiently.
4. Insulin elevation is a potent growth factor and can also accelerate telomere shortening, which is an essential factor in premature cellular aging.

In addition to preventing the burning of stored body fat, elevated insulin levels lead to low blood glucose levels **(hypoglycemia)**. Low blood glucose levels cause such symptoms such as loss of concentration, mood swings, fatigue, low energy, and uncontrollable hunger and sugar cravings.

The fat-storage, sugar-burning cycle is initiated over and over again with this scenario of rapidly elevated insulin levels. The brain counters a drop in blood sugar with signals to eat more carbohydrates. The transient elevation of blood sugar then triggers more insulin release from the pancreas, and the cycle continues. Unabated, this cycle produces hypoglycemia and, eventually, **diabetes**. During this progression, one's body reacts to food with a hormonal *alarm response* that is usually reserved for situations such as injury or impending danger. This has been termed the **flight or fight response**.

Glucagon Glucagon is a fat-burning hormone. The pancreas secretes glucagon in place of insulin when a meal contains a higher quantity of protein. In the "checks and balances" design of the endocrine system, glucagon and insulin work together to keep blood sugar levels stable. Glucagon is responsible for mobilizing stored body fat as the key energy source. Therefore, excess fat in our hips, waist, thighs and buttocks is used to create energy for the body in general.

A *high-carbohydrate* diet means high insulin and low glucagon. This is the "wrong" hormonal profile for burning body fat. The "correct" profile is high glucagon and low insulin. The best way to stimulate glucagon is to eat the right balance of carbohydrates, proteins and fats.

When we eat a protein-rich meal, blood glucose fluctuations are minimized and glucagon is secreted appropriately. Glucagon has been termed the "magic hormone" for burning fat and for improving body composition. It directs the body to use stored fat to fuel energy requirements. Using fat as an energy source is a hallmark of youth. The anti-aging diet and body composition plan described here restores these key tendencies.

Just like insulin, glucagon also stimulates eicosanoid production; however, these eicosanoids are of the "good" variety. These *good* hormones cause the diameter of the blood vessels to increase, or dilate **(vasodilation)**. This brings more oxygen into the cells, resulting in **aerobic metabolism**, the most *efficient* mechanism by which the body burns fat.

From this discussion, one might conclude that the key to losing weight is to increase protein consumption. Although this is true to a point, protein in excess results in a condition known as **ketosis**. This is a common consequence of the high-protein diets so prevalent today. Although this diet works temporarily, it results in acidic pH levels in our "*cellular soup*" over a period of time. This *accelerates* aging and the appearance of age-related diseases. The key is to *balance* the ratio of protein and carbohydrates. The anti-aging diet and body composition plan focuses on balancing this appropriate ratio.

In summary, elevated glucagon levels can be accomplished by:

1. Eating smaller meals more frequently spaced.
2. Eating meals and snacks with moderate to low-glycemic carbohydrates (Diagram V-5).
3. Eating meals and snacks with higher protein concentrations.
4. Eating meals with fewer carbohydrates later in the day.

Elevated glucagon levels help to burn body fat because:

1. Stored fat instead of stored glucose is burned for energy.
2. An elevated glucagon level means that excess blood glucose is not present.
3. Elevated glucagon induces the "good" eicosanoids to dilate the blood vessels and make fat burn aerobically.

When the blood glucose level is stable, higher amounts of oxygen are available. This allows the body to use stored body fat for energy. The anti-aging diet and body composition plan can help keep blood glucose levels stable throughout the day. This is the physiological response that is found in a younger person, which helps to maintain a favorable balance between insulin and glucagon. That means whether we are sleeping, reading, walking or exercising, we will be using fat to satisfy energy requirements.

Cortisol Cortisol is termed the "alarm response hormone" and is the primary culprit in the acceleration of aging. Elevations in cortisol are *induced* by elevations in

Diagram V-5

Low Glycemic Foods

Fruits	Apples, apricots, cherries, grapefruits, oranges, peaches, and plums
Vegetables	Artichokes, asparagus, broccoli, cauliflower, and green beans
Grains	Oatmeal, rye and wild rice
Legumes	Black beans, chick peas, kidney beans and lentils
Starches	Sweet potatoes, whole grain pasta and yams

insulin. They are directly related to each other; as insulin levels increase, so do cortisol levels. Therefore, the "*flight or fight*" response associated with cortisol is directly related to blood sugar elevation.

An elevated cortisol level is a biomarker of advanced aging.[31–39] This elevation not only triggers the inflammatory response associated with the bad eicosanoids but also sets the stage for age-related degenerative diseases. One of the most *detrimental* effects of elevated cortisol is destruction of nerve and brain tissue, causing memory loss and destroying the central homeostatic mechanisms responsible for maintaining hormonal levels.[40] The destruction of collagen and elastin fibers also frequently affects our cosmetic surgery patients directly. This hormonal profile of high cortisol, and high insulin levels leads to a drop in human growth hormone levels (HGH) and IGF-1 production. Decreased growth hormone levels are associated with increased body fat and decreased muscle mass.

DHEA DHEA is secreted by the adrenal glands. **DHEA** is known as the "mother," or *precursor hormone* because it can be converted into many other hormones, which include estradiol, estrone, estriol, progesterone and testosterone. Because of its centrality, DHEA plays a role in enhancing the immune function as well. It also helps to optimize body composition and inhibits the detrimental effects of the elevated cortisol.[41–49] The quickest and easiest way to balance elevated cortisol levels is to prescribe an oral dose of DHEA.

Human Growth Hormone Human growth hormone (HGH) is the master hormone that *controls* the rate of cell division in both young and old cells. It has a direct effect on the cell cycle of each cell in the body, especially as we age. It can stimulate senescent cells from G0 (zero) into the active "S" phase of the cell cycle.[50] It is converted into the more active insulin-like growth factor-1 (IGF-1) mainly in the liver. Therefore, optimal liver function is an essential component to optimal aging.

Optimal levels of HGH lead to increased muscle mass and decreased body fat. By accomplishing this, HGH helps burn calories more efficiently and helps maintain ideal body composition.[51–58] The bad eicosanoids lower HGH levels, and the good eicosanoids increase them.

Human Growth Hormone Replacement Therapy HGH has not been shown to be related to an increase in cancer formation when used in vivo.[59–61] Growth hormone therapy in vivo also elevates IGF-BP-3 (binding protein-3), which has been shown to be protective against cancer. It appears that IGF-BP-3 has inhibitory effects on cell growth.

In summary, growth hormone replacement is not associated with an increased risk of cancer, increasesd cancer recurrence, or de novo cancer or leukemia.[61–62]

Eicosanoids This family of hormones is derived mainly from dietary fat. Eicosanoids actually constitute the "letters of the alphabet" in the language of **cellular communication**. They are responsible for the "chatter" that occurs from cell to cell. The anti-aging diet and body composition plan focus on balancing the production and ratio of these good messages with the bad ones. This is accomplished by *balancing* the micronutrient component responsible for the production of all the eicosanoids — omega-3 and omega-6 fatty acids.[62–68] In addition, the balance of carbohydrates and protein in a meal determines the *insulin-glucagon ratio*, which directly affects eicosanoid production.

In general, omega-6 fatty acids produce pro-inflammatory, or fatty, eicosanoids; omega-3 fatty acids produce anti-inflammatory, or good, eicosanoids. Bad eicosanoids are also part of the alarm response, which results from chronic insulin elevation. The standard American diet contains 20 times more omega-6 fatty acids than omega-3 fatty acids. High levels of substances derived from omega-6 fatty acids are responsible for and characterize obesity, accelerated aging, diabetes and cardiovascular disease.

These important facts illustrate that achieving a balance of fatty acid intake is a crucial ingredient for maintaining optimal good health and optimal aging.[69–76]

Thyroid Hormone A thyroid deficiency, frequently undetectable by a standard serum thyroid T_3 panel, can directly lower the production of energy, adenosine triphosphate (ATP), in each one of the hundred trillion cells in the human body.[77] Energy production is actually controlled by the microscopic structures in the cells called **mitochondria**. Known as the "powerhouses" of the cell, the mitochondria require ample supplies of T_3, the active thyroid hormone form. In the mitochondrion, T_3 works with oxygen to transform food into the cellular fuel known as ATP. It is ATP, produced in the mitochondria, that fuels the nuclear genetic machinery responsible for cell reproduction and protein and enzyme formation.

The thyroid gland is also *responsible* for establishing the basal metabolic rate. This gland *regulates* body temperature and modulates oxygen consumption, which is directly related to body composition.[77]

The thyroid gland also secretes a simple pro-hormone composed of two amino acids and four iodine residues called thyroxine (T_4). An enzyme clips off one of the iodine residues that create the active hormone called T_3. A deficiency of **selenium** prevents T_4 from being efficiently converted into T_3, which is one of the most dietary common causes of hypothyroidism as we age. In cases of accelerated aging, when liver pathways are compromised, the body may not convert T_4 to the active hormone of T_3 as well. Indeed, our body produces reverse T_3 (rT_3), which slows enzyme activity, lowers body temperature, causes free radical damage and alters genetic transcription. rT_3 accomplishes this by occupying receptor sites for normally active and positive T_3 molecules.

The anti-aging diet and body composition plan emphasize nutrients that support proper thyroid function. Alkaline minerals are important to help the body break down protein without altering pH. If alkaline minerals are absent from the diet, the body leaks calcium from the bone to help control its acidic pH environment. Thus, *a deficient thyroid state* also directly contributes to osteoporosis, another acute finding in the aging process.[72–83]

What Happens to Insulin Levels as We Age?

In order for insulin to perform its main function, which is to allow glucose to enter the cell for energy, it must fit snugly into a cell surface receptor. As we age, these receptor sites are constantly becoming less sensitive as a result of the excessive buildup of sugar residues on protein complexes that must fit into the sites on cell membranes. This alters the three-dimensional (3D) configuration of the receptor site. This process is termed "glycation." As a result, the pancreas overworks itself and produces excess insulin to compensate for the decreased receptor-binding ability of insulin.

It has become increasingly apparent that **insulin resistance** is a hallmark of aging. With insulin resistance, a decrease in insulin receptor sensitivity leads to a clustering of risk factors. These risk factors may include elevated insulin, excessive sympathetic tone (vasoconstriction), high blood pressure, elevated LDL cholesterol, and elevated triglycerides (see Diagram V-4). With this condition, the environment is set for obesity, chronic inflammation, arteriosclerosis, and adult-onset diabetes — all hallmarks of degenerative diseases of aging. In its early stages, insulin resistance can also be responsible for cellulite and even localized lipodystrophy.

In addition, high levels of oxidative stress (free radicals), arthritis, altered gene expression and chronic fatigue syndrome have all been linked to insulin resistance. As we shall see in later chapters, this constellation of symptoms has been called **syndrome X**.

Other important factors in maintaining insulin receptor sensitivity include:

1. Correcting deficiencies of essential fatty acids.
2. Adding phytonutrients (plant nutrients) to the diet.
3. Achieving the proper ratios of vanadium, chromium and magnesium.
4. Normalizing pH levels in the extracellular and intracellular environments (the cellular soup).

All of these factors are taken into account when an anti-aging **nutraceutical supplement** is recommended and is designed to be utilized with an anti-aging diet and body composition program.

Clinical studies[84–85] have documented that a nutraceutical supplement can accomplish these goals as well as limit DNA damage, improve DNA repair, and decrease free radical levels. Recent clinical trials measuring these parameters at the University of Lund in Sweden[86] have documented that the appropriate quantity and combination of antioxidants and phytochemicals can be extremely effective in controlling these aspects of aging (Diagrams V-6 and V-7). This is proof that we as anti-aging physicians and surgeons *can* alter the aging process.

The Primary Role of Insulin

The primary role of insulin is to transport glucose into the cell. This process occurs most efficiently at an alkaline pH of 7.8 to 8.0. Today's acidic fast foods work to suppress insulin's function, also contributing to elevated insulin, cortisol and glucose levels. As we age, we absorb fewer nutrients because of a loss of gastrointestinal function. Our needs for the type and quantity of nutrients change as well (Diagrams V-8 and V-9a, 9b). When nutritional deficiencies are chronic, pH becomes acidic. This can rob insulin and its receptors of their ability to function. In this case, fat stores instead of glycogen build up because elevated blood sugar levels are fueling fat deposition adipose tissue rather than glycogen deposition in the muscle and liver.

The vicious circle ensues as other nutrient deficiencies trigger an immense hunger for vital substances, which can be satisfied only with large quantities of carbohydrates. This, in turn, elevates insulin throughout the body, signaling us to eat past satiety (see Diagram V-10). In its mild form, this cycle leads to adult weight

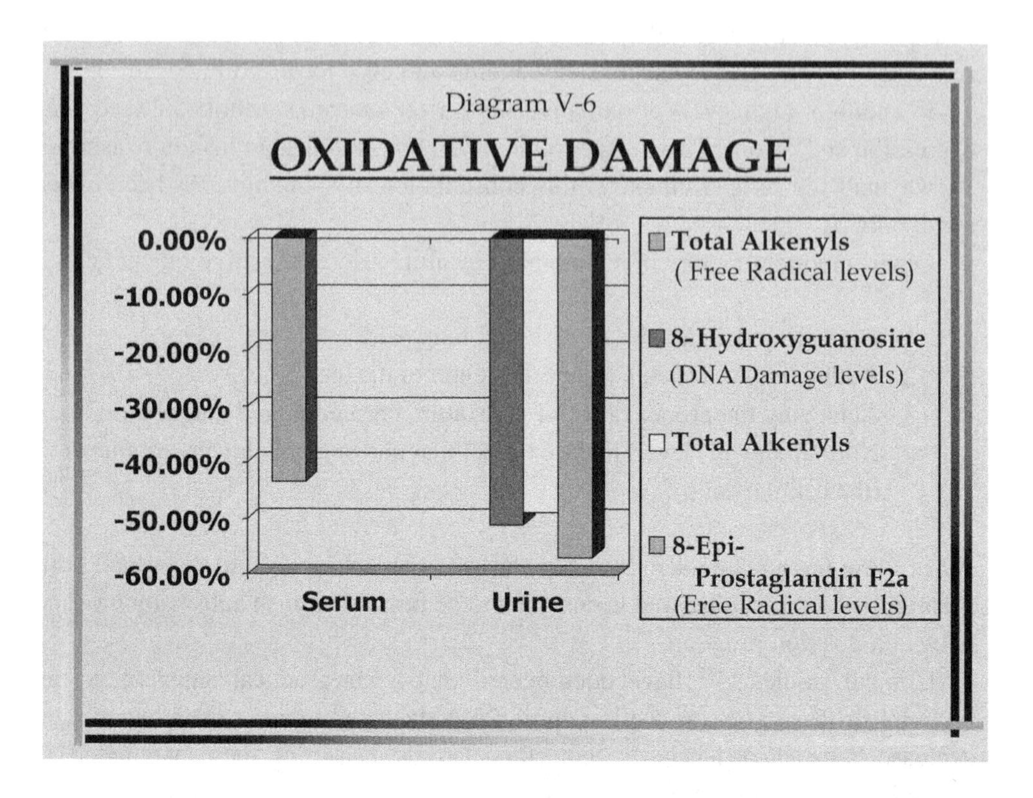
Diagram V-6
OXIDATIVE DAMAGE
0.00%
-10.00%
-20.00%
-30.00%
-40.00%
-50.00%
-60.00%
Serum
Urine
Total Alkenyls
(Free Radical levels)
8-Hydroxyguanosine
(DNA Damage levels)
Total Alkenyls
8-Epi-
Prostaglandin F2a
(Free Radical levels)

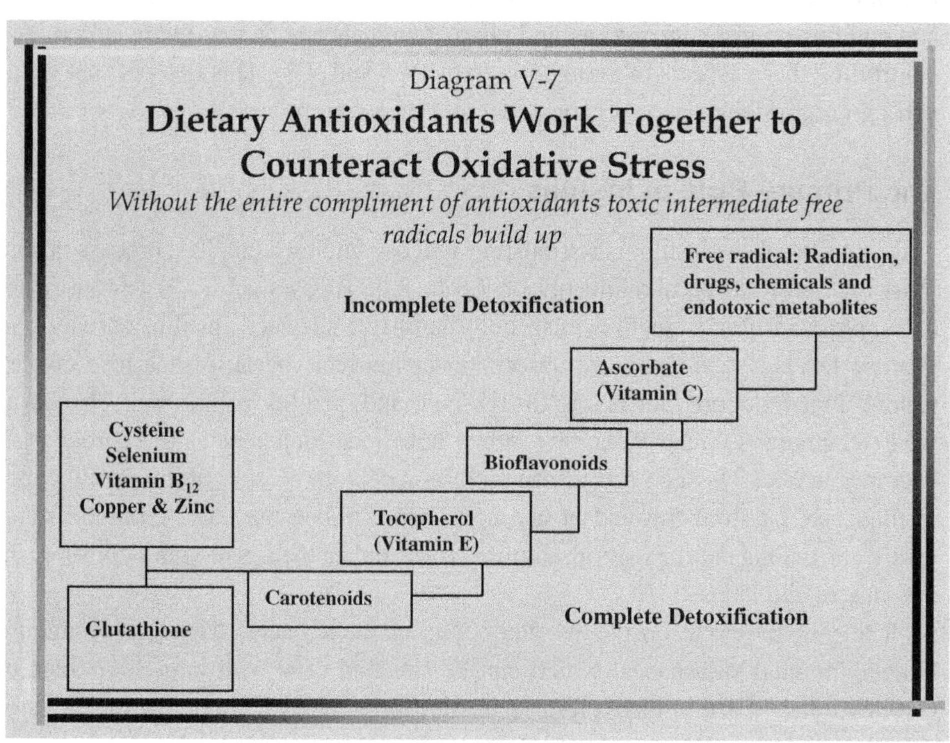
Diagram V-7
Dietary Antioxidants Work Together to Counteract Oxidative Stress
Without the entire compliment of antioxidants toxic intermediate free radicals build up
Free radical: Radiation, drugs, chemicals and endotoxic metabolites
Incomplete Detoxification
Ascorbate (Vitamin C)
Bioflavonoids
Tocopherol (Vitamin E)
Carotenoids
Cysteine
Selenium
Vitamin B12
Copper & Zinc
Glutathione
Complete Detoxification

Diagram V-8

ANTI-AGING NUTRIENTS THE KEY CONCEPTS

- As we age so do our needs for types and quantities of nutrients change. These needs and changes have been just recently described.

Diagram V-9a

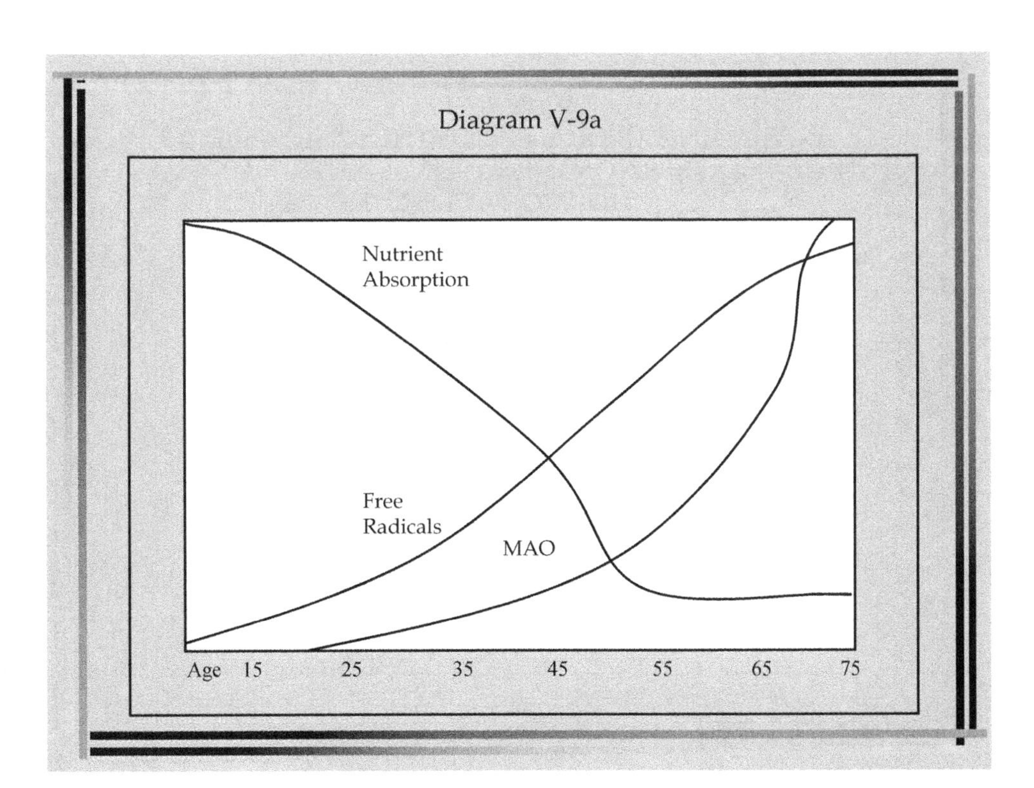

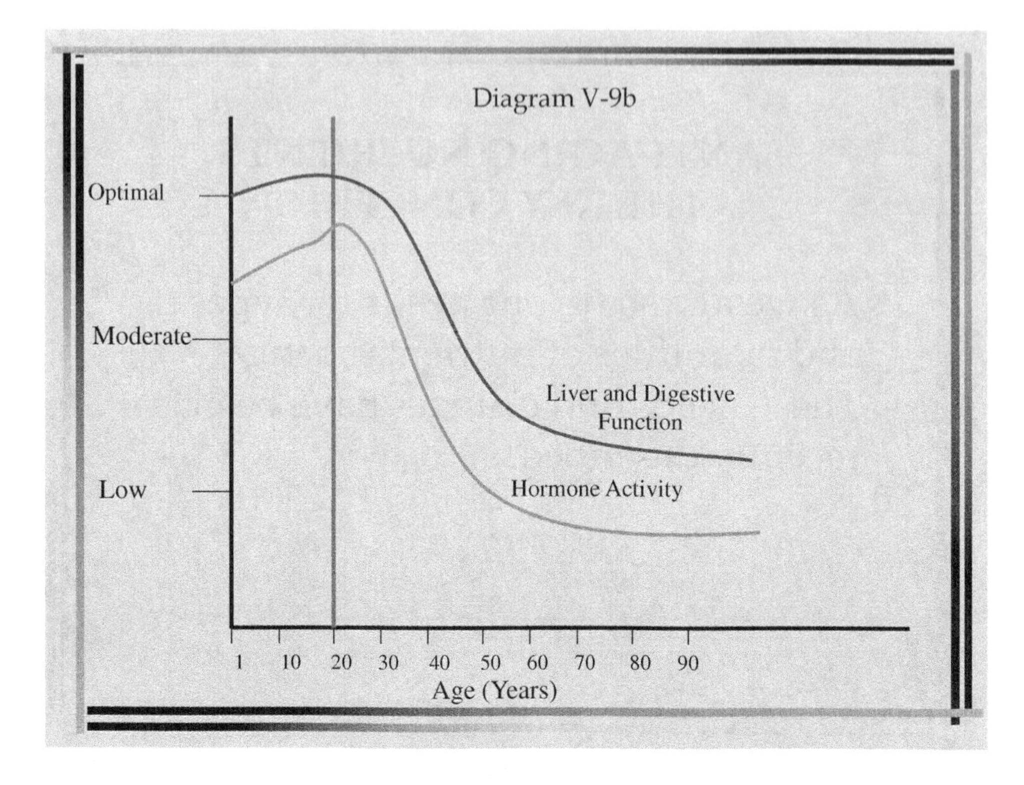

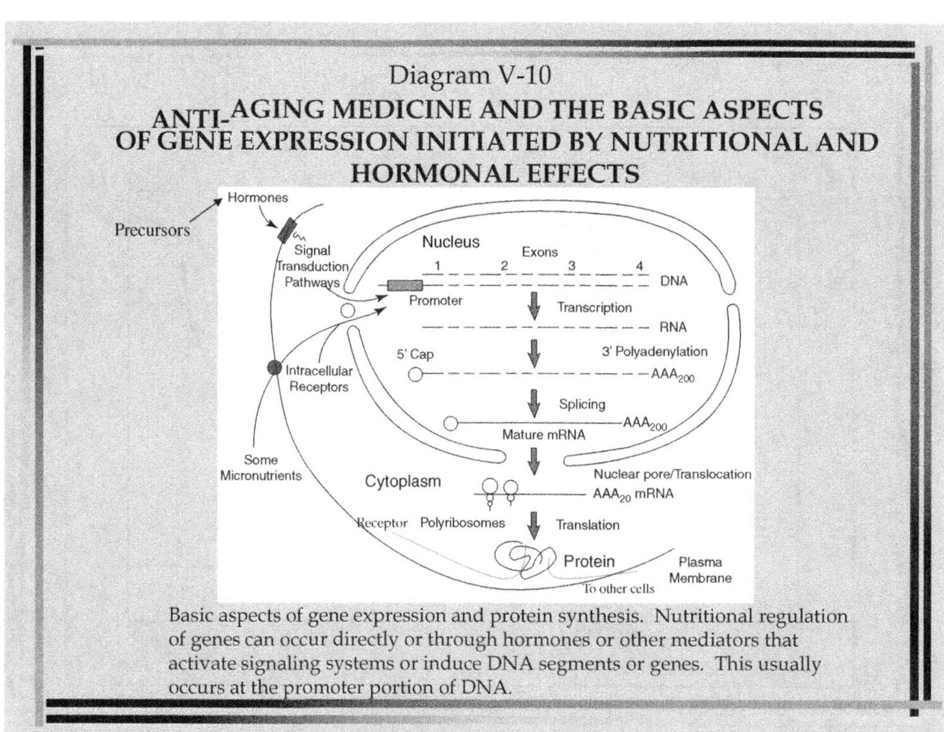

Basic aspects of gene expression and protein synthesis. Nutritional regulation of genes can occur directly or through hormones or other mediators that activate signaling systems or induce DNA segments or genes. This usually occurs at the promoter portion of DNA.

gain and generalized lipodystrophy; in its most severe form, it leads to adult-onset diabetes!

In summary, we all tend to become diabetic-like as we grow older. Supplements such as chromium and vanadium can help improve insulin sensitivity. The natural herbs fenugreek and gymnema also help to maintain insulin receptor sensitivity.

Prescription medication such as metformin can improve insulin receptor sensitivity dramatically with minimal side effects, especially when used in microdoses of approximately 125 mg in the afternoon and evening with meals. This works clinically in humans and has been shown to dramatically improve life span in animals. Metformin is as close as any other compound we presently have to a caloric-mimetic drug. No doubt in the near future caloric-mimetics will occupy an important place in age management.

The Mind and Diet

A direct connection exists between mood and food. **Stress** can have very negative effects on one's diet. When intense unresolved and unrelenting stress weakens a persons self-control, there is a tendency to binge on carbohydrates. Along with the regulation of key hormones and control of pH, research in chemistry and weight gain is also incorporated into the anti-aging diet and body composition plan.

New research has shown that people who are chronically overweight have elevated monoamine oxidase (MAO) levels.[87] This enzyme inhibits one of the key neurotransmitters in the brain, **serotonin**. *Low serotonin* levels are not only associated with depression but are also linked to overeating and binging.

With *decreased* MAO levels and *elevated* serotonin levels, many people notice a decrease in appetite. Thus, incorporating key supplements into the anti-aging body composition plan that help accomplish this state can have a positive effect on neurotransmitter levels. One of the most effective and easiest ways to elevate serotonin levels is to take tryptophan, 500 to 2000 mg orally, before sleep. This regimen not only helps initiate the onset of sleep but also results in an elevated serotonin level throughout the day.

Appropriate serotonin levels help curb the desire for food and actually alter the quantity of food required to achieve a feeling of satiety. More important, with higher serotonin levels, patients feel less angry and upset, more relaxed and better able to cope with the stress of life in general. The serotonin level is also a key biomarker of brain function when the physician is completing a comprehensive anti-aging evaluation.

The Mind-Mood Connection

The most common cause of overeating is the fact that "simple" carbohydrates elevate serotonin levels. A high serotonin level makes people feel good. Therefore, many people eat to feel better — not because they are hungry. In this scenario, carbohydrates have become a "drug."

This results in a conditioned response and a cyclic behavior pattern. People desire to binge in order (unconsciously) to increase serotonin levels. This desire can be suppressed with supplements that elevate and maintain serotonin levels evenly throughout the day, such as tryptophan and prescription medications. If the intensity of sugar cravings is also diminished, the discipline necessary to conform to a change in dietary habits and patterns may be easier to maintain.

Anti-aging physicians may even choose to elevate serotonin levels by using low doses of a prescription medication called Selegiline (5 mg once to three times per week). Selegiline may also upregulate the **intracellular intrinsic antioxidants** superoxide dismutase and catalase, which are formed in the mitochondria. This helps to limit free radical attacks on molecules such as DNA both within the nucleus and mitochondrion of the cells.

Stress: Two Varieties

Stress can be induced in two fashions. First, stress can be defined, biochemically, by an elevated level of cortisol.

Cortisol has been shown to be responsible for wreaking havoc on the *immune system* and on *brain function* and for having a negative impact on many of the other anti-aging hormones.[40,88–93]

The most common cause of stress that we encounter is environmental. This can be termed "*time urgency*." In the 21st century the most common causes of "time urgency" are an overcrowded schedule and constant stimuli and interruptions of our thoughts throughout the day. This process, by which our mind jumps from thought to thought and situation to situation, without being able to satisfactorily focus on the present moment "the now," can be referred to as time "urgency." This constant mental stress results over time in a conditioned mental response, which causes subtle but chronic elevations of cortisol — the age-accelerating hormone.[86,94]

The second, and a very common form of stress is initiated by our poor dietary habits. **Dietary stress** is induced by meals that contain large quantities of simple carbohydrates, which cause a rapid elevation of insulin and, subsequently, cortisol as well.

In the 21st century, we find ourselves in a situation **in which both forms of stress become chronically induced and conditioned** as to create a state of constant accelerated aging.

Macronutrients: The Constituents of Food

What is food? We should think of food as our primary means of balancing our hormones. It is also a major source of energy, which ultimately originates from the production of ATP from the utilization of the macronutrients contained in each meal. Macronutrients are divided into three main categories: carbohydrates, proteins and fats.

In essence, the foods we eat *control* which genes are expressed.[16] Our genes are responsible for making hormones and for helping to transmit their signals. Therefore, the micronutrients and macronutrients in the diet are our primary means of upgrading our genetic potential (Diagram V-10).

The anti-aging diet and composition plan focuses on the ratio of three macronutrient food groups:

1. Carbohydrates are mainly responsible for energy.
2. Proteins are mainly responsible for the building blocks of DNA.
3. Fats contribute mainly to the production and stability of cell membranes and cell communication.

Calculating macronutrient ratios for an individual is easily accomplished (Diagram V-11).

The micronutrient side of the equation is where supplementation with key vitamins and minerals cofactors and phytochemical components come into play. As previously mentioned, recent clinical experiments have documented a tremendous

Diagram V-11

Calculation Sheet:

TURNING THEORY INTO REALITY

Calculating Macronutrient Ratios for Individuals **Carbohydrates:** Should be in a 1.3 to 1 ratio with protein

Protein: For each pound of body weight eat between ½ gram and 1 gram of protein

Fat: Daily intake should be approximately half the weight of protein

Macronutrient Ratios for 150 Pound Person

Carbohydrate = 200grams

Protein = 150 grams

Fat = 75 grams

Converting Grams to Calories

1 gram of carbohydrate = 4 calories

1 gram of protein 4 calories

1 gram of fat = 9 calories

2,075 daily calories for a 150lb person

Carbohydrate 40% - 200 grams x 4 = 800 calories

Protein 30% - 150 grams x 4 = 600 calories

Fat 30% - 75 grams x 9 =675 calories

ability of these natural compounds to improve gene expression, decrease DNA damage, improve DNA repair rates, and limit free radical levels.[95]

Although both macronutrients and micronutrients play a role in determining an individual's hormone profile, *balancing* the macronutrients in the diet can be very difficult to achieve. While we can easily supplement our diets with micronutrients in the form of supplements, it takes discipline and knowledge of what to eat in order to balance the proper macronutrient ratio. Therefore, it is important to see how and what you eat affects the hormones that your body produces.

Thus, what is the best ratio of carbohydrates, proteins and fats to optimize hormonal responses and reactivate optimal genetic expression? A large series of scientific studies[94] make it clear that in most people, a diet consisting of 40% carbohydrates, 30% proteins and 30% fats is optimal. In these studies, the 40/30/30 ratio has been demonstrated to promote fat loss and to enhance muscle mass at the same time. Some individuals may require slightly different ratios of macronutrients to function optimally according to their hormonal levels and body composition. These combinations have included ratios between 50/20/20 and 50/30/20, which represent a smaller percentage of the overall population. These people are easily identified after initial trials with the 40/30/30 diet. The basis of this difference has to do with the *genetic inheritance* of how well we tolerate insulin and receptor sensitivity.

The marked improvement of body composition that follows when one adheres to a ratio of this degree is due to enhanced hormonal responses. This is accomplished by creating conditions that allow the body to utilize its inherited genetic potential more efficiently.

Important Facts About Macronutrients

Carbohydrates Forty percent of the calories in the anti-aging diet and body composition plan come from carbohydrates. In the blood, carbohydrates are converted into glucose, which is ultimately stored in the liver and muscle cells as glycogen. Glycogen is a *primary source* of energy for the body.

The best carbohydrate foods are those high in fiber, low in starch and low in simple sugars. These include certain fruits, vegetables, grains, legumes and starches (see Diagram V-5). Carbohydrates may also be rated according to their **glycemic index**. The lower the rating, the better. A low glycemic index means that it takes the body a long time to break the carbohydrate down into simple sugars. This results in lower blood sugar levels in general and in lower insulin levels over a longer period of time.

Proteins Thirty percent of the calories in the anti-aging diet and body composition plan come from protein. As stated previously, protein stimulates the release of glucagon, which is the key fat-burning hormone. Proteins also supply the key building blocks for DNA and enzymes that are required to run the essential cellular processes. Excellent sources of protein include lean white meat such as turkey, chicken and fish. Fish is also an excellent combined source of proteins and fats in the form of omega-3 and omega-6 acids, especially from cold water North Atlantic fish.

Vegetable proteins, such as spinach, green beans, legumes, peas, beans, and soy products like tofu (bean curd) are an excellent source of protein as well and should be utilized as the primary source of protein in order to avoid excess meat, especially red meat such as beef.

Three basic types of proteins are produced from the information that is copied from DNA to ribonucleic acid (RNA). These three types are formed from information that has been stored on RNA and that is then delivered to tiny cellular factories within the cell, called **ribosomes**. These ribosomes are then instructed to build these three types of proteins:

1. **Structural proteins** are like the framework of a house for cellular and organ function.
2. **Signaling proteins** act like a telephone; they are needed to contact cell to cell and are responsible for communications.
3. **Enzymes** act like the electrical system in the home; they provide the energy to spark changes in molecules and rebuild other damaged proteins.

Fats Thirty percent of the calories in the anti-aging diet and body composition plan come from fat. "Good" fat can slow down the digestion and absorption of carbohydrates. In essence, fat can act as a control rod[96] slowing the reaction and absorption of carbohydrates in order to keep blood glucose low. Fats are the essential components of all cell membranes and cell receptors as well and are required in appropriate quantities to optimize hormonal signaling and hormonal responses both on the surface and *within* the cell.

Unsaturated fats are better than *saturated* fats. Extra virgin olive oil is an ideal source of fats. Among unsaturated fats, there are two major types. *Polyunsaturated* oils (corn oil, safflower oil, flaxseed oil and primrose oil) are excellent sources. *Monosaturated* oils come in two main types, omega-3 and omega-6. Although both are important, most Americans consume too much of the omega-6 variety.

The appropriate balance of omega-3 and omega-6 fatty acids has a direct effect and response on which eicosanoids, the good or the bad, are produced. If the ratio of omega-3 and omega-6 acids is abnormally high in favor of omega-6 fatty acids, increased inflammation and even bone and joint pain can occur.

How Does the Glycemic Index Relate to Hormones and Body Composition?

The **glycemic index** is a measure of how quickly a food increases the blood sugar level. Simple sugars, like high-fructose corn syrup, quickly flood the blood with glucose. This puts stress on the body in the form of elevated cortisol. The body then attempts to bring its blood sugar level back to equilibrium, which requires extra effort and energy in general. The *lower* the glycemic index, the *slower* the blood sugar level rises. This is because carbohydrates, that are complex (from vegetable sources) take a longer time for the body to break down and digest.

The glycemic index is also related to fiber, protein and fat content as well as cooking technique. The higher the glycemic index in food, the less the food can maintain optimal levels of anti-aging hormones, which are directly related to stable levels of blood sugar. In addition, high glycemic foods lead to an increase in fat deposition, which is a direct result of higher insulin levels.

Other benefits of the anti-aging diet and body composition plan *directly* affect many aspects of health in general, for instance, **mental performance and aging of the brain**. The brain needs glucose (blood sugar) for energy. If we consume a high-carbohydrate meal, the pancreas releases insulin in an attempt to stabilize blood sugar. In many cases, blood sugar drops far too low. This "short-circuits" the brain's supply of glucose, which is needed for energy and optimal brain function, and the ability to concentrate begins to deteriorate. *High* levels of cortisol, which are released in response to blood sugar spikes, create *massive* damage to the brain cells and central nervous system tissue.[86]

Diabetes

Type II diabetes, or insulin resistance diabetes, is also known as hyperinsulinemia. High levels of insulin occur because the insulin receptors located in cell membranes become less sensitive to the insulin molecules. That is, when insulin binds to a receptor, little happens and blood glucose, levels remain elevated. It is this constant elevation of blood glucose, which creates glycated, or "distorted," proteins. These proteins then become fixed to these receptor sites and inhibit insulin from three-dimensionally "fitting" into its receptor site to do its job — which is to allow glucose into the cell!

Insulin resistance is a predominant type of adult-onset diabetes in modern countries.

Hypoglycemia is another major problem that results from eating too many carbohydrates. Initially, the body responds to high blood sugar with an insulin spike and a concomitant cortisol elevation. When blood sugar rises too much or too fast, however, insulin is released to lower it. Hypoglycemia occurs when insulin drives blood sugar levels too low, frequently below the level before the meal! The brain is deprived of its sole source of fuel, and this induces symptoms such as fatigue, rapid heart beat, sweating, hunger, trembling, headache, mental dullness and confusion as well as poor memory.

Elevated insulin also triggers bad eicosanoid production, resulting in blood vessel constriction (vasoconstriction) and thus increased blood pressure. Elevated insulin levels also activate the enzyme hydroxymethylglutaryl coenzyme A (HMG CoA) reductase, located in the liver, and causes the liver to make more cholesterol. No matter how low the diet is in cholesterol, elevated insulin levels can keep cholesterol levels higher than desired.

On the other hand, glucagon inhibits the activation of HMG CoA, therefore *reducing* the amount of cholesterol made by the liver. Appropriate nutrition can help keep blood sugar levels stable, elevate glucagon to help control cholesterol synthesis and provide nutritional support needed to maintain optimal cholesterol levels.

One of the body's *main* defenses against cancer is the immune system. Elevated insulin levels suppress the immune system by stimulating production of bad eicosanoid hormones, mainly prostaglandin E_2 (PGE-2) (Diagram V-12). This hormone inhibits the activation of natural killer cells and decreases oxygen transfer. Elevated insulin also accelerates telomere turnover at the microcellular (nucleus) level and may be directly responsible for decreasing stem cell populations in general.

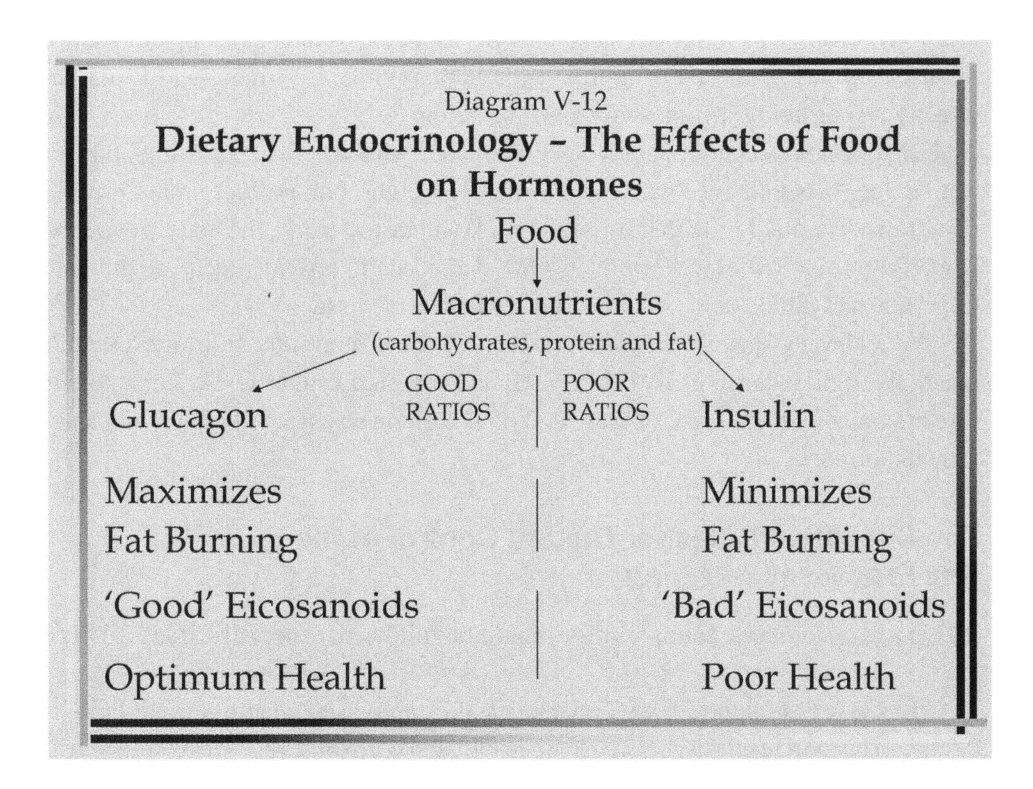

Many of the symptoms associated with premenstrual syndrome (PMS) are directly related to elevated blood glucose levels as well. Controlling and maintaining a regular blood sugar level can decrease or eliminate the monthly episodes of PMS and its related symptoms.

High-carbohydrate diets can lead to protein and amino acid deficiencies as well. These deficiencies can decrease neurotransmitter production and elevate insulin level production. This ultimately creates a cascade of hormone responses responsible for the uptake and release of neurotransmitters. If the brain has low levels of neurotransmitters or poor uptake in releasing neurotransmitters, depression quickly sets in.

Optimal nutrition and ratios of macronutrients, as well as adequate supplies of good quality protein, are the key sources of supplying the precursors for brain neurotransmitters.

Sleep disorders can occur after we eat a high-carbohydrate snack before bedtime. Carbohydrates can cause blood sugar and insulin levels to surge. Elevated insulin during sleep not only blocks release of growth hormone, but also inhibits proper repair and recovery of the body at the cellular level. This elevation also causes us to feel groggy upon awakening. We might also experience a low blood sugar level, which can disrupt sleep patterns, causing irregular sleep patterns. An appropriate snack before bed (proteins) can *stabilize* blood sugar levels and can *trigger* the hormonal responses to help release maximum levels of growth hormone for cellular and body repair. This improves sleep patterns as well. Amino acid supplementation or snacks with high doses of amino acids like arginine, ornithine, and lysine are directly responsible for improving HGH secretion.

For people who cannot gain weight, no matter what they eat, adjusting nutrition can be very beneficial. To gain mainly muscle weight, one needs to maximize the body's production of growth hormone (HGH) and testosterone, the body's two most powerful muscle-building hormones. One also needs to provide adequate amounts of protein and amino acids to build and repair muscle tissue.

Since elevated insulin *inhibits* release of growth hormone from the pituitary gland, the right balance of nutrition can help keep blood sugar levels stable, therefore maximizing the release of these essential hormones, especially glutathione and growth hormone.

The Simplified Rules for Dietary Control of the Key Processes of Aging

As we age, a number of simple rules should be followed, especially after age 40:

1. Eat lesser quantities of food, and eat the appropriate macronutrient ratio. This will result in a decrease in insulin and cortisol levels. Basic research

has shown that decreasing calories and increasing micronutrients is the only proven way to *increase* longevity and decrease age-related diseases (Diagrams V-13 and V-14).

2. Eat more frequent, smaller meals. As we age, because of a decrease in gastrointestinal function and digestive enzymes, it is much easier to tolerate smaller meals more often than large meals episodically. *Better absorption* of food content occurs with this regimen.
3. Eat low-glycemic carbohydrates, which helps stabilize insulin and elevate glucagon levels. A greater number of carbohydrates should be eaten in the morning and a decreasing amount eaten as the day goes on. Large amounts of carbohydrates in the evening meal are the *worst* approach because they cause marked insulin elevation and fat deposition will occur during sleep.
4. Eat more protein. In general, slowly shift the ratio of carbohydrates, proteins and fats to more protein, especially a vegetable source, and eat fewer high-glycemic carbohydrates as you age. This habit helps supply a greater source of building blocks of DNA; it helps *augment* growth hormone levels and growth hormone release later on in the day.
5. Avoid drinking large quantities of liquids while eating meals, especially carbonated and acidic drinks such as soft drinks. This habit dilutes

Diagram V-13

THE APPROACH:
A BASIC PROGRAM

- DECREASE CALORIES BY APPROXIMATELY 30% DAILY
- ADD HIGH DENSITY NUTRIENTS
- IMPROVE DIGESTION AND GUT FUNCTION

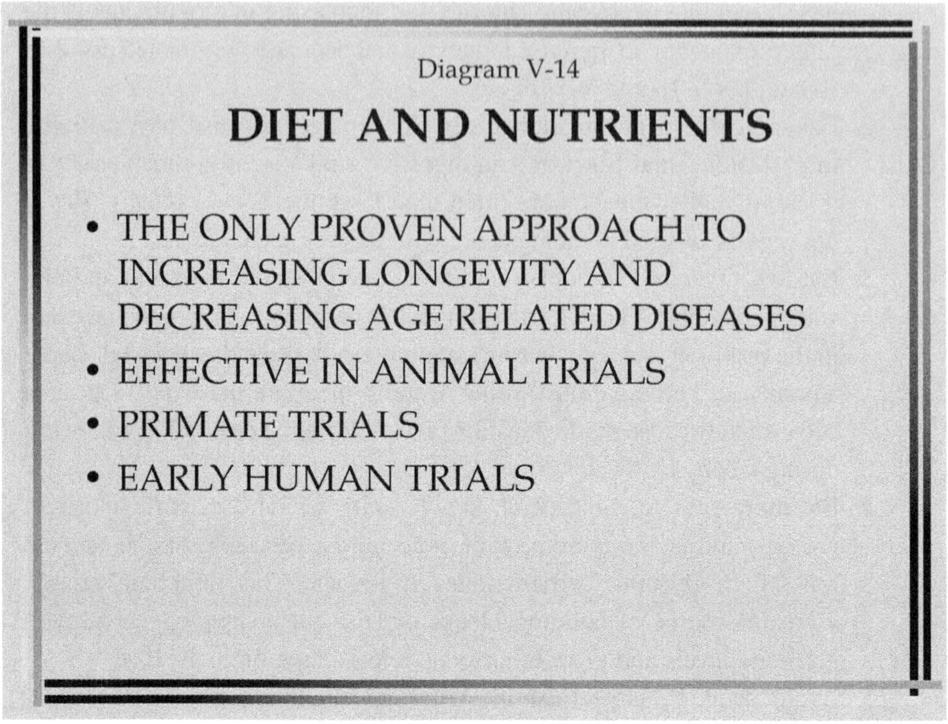

Diagram V-14

DIET AND NUTRIENTS

- THE ONLY PROVEN APPROACH TO INCREASING LONGEVITY AND DECREASING AGE RELATED DISEASES
- EFFECTIVE IN ANIMAL TRIALS
- PRIMATE TRIALS
- EARLY HUMAN TRIALS

digestive enzymes and gastric juices, which are already on the decline and are essential for appropriate food digestion. This habit also *acidifies* the intracellular fluids and decreases cellular efficiency and biochemistry throughout the trillions of cells in the body!

6. Eat protein before eating carbohydrates in order to maximize stimulation of the fat-burning hormone glucagon. This step helps to slow digestion and absorption of the carbohydrates in that meal, resulting in *better stabilization* of blood sugar levels.
7. Drink at least six to eight glasses of water or appropriate beverages, such as natural nonacidic beverages each day *between meals*.
8. Avoid alcoholic beverages, which elevate blood sugar levels. If you do drink these beverages, have them with your meal or with a higher-protein appetizer, which will *control* blood sugar levels and keep them lower.

Summary

The anti-aging diet and body composition plan is designed to keep the body at near-peak performance during the advancing years, when it normally begins to decline. This is accomplished by:

1. Maintaining and controlling pH levels.

2. Regulating key hormones of aging and health, which include insulin, cortisol, growth hormone and the eicosanoids.
3. Regulating and maintaining key neurotransmitters such as serotonin, thereby reducing sugar cravings throughout the day.

After reading the information presented in this chapter, you should find it relatively easy to establish new habits to optimize your hormonal levels and body composition. *It is essential to keep in mind that good habits, once established, are just as easy to maintain as bad habits*. The difficulty is transitioning from bad to good habits. It is hoped that this chapter will help readers and their patients accomplish the transition efficiently.

The next chapter discusses the importance of exercise and how the combination of the appropriate amount of exercise and diet further enhances hormonal production, body composition and optimal aging.

References

[1] Bodkin NL, Ortmeyer HK, Hansen BC. Long-term dietary restriction in older-aged rhesus monkeys: effects on insulin resistance. *J Gerontol Biol Sci Med Sci*. 1995; 50: B142–B147.
[2] Cefalu WT, Wagner JD, Wang ZQ, et al. A study of caloric restriction and cardiovascular aging in cynomolgus monkeys: a potential model for aging research. *J Gerontol Biol Sci Med Sci*. 1997; 52: B98–102.
[3] Cerami A. Hypothesis: Glucose as mediator of aging. *J Am Gerontol Soc*. 1985; 33: 626–634.
[4] Duffy PH, Reuers RJ, Leakey JA, et al. Effect of chronic caloric restriction on physiological variables related to energy metabolism in male Fischer 344 rat. *Mech Ageing Dev*. 1989; 48: 117–133.
[5] Fernades G, Friend P, Yunis EJ, Good RA. Influence of dietary restriction on immunologic function and renal disease in (NZB X NZW) F1 mice. *Proc Natl Acad Sci U S A*. 1978; 75: 1500–1504.
[6] Holehan AM, Merry BJ. The experimental manipulation of ageing by diet. *Biol Rev*. 1986; 61: 329–368.
[7] Kim MJ, Roecher EB, Weindruch R. Influences of aging and dietary restriction on red blood cell density profiles and antioxidant enzyme activities in rhesus monkey. *Exp Gerontol*. 1993; 28: 515–527.
[8] Lee DW, Yu BP. Modulation of free radicals and superoxide dismutase by age and dietary restriction. *Aging*. 1991; 2: 357–362.
[9] Parr T. Insulin exposure controls the rate of mammalian aging. *Mech Ageing Dev*. 1996; 88: 75–82.
[10] Venkatraman JT, Fernades G. Mechanisms of Delayed Autoimmune Disease in B/W Mice by Omega-3 Lipids and Food Restriction. In: Chandra RK, ed. *Nutrition and Immunology*. St. John's, Newfoundland: ARTS; 1992: 309–323.
[11] Wolff SP, Bascal ZA, Hunt JV. Autooxidative Glycosylation: Free Radicals and Glycation. In: Baynes JW, Monner VM, eds. *The Maillard Reaction in Aging Diabetes and Nutrition*. 1989: 259–273.
[12] Yu BP. Food restriction research: past and present status. *Rev Biol Res Aging*. 1990; 4: 349–371.
[13] Yu BP, Lee DW, Marler CG, Choi J-H. Mechanism of food restriction: protection of cellular homeostasis. *Proc Soc Exp Biol Med*. 1990; 193: 13–15.
[14] Cerami, A. Hypothesis: Glucose as mediator of aging. *J Am Gerontol Soc*. 1985; 33: 626–634.

[15] Parr T. Insulin exposure controls the rate of mammalian aging. *Mech Aging and Dev.* 1996; 8: 75–82.

[16] Cousins RJ. Nutritional Regulation of Gene Expression. In: Shils ME, Olson JA, Shike M, Ross AC. *Modern Nutrition in Health and Disease*, ed 9. Baltimore: Lippincott Williams & Wilkins; 1999: 573–603.

[17] Austin MA, Breslow JL, Hennekens CH, et al. Low density lipoprotein subclass patterns and risk of myocardial infarction. *JAMA.* 1988; 260: 1917–1920.

[18] Brandes J. Insulin induced overeating in the rat. *Physiol Rev.* 1977; 18: 1095–1102.

[19] Bruning PF, Bonfrer JMG, van Noord PAH, et al. Insulin resistance and breast cancer risk. *Int J Cancer.* 1992; 52: 511–516.

[20] Dek SB, Walsh MF. Leukotrienes stimulate insulin release from rat pancreas. *Proc Natl Acad Sci U S A.* 1985; 81: 2199–2202.

[21] Ducimetiere P, Richard JL, Cambrien I. The pattern of subcutaneous fat distribution in middle-aged men and risk of coronary heart disease. *Int J Obes.* 1986; 10: 229–240.

[22] Fanaian M, Szilasi J, Storlien L, Calvert GD. The effect of modified fat diet on insulin resistance and metabolic parameters in type II diabetes. *Diabetologia.* 1996; 39: A7.

[23] Giovannucci E. Insulin and colon cancer. *Cancer Causes Control.* 1995; 6: 164–179.

[24] Job FP, Wolfertz J, Meyer R, et al. Hyperinsulinism in patients with coronary artery disease. *Coronary Artery Dis.* 1994; 5: 487–492.

[25] Kaplan N. The deadly quartet: upper body obesity, glucose intolerance, hypertriglyceridemia, and hypertension. *Arch Intern Med.* 1989; 149: 1514–1520.

[26] Kern PA, Ong JM, Soffan B, Carty J. The effects of weight loss on the activity and expression of adipose-tissue lipoprotein lipase in very obese individuals. *N Engl J Med.* 1990; 322: 1053–1059.

[27] Nestler JE, Beer NA, Jakubowicz DJ, et al. Effects of insulin reduction with benfluorex on serum dehydroepiandrosterone (DHEA), DHEA sulfate, and blood pressure in hypertensive middle-aged and elderly men. *J Clin Endocrinol Metab.* 1995; 80: 700–706.

[28] Pek SB, Walsh MF. Leukotrienes stimulate insulin release from rat pancreas. *Proc Natl Acad Sci U S A.* 1984; 82: 2199–2202.

[29] Reaven GM. Role of insulin resistance in human disease. *Diabetes.* 1989; 37: 1595–1607.

[30] Reaven GM, Hoffman B. Abnormalities of carbohydrate metabolism may play a role in the etiology and clinical course of hypertension. *Trends Pharmacol Sci.* 1988; 9: 78–79.

[31] Cupps TR, Fauci AS. Corticosteroid-mediated immunoregulation in man. *Immunol Rev.* 1982; 65: 133–155.

[32] Fauci AS, Dale DC. The effect of in vivo hydrocortisone on subpopulation of human lymphocytes. *J Clin Invest.* 1974; 53: 240–246.

[33] Haynes BF, Fauci AS. The differential effects of in vivo hydrocortisone on kinetics of subpopulations of human peripheral blood thymus-derived lymphocytes. *J Clin Invest.* 1978; 61: 703–707.

[34] Munch A, Crabtree GR. Glucocorticoid-Induced Lymphocyte Death. In: Bower ID, Lockskin RA, eds. Cell Death in Biology and Pathology. New York: Chapman and Hall; 1981: 329–357.

[35] Norman AW, Litwack G. *Hormones*, ed 2. New York: Academic Press; 1997.

[36] Orth DN. Cushing's syndrome. *N Engl J Med.* 1995; 332: 791–803.

[37] Romero LM, Raley-Susman KM, Redish DM, et al. Possible mechanism by which stress accelerates growth of virally derived tumors. *Proc Natl Acad Sci U S A.* 1992; 89: 11084–11087.

[38] Sapolsky RM, Krey L, McEwen BS. Prolonged glucocorticoid exposure reduces hippocampal neuron number: implications for aging. *J Neurosci.* 1985; 5: 1222–1227.

[39] Sapolsky RM, Uno H, Rebert CS, Finch CE. Hippocampal damage associated with prolonged glucocorticoid exposure in primates. *J Neurosci.* 1990; 10: 2897–2902.

[40] Sapolsky RM, Uno H, Rebert CS, Finch CE. Hippocampal damage associated with prolonged glucocorticoid exposure in primates. *J Neurosci.* 1990; 10: 2897–2902.

[41] Field AER, Colditz GA, Willett WC, et al. The relation of smoking, age, relative weight, and dietary intake to serum adrenal steroids, sex hormones, and sex hormone–binding globulin in middle-aged men. *J Clin Endocrinol Metab.* 1994; 79: 1310–1316.

[42] Fleshner M, Pugh CR, Tremblay D, Rudy JW. DHEA-S selectively impairs contextual-fear conditioning: support for the antiglucocorticoid hypothesis. *Behav Neurosci.* 1997; 111: 512–517.

[43] Haffner SM, Valdez RA, Mykkanen L, et al. Decreased testosterone and DHEA sulfate concentrations are associated with increased insulin and glucose concentration in nondiabetic men. *Metabolism.* 1994; 43: 599–603.

[44] Kalimi M, Shafagoj Y, Loria R, et al. Anti-glucocorticoid effects of dehydroepian-drosterone (DHEA). *Mol Cell Biochem.* 1994; 131: 99–104.

[45] Labrie F, Belanger A, Cusa L, Candas B. Physiological changes in dehydroepiandrosterone are not reflected by serum levels of active androgens and estrogens, but of their metabolites: intracrinology. *J Clin Endocrinol Metab.* 1997; 82: 2403–2409.

[46] Lane MA, Ingram DK, Ball SS, Roth GS. Dehydroepiandrosterone sulfate: a biomarker of primate aging slowed by caloric restriction. *J Clin Endocrinol Metab.* 1997; 82: 2093–2096.

[47] May ME, Hollmes E, Rogers W, Poth M. Protection from glucocorticoid induced thymic involution by dehydroepiandrosterone. *Life Sci.* 1990; 46: 1627–1631.

[48] Nestler JE, Clore JN, Blackard WG. Dehydroepiandrosterone: the "missing link" between hyperinsulinemia and arteriosclerosis. *FASEB J.* 1992; 6: 3073–3075.

[49] Academy of Pharmaceutical Research and Science, Washington, DC.

[50] Daughaday WH, Harvey S. Growth Hormone Action: Clinical Significance. In: Harvey S, Scanes CG, Daughaday WH, eds. *Growth Hormone.* Boca Raton, FL: CRC Press; 1995.

[51] Rudman, D. Growth hormone, body composition, and aging. *J Am Geriatr Soc.* 1985; 33: 800–807.

[52] Rudman D. Impaired growth hormone secretion in the adult population: relation to age and adiposity. *J Clin Invest.* 1981; 67: 1361–1369.

[53] Rudman D, Feller A, Cohn L, et al. Effects of human growth hormone on body composition in elderly men. *Horm Res.* 1991; 36(Suppl. 1): 73–81.

[54] Rudman D, Feller A, Nagrag GA, et al. Effects of human growth hormone in men over 60 years old. *N Engl J Med.* 1990; 323: 1–6.

[55] Rudman D, et al. Relations of endogenous anabolic hormones and physical activity to bone mineral density and lean body mass in elderly men. *Clin Endocrinol.* 1994; 40: 653–661.

[56] Russell-Jones DL. The effects of growth hormone on protein metabolism in adult growth hormone deficient patients. *Clin Endocrinol.* 1993; 38: 427–431.

[57] Salomon F, Cuneo RC, Hesp R, Sonksen PH. The effects of treatment with recombinant human growth hormone on body composition and metabolism in adults with growth hormone deficiency. *N Engl J Med.* 1989; 321: 1797–1803.

[58] Whitehead J, et al. Growth hormone treatment of adults with growth hormone deficiency: results of a 13-month placebo controlled cross-over study. *Clin Endocrinol.* 1992; 36: 45–52.

[59] Giovannucci E, et al. Insulin-like growth factor-1 and binding protein-3 and risk of cancer. *Horm Res.* 1999; 51(Suppl. S3): 34–41.

[60] Circulating concentration of IGF-1 and risk of breast cancer. *Lancet.* 1995; 351(9113): 1393–1396.

[61] Adam O. Polyenoic fatty acid metabolism and effects on prostaglandin biosynthesis in adults and aged persons. In: *Polyunsaturated Fatty Acids and Eicosanoids.* IL: American Oil Chemical Society Press; 1987: 213–219.

[62] Bergstrom S, Rhyhage R, Samuelsson B, Sorval J. The structure of prostaglandins E_1, E_{1a}, and F_{1B}. *Biol Chem.* 1963; 238: 3555–3565.

[63] Burr GO, Burr MR. On the nature and role of the fatty acids essential in nutrition. *J Biol Chem.* 1930; 86: 587–621.

[64] Chatzipanteli K, Rudolph S, Axelrod L. Coordinate control of lipolysis by prosta-glandin E_2 and prostacyclin in rat adipose tissue. *Diabetes.* 1992; 41: 927–935.

[65] Eaton SB. Humans, lipids and evolution. *Lipids.* 1992; 27: 814–820.

[66] Enders S, Ghorbani R, Kelly VE, et al. The effect of dietary supplementation with n-3 polyunsaturated fatty acids on the synthesis of interleukin-1 and tumor necrosis factor by mononuclear cells. *N Engl J Med.* 1989; 320: 265–271.

[67] Giron DJ. Inhibition of viral replication in cell cultures treated with prostaglandin E_1. *Proc Soc Exp Biol Med.* 1982; 170: 25–28.
[68] Hamberg M, Svensson J, Samuelsson B. Thromboxanes: a new group of biologically active compounds derived from prostaglandin endoperoxides. *Proc Natl Acad Sci U S A.* 1975; 72: 2994–2998.
[69] Horrobin DF. Loss of delta-6 desaturase activity as a key factor in aging. *Med Hypotheses.* 1981; 7: 1211–1220.
[70] Horrobin DF, ed. *Omega-6 Essential Fatty Acids.* New York: Wiley-Liss; 1990.
[71] Kirtland SJ. Prostaglandin E_1: a review. *Prostaglandins Leukot and Essentl Fatty Acids.* 1988; 32: 165–174.
[72] Kunkel SL, Fantone JC, Ward PA, Zurier RB. Modulation of inflammatory reaction by prostaglandins. *Prog Lipid Res.* 1982; 20: 633–640.
[73] Chakrin LW, Bailey DM, eds. *The Leukotrienes.* New York: Academic Press; 1984.
[74] Metz S, Fukimoto W, Robertson RO. Modulation of insulin secretion by cyclic AMP and prostaglandin E. *Metabolism.* 1982; 31: 1014–1033.
[75] Oates JA, FitzGerald GA, Branch RA, et al. Clinical implications of prostaglandin and thromboxane A2 formation. *N Engl J Med.* 1988; 319: 689–698.
[76] Raheja BS, Sakidot SM, Phatak RB, Rao MB. Significance of the n-6/n-3 ratio for insulin action in diabetics. *Ann NY Acad Sci.* 1993; 983: 258–271.
[77] Brent GA. The molecular basis of thyroid action. *N Engl J Med.* 1994; 331: 847–853.
[78] Giustna A, Wehrenberg WB. Influence of thyroid hormones on the regulation of growth hormone secretion. *J Endocrinol.* 1995; 133: 646–653.
[79] Friedland IB. Investigations on the influence of thyroid preparations on experimental hypercholesterolemia and atherosclerosis. *Z Ges Exp Med.* 1933; 87: 683–695.
[80] Greenspan SL, Klibanski A, Rowe JR, Elahi D. Age-related alterations in pulsatile secretion of TSH: role of dopaminergic regulation. *Am J Physiol.* 1991; 260: E486–E491.
[81] Greer MA, ed. *The Thyroid Gland.* New York: Raven Press; 1990.
[82] Pasquini AM, Adamo AM. Thyroid hormones and the central nervous system. "*Dev Neurosci.* 1994; 16: 161–168.
[83] Rosenthal MS. *The Thyroid Sourcebook.* Los Angeles: Lowell House; 1996.
[84] Sheng Y, Pero RW, Olsson AR. DNA repair enhancement by a combined supplement of carotenoids, nicotinamide and zinc. *Cancer Detect Prev.* 1998; 22(4): 284–292.
[85] Sheng Y, Brygelsson C, Pero RW. Enhanced DNA repair, immune function and reduced toxicity of C-MED-100, a novel aqueous extract of *Uncaria tomentosa. J Ethnopharmacol.* 1999; 69(2): 115–126.
[86] Sapolsky RM. *Stress, The Aging Brain and the Mechanisms of Neuronal Death.* Cambridge, MA: MIT Press; 1992.
[87] Wurtman JJ. *The Serotonin Solution.* New York: Balantine Books; 1996.
[88] Sapolsky RM, Krey L, McEwen BS. Stress down-regulates corticosterone receptors in a site-specific manner in the brain. *Endocrinology.* 1984; 114: 287–292.
[89] Sapolsky RM, Krey L, McEwen BS. Glucocorticoid-sensitive hippocampal neurons are involved in terminating the adrenocortical stress response. *Proc Natl Acad Sci USA.* 1984; 81: 6174–6177.
[90] Sapolsky RM, Krey L, McEwen BS. Prolonged glucocorticoid exposure reduces hippocampal neuron number: implications for aging. *J Neurosci.* 1985; 5: 1222–1227.
[91] Sapolsky RM, Packan DR, Vale WW. Glucocorticoid toxicity in the hippocampus: in vitro demonstration. *Brain Res.* 1998; 453: 367–371.
[92] Blumenkrantz N, Sondergaard J. Effect of prostaglandin E_1 and F_{1a} on biosynthesis of collagen. *Nature.* 1972; 239: 246–247.
[93] Wertz PW, Abraham W, Landmann L, Downing DT. Preparation of liposomes from stratum corneum lipids. *J Invest Dermatol.* 1986; 87: 582–584.
[94] Sears B. *The Anti-Aging Zone,* 2nd ed. New York: Regan Books, HarperCollins; 1999.
[95] Giampapa VC, Pero R. Optigene Study. *J Anti-Aging Med.* (pub pending 2002).
[96] Sears B. *The Anti-Aging Zone.* New York: Regan Books, HarperCollins; 1995.

6

Exercise and Aging

Vincent C. Giampapa, M.D., F.A.C.S.

> If I knew I would live as long as I have, I would have taken better care of myself.
>
> Comedian George Burns at age 101

Overview

Chapter V discussed the effects of diet on aging and the powerful effects of food on aging. Exercise, like food, is an essential way of controlling essential hormonal levels. It is also one of the primary means of controlling body composition and creating positive changes in **quality of life**, or what we call "functionality." A number of key biomarkers of aging are improved with exercise, and although they may not increase longevity, they improve our **quality of time** as we age.

Exercise speeds up the fat-burning process by burning calories more efficiently and improving our metabolic rate. It also lowers insulin levels. When the insulin level is lowered, glucagon is elevated and exercise really starts to pay off by burning maximum levels of stored body fat for energy. Through control of the effects of insulin on aging mechanisms, a large cascade of events is set into place that affects everything from nuclear deoxyribonucleic acid (DNA) to hormonal signaling and the key components of the aging equation itself (Diagram VI-1).

Recent studies have shown that fat loss can result simply through a long-term walking program without any dieting. Vigorous walking has resulted in reduced fat body stores and decreased insulin requirements in both animal and human studies (see Diagram VI-1).[1–6]

The Principles and Practice of Antiaging Medicine for the Clinical Physician, 93–107.

Diagram VI-1

The Insulin - Exercise Connection

- Fat Loss -could be produced simply through a regular (i.e. at least four days a week), long-term walking program without any dieting.

- Vigorous regular walking has resulted in reduced body fat stores, reduced…insulin requirements (a 36% decrease in the ratio of insulin/glucose concentration occurred), and spontaneous reduced food intake.

1. J. Thompson Et. AL "Exercise and Obesity: Etiology, Physiology, and Intervention" Psych Bull 1982, 91:55-79
2. G. Gwinup: "Effect of Exercise Along on the Weight of Obese Women" Arch in Med 1975, 135: 676 -80
3. A.Leon Et.Al: "Effects of a Vigorous Walking Program on Body Composition…"J.Clin Nutr 1979, 32:17787.

In fact, large changes in insulin resistance, which normally occur as we age, have tremendous effects on aging as well (Diagram VI-2).[7] The effects of increased insulin resistance as we age include decreased longevity, hypertension, arteriosclerosis, blood lipid abnormalities, glucose fluctuations, osteoporosis, obesity and even tumor growth.

It has also been shown[8] that burning more than 2,000 calories per week on an exercise program does not improve longevity because it results in excess production of (1) free radicals and (2) cortisol.[9,10] Therefore, it becomes essential to understand, from a longevity and aging point of view, that exercise should be kept *moderate* and not *intense*.

Intense exercise, although not related to longevity, can improve functionality, thus resulting in extra muscle mass, strength, and improved quality of life (Diagram VI-3). Basically, exercise is the easiest way to control a number of the essential hormones and their overall physiological effects.

In general, exercise can:

1. Lower excess blood glucose.
2. Lower excess insulin.
3. Elevate growth hormone.
4. Elevate testosterone.

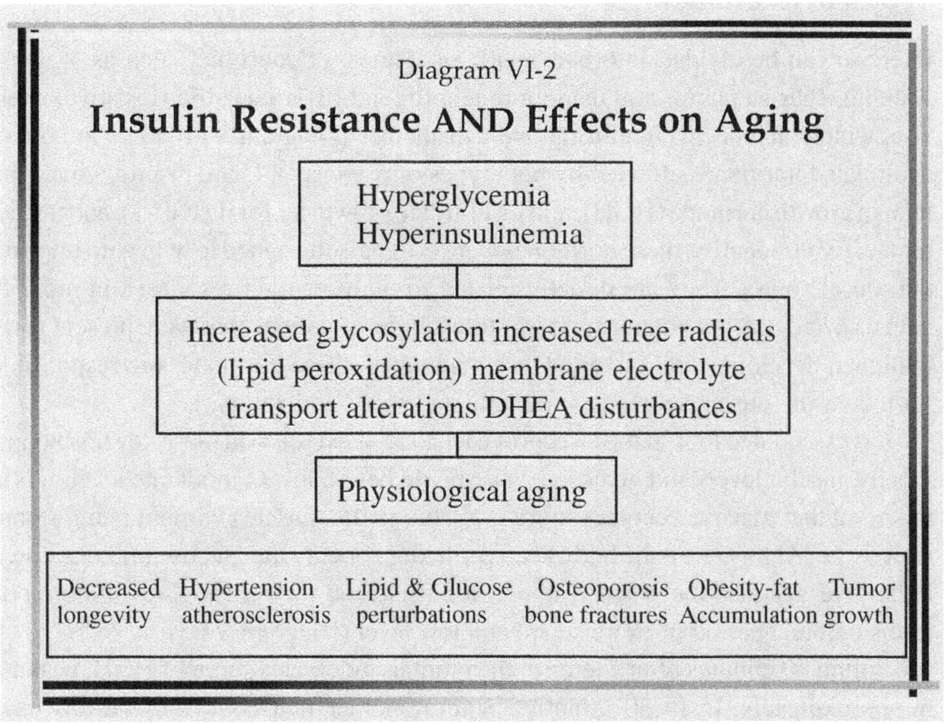

Diagram VI-2

Insulin Resistance AND Effects on Aging

Hyperglycemia
Hyperinsulinemia

Increased glycosylation increased free radicals (lipid peroxidation) membrane electrolyte transport alterations DHEA disturbances

Physiological aging

Decreased longevity	Hypertension atherosclerosis	Lipid & Glucose perturbations	Osteoporosis bone fractures	Obesity-fat Accumulation	Tumor growth

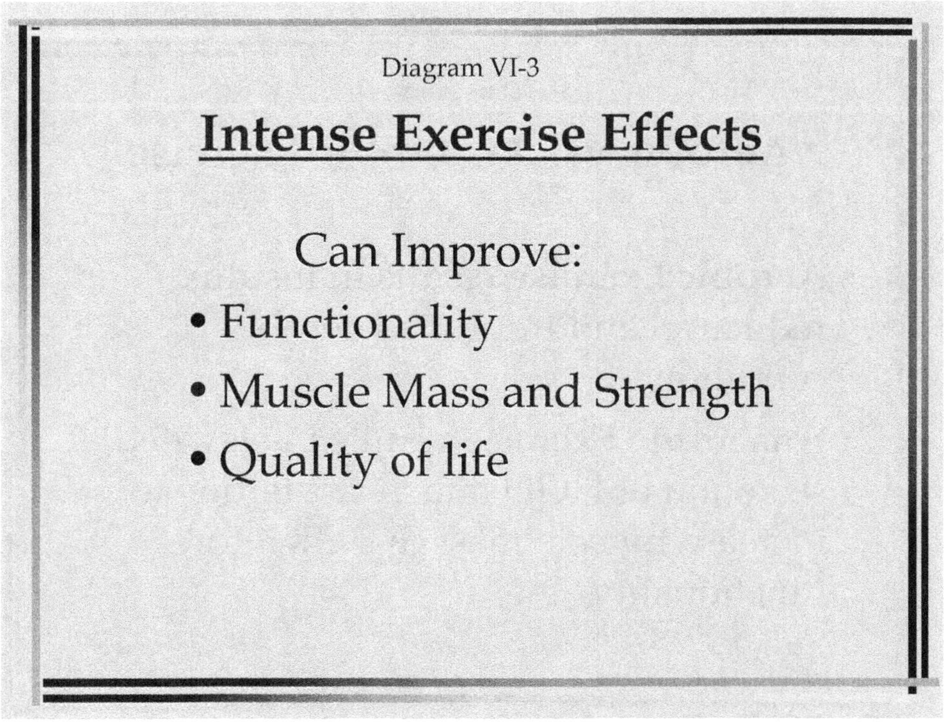

Diagram VI-3

Intense Exercise Effects

Can Improve:

- Functionality
- Muscle Mass and Strength
- Quality of life

Types of Exercises

Exercise can be divided into two major categories: (1) **aerobic**, such as jogging, walking, stair-stepping, and using a treadmill, and (2) **anaerobic** (resistive) exercise, which includes weightlifting and calisthenics (Diagram VI-4). The important age-related hormones affected by both types of exercise[11–13] are insulin, glucagon, human growth hormone (HGH), and insulin-like growth factor-1 (IGF-1), and testosterone. Evolutionarily, these hormones were originally designed to help with improving muscle mass. They are directly related to flight or fight responses in order to help us avoid potential danger and life-threatening situations that were present early in human developmental history. Once stimulated, these hormones are responsible for increasing our muscle mass, strength, and speed and agility.

It is essential to look at the effects of each type of exercise on these key hormones. Ideally, insulin levels and glucose levels should be at a low to moderate level. It is at this level that glucose becomes more available to the muscles without being stored as body fat. At this level, the body keeps utilizing not only the glucose initially stored in the liver but also the glucose stored in muscle and fat tissues. Growth hormone levels before a period of exercise are at a low level (Diagram VI-5).

Within 30 minutes after exercise, the pituitary releases a surge of HGH, peaking in approximately 15 to 20 minutes. Within another half-hour, these levels have

Diagram VI-4

Aerobic vs. Anaerobic Exercise

- **Aerobic Exercise** results in insulin reduction and improved aerobic condition
- **Anaerobic Exercise** results in elevated secretion of HGH and Testosterone to increase muscle mass and therefore functionality.

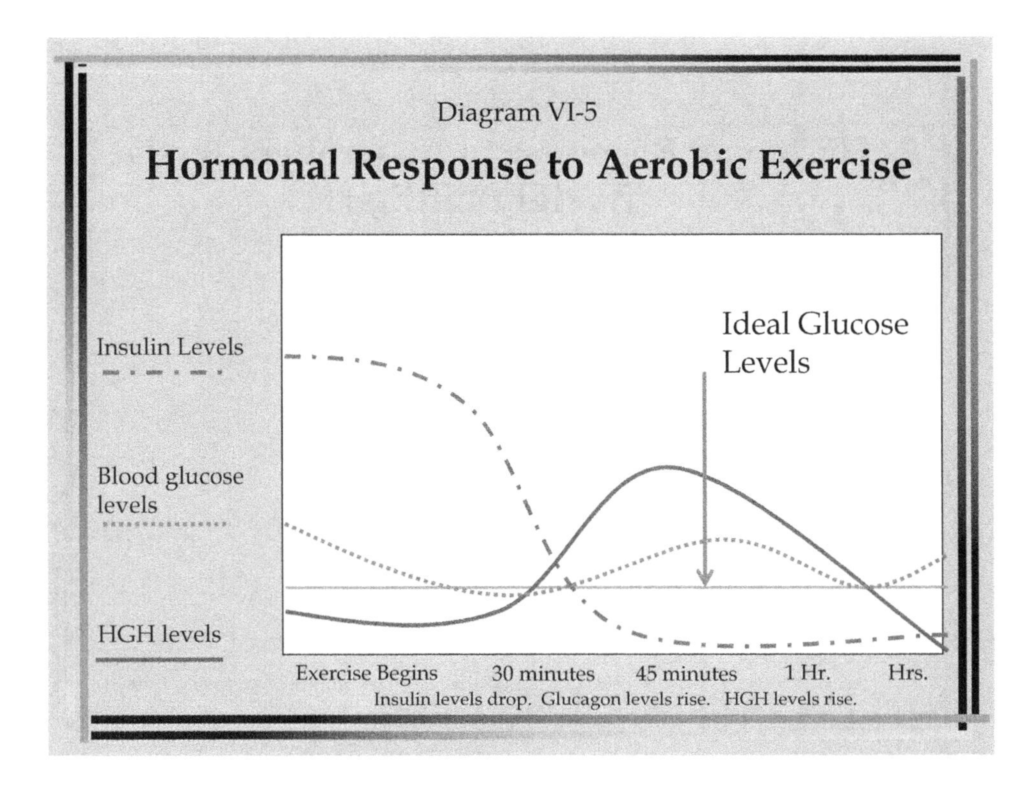

dropped back to baseline values. It is this surge of hGH that helps repair damaged muscle mass and increases muscle mass while decreasing body fat. In an earlier hostile environment, these were all protective responses for humans (e.g., cavemen). Today, the environment remains even more hostile, although in different ways. As growth hormone levels increase with the initiation of exercise, insulin levels begin to drop and glucagon levels begin to rise. This drop in insulin levels and the rise in glucagon and glucose levels helps to further stimulate the release of growth hormone.[14,15]

Aerobic Exercise These responses just described occur during **aerobic exercise**. For these reasons, it is more beneficial to begin aerobic exercises *before* a resistive exercise workout. In this state, the release of glucagon and insulin helps to stimulate the release of growth hormone. This effect is due to the aerobic exercise itself and also helps to improve the release of growth hormone and testosterone when resistive exercises or weightlifting has begun. The hormonal response to anaerobic exercise or weightlifting differs from the aerobic response just described.[16]

In summary, aerobic exercise results in improved cardiac conditioning and pulmonary function, which are some of the most important and earliest reversible biomarkers that *decrease* with aging. These biomarkers improve functionality, or quality of life.

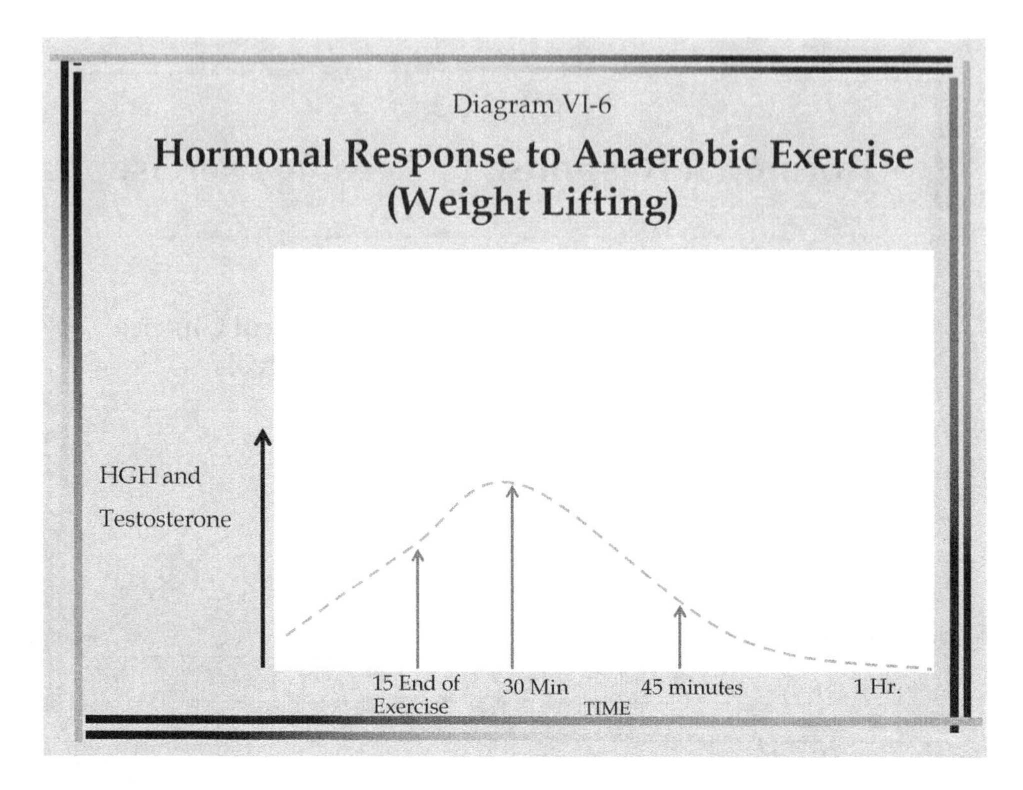

Anaerobic Exercise Anaerobic (resistive) exercise (Diagram VI-6) results in increased secretion of growth hormone,[17,18] testosterone and a marked improvement in muscle mass and body composition as well.

At the end of the weightlifting session, within minutes, both testosterone and growth hormone are released; in 30 minutes, they have hit their maximum peak. The effects of growth hormone and testosterone continue for approximately another 45 to 60 minutes. It is during this period, again, that weightlifting, which improves muscle tone, muscle mass, and tendon and ligament strength, has its effect.

Diagram VI-7 illustrates the fact that weightlifting has a greater effect than aerobic exercise in releasing more growth hormone and testosterone. In Diagrams VI-6 and VI-7, we see that the ideal sequence of exercise for anti-aging purposes would include a period of approximately 30 minutes of aerobic exercise (e.g., jogging) before anaerobic exercises (e.g., weightlifting). In this type of exercise plan, with an initial drop in insulin levels and a rise in glucagon levels, growth hormone is stimulated and released in greater quantities. With the glucose levels present under control and the glucose levels kept lower, growth hormone is also stimulated more effectively.

The stimulation of growth hormone occurs both aerobically and anaerobically.[18] In effect, the two stimulatory peaks can be added together to induce a greater stimulus to the pituitary to release its own growth hormone naturally. This allows the body to experience the benefit of both improved muscle mass and tissue repair.

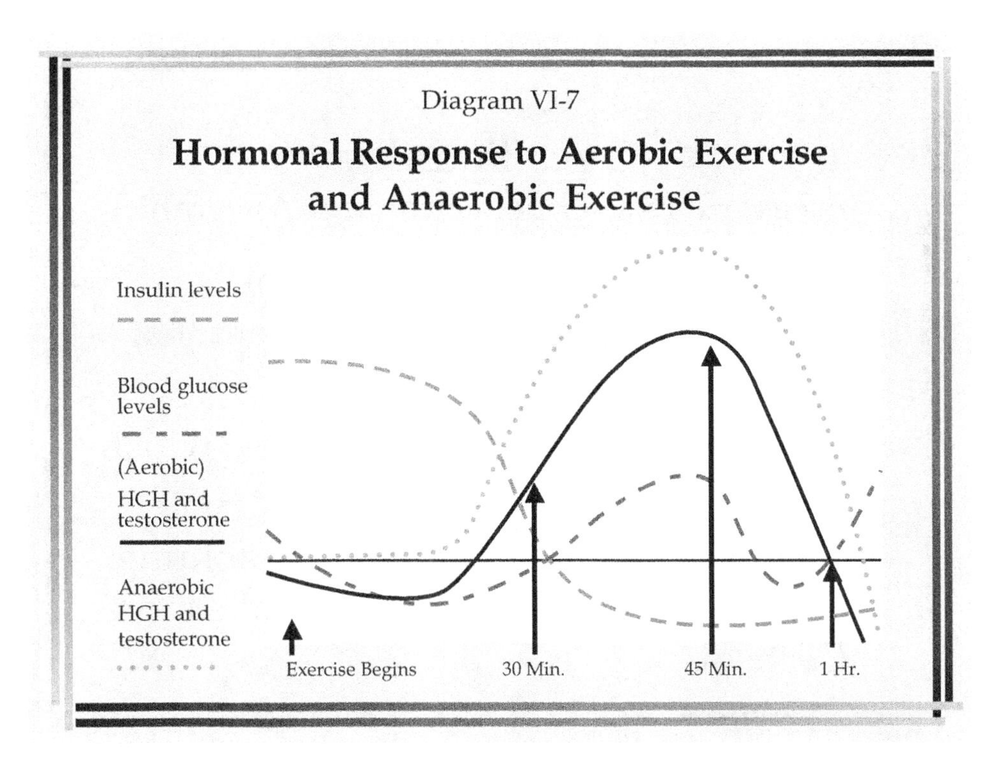

Taking certain supplements, or nutraceuticals, at a specific time can also help enhance both growth hormone and testosterone release. Supplements that help stimulate growth hormone release (e.g., arginine, ornithine or glutamine) ideally should be taken 30 minutes before weightlifting or anaerobic activity. Supplements that improve glucose levels (e.g., products containing vanadium, chromium, fenugreek, gymnema or carnosine) should be taken 30 minutes before the initial aerobic exercise. These supplements help stabilize glucose levels and therefore may help to improve glucagon response.

Testosterone-releasing compounds should be taken 30 to 45 minutes before the anaerobic or weightlifting activity. These supplements usually contain androstenedione, yohimbine extracts and other compounds, which help to stimulate testosterone release naturally.

Compounds that contain creatine enhance hormonal response and improve aerobic and anaerobic exercise capacity. This effect is due to improved energy response in the muscles because of the increased production of adenosine triphosphate (ATP), allowing greater endurance in both types of exercise responses.

Note: Too much exercise (more than 45 minutes) can create *negative* effects as far as anti-aging parameters are concerned (Diagram VI-8). These effects occur whether the exercise is aerobic or anaerobic.

Diagram VI-8

Over Training

(Excess exercise either Aerobic or Anaerobic)

- After 45 minutes excess levels of Cortisol increase to such an extent that other hormones begin to suffer
- > Cortisol causes > blood glucose which causes > insulin production
- Testosterone levels drop due to shifting of testosterone precursors to make more Cortisol

After 45 minutes, excess levels of cortisol begin to increase to such an extent that other hormone functions begin to become impaired. Free radical production also accelerates! Burning more than 2000 calories per week by exercise does not improve longevity at all but may improve strength and functionality (Diagram VI-9).

This elevation of cortisol causes an increase in blood glucose, which causes a stimulation of insulin production. We have already discussed how an elevated insulin level has detrimental effects on aging in general. In addition, at this time, testosterone levels drop because of the shifting of testosterone precursors, or dehydroepiandrosterone (DHEA), to make more cortisol.

From an anti-aging point of view, aerobic exercise, which affects mainly cardiovascular status, should be accomplished at least five times a week for a minimum of 20 minutes.

For heart and pulmonary biomarker improvement, cardiovascular activity at 55% to 65% the maximum heart rate creates the ideal level of exercise. (To calculate the maximum heart rate, simply deduct your age from 220.)

In summary, aerobic exercise has the following benefits (Diagram VI-10):

1. It enhances circulation and blood vessel elasticity, especially in the heart, as the heart muscle strength and cardiac mitochondrial density increases.

Diagram VI-9

Exercise and Longevity

- Burning greater than 2000 Calories per week does not improve longevity due to:

1 – Excess Free Radical Production
2 – Excess Cortisol Production

* Keep exercise moderate **Not** intense to improve longevity

Diagram VI-10

Exercise and Aging Benefits

Aerobic (Cardiovascular Exercise)

- Heart vascular benefits
- Fat Burning benefits

TYPE

- Biking
- Walking
- Running
- Jogging

2. It burns mostly fat for fuel, especially when we exercise on an empty stomach, or when we exercise 2 to 3 hours after a balanced meal with a minimum content of carbohydrates.
3. It increases the serotonin levels, if the duration is more than 45 minutes.
4. It improves body mass index (BMI), which is directly related to improved body composition.[16] A BMI greater than 27 kg/m^2 is related to an increase in age-dependent diseases (Diagram VI-11).

Anaerobic, or weight-resistive, exercise should be accomplished three to four times a week for no more than 45 minutes. This type of exercise benefits the body by increasing lean body mass (muscle), which increases the basic metabolic rate. It prevents bone loss by stimulating and strengthening the tendons and ligaments through the piezoelectric effect on osteoblasts in bone and may even increase bone density. Anaerobic activity also helps to maintain optimal joint health and strength (Diagram VI-12).

Stretching and Breathing Two other types of exercise are part of an age-management program. **Stretching** is important in parasympathetic and sympathetic (autonomic nervous system) regulation and in normalizing the essential nervous system neurotransmitters through the motion generated by inner ear stimulation (Diagram VI-13).[19] This relatively new finding has been used to benefit severely

Diagram VI-11

A BMI Greater Than or Equal to 27kg/m2 Equals High Risk for:

- Hypertension (> 140 systolic/ >90 diastolic mm hg
- Hyperlipidemia (cholesterol > 200mg dl triglycerides >225mg/dl
- Noninsulin-dependent (type II) diabetes mellitus
- Osteoarthritis
- Sleep apnea

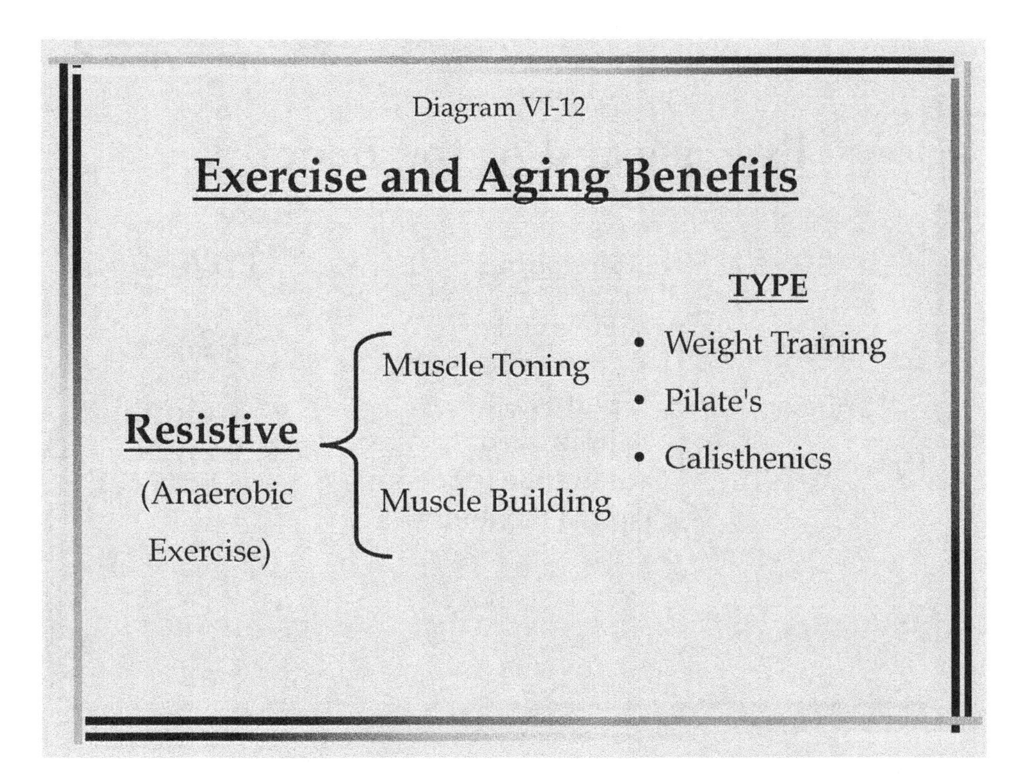
Diagram VI-12
Exercise and Aging Benefits
TYPE
Muscle Toning
Weight Training
Pilate's
Resistive
Calisthenics
(Anaerobic
Exercise)
Muscle Building

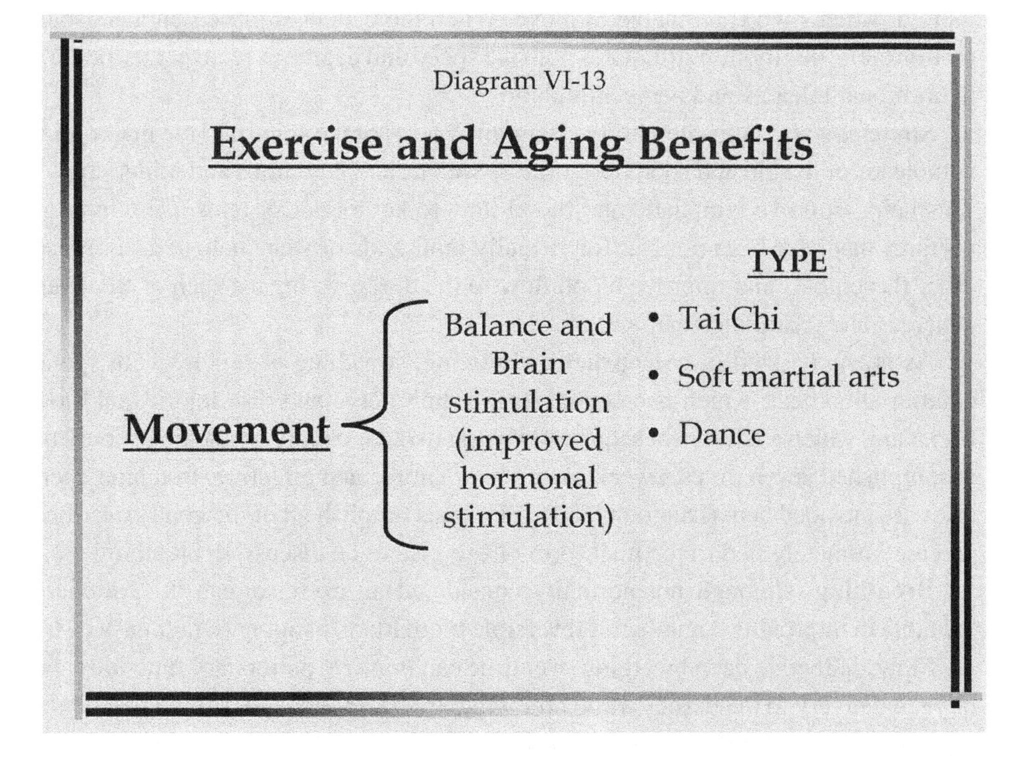
Diagram VI-13
Exercise and Aging Benefits
TYPE
Balance and
Brain
stimulation
(improved
hormonal
stimulation)
Tai Chi
Soft martial arts
Dance
Movement

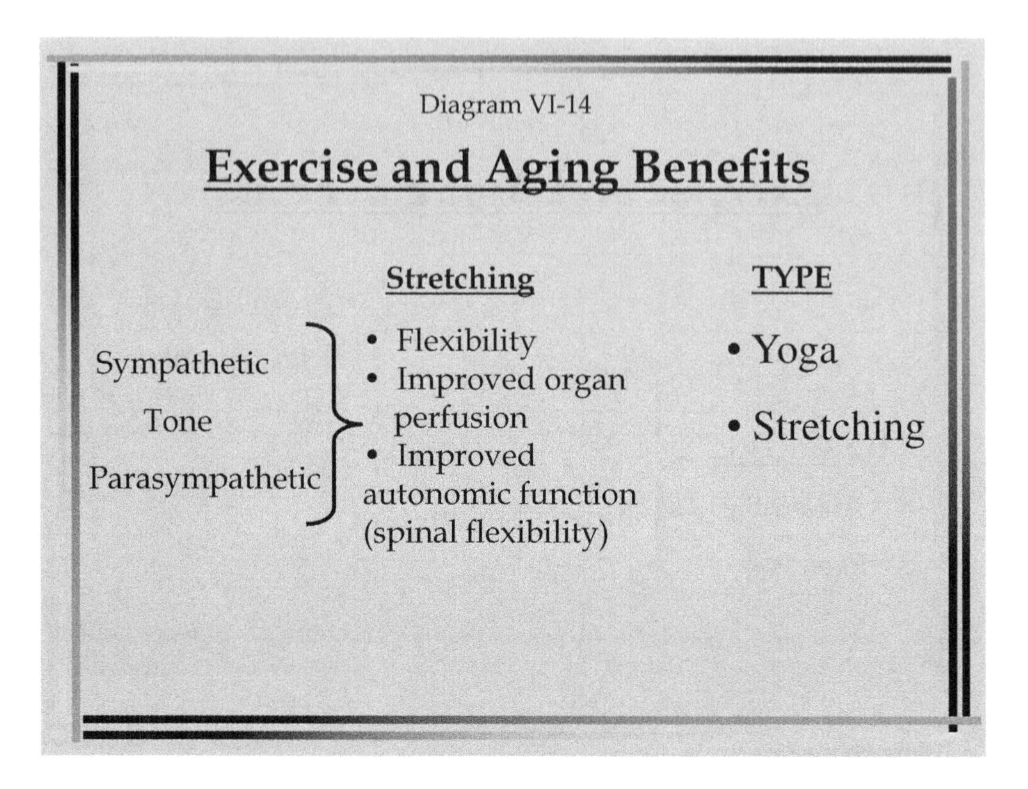

injured patients who are unable to move. When the patient is placed on a constant motion table, this motion stimulates both the speed and degree of recovery secondary to hormonal releases and nerve stimulation.

Stretching also helps flexibility (Diagram VI-14) of the spine and the major joint complexes of the hip and shoulder girdle as well as keeping limbs and joints supple. This helps improve lymphatic and blood flow to key organ systems. Certain yoga postures that have been popular for virtually thousands of years help to accomplish all of these goals and improve blood flow to the digestive organs such as the liver, pancreas and gastrointestinal system.

As is emphasized in **chiropractic medicine**, stretching also helps with spinal column alignment, which is essential in avoiding nerve and disc injuries and low back pain, which are frequent sequelae to aging in older people. Stretching should be accomplished seven times a week. A series of simple and effective stretching exercises are included here (Diagram VI-15). They accomplish all of the goals described and are intimately tied to the final form of exercise to be discussed, breathing.

Breathing, although not normally considered an exercise, can be extremely valuable in improving a number of reversible biomarkers of aging (Diagram VI-16).

Slow, deliberate deep breathing over time can improve pulmonary functions and oxygen delivery to the tissues while removing excess acidity and helping to balance not only the blood pH but also the cellular fluids (our "cellular soup"). There is

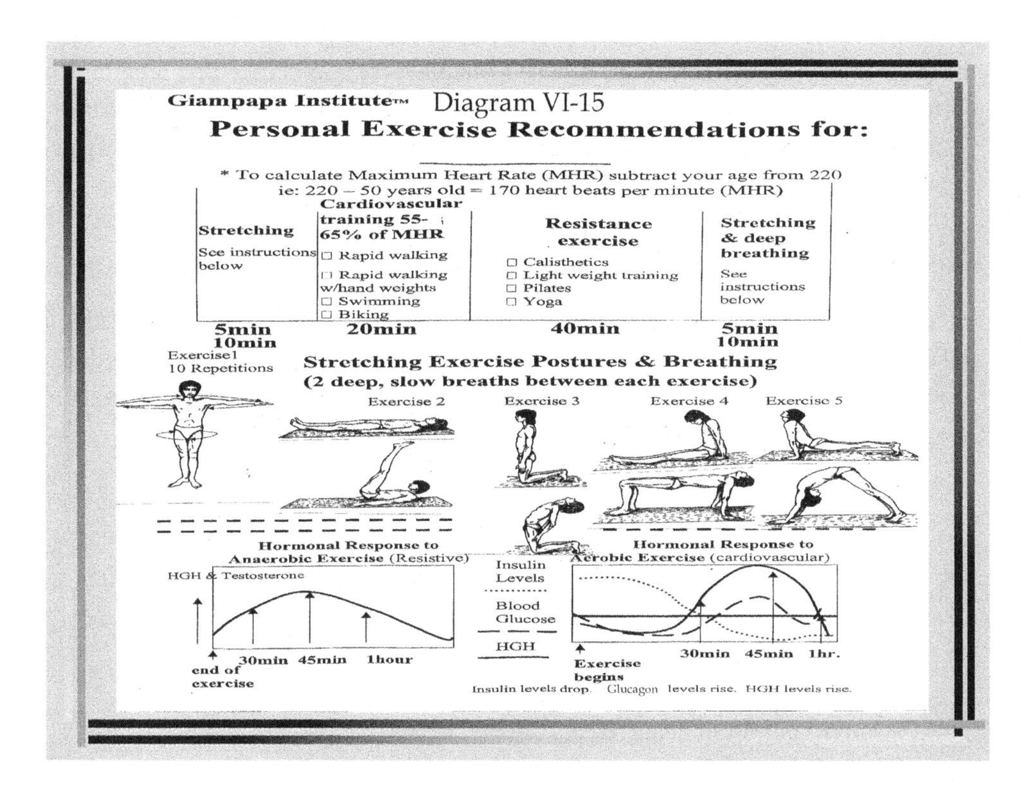

Giampapa Institute™ Diagram VI-15
Personal Exercise Recommendations for:
* To calculate Maximum Heart Rate (MHR) subtract your age from 220
ie: 220 – 50 years old = 170 heart beats per minute (MHR)
Stretching
See instructions below
Cardiovascular training 55-65% of MHR
Rapid walking
Rapid walking w/hand weights
Swimming
Biking
Resistance exercise
Calisthetics
Light weight training
Pilates
Yoga
Stretching & deep breathing
See instructions below
5min 10min
20min
40min
5min 10min
Exercise1
10 Repetitions
Stretching Exercise Postures & Breathing
(2 deep, slow breaths between each exercise)
Exercise 2
Exercise 3
Exercise 4
Exercise 5
Hormonal Response to Anaerobic Exercise (Resistive)
HGH & Testosterone
end of exercise
30min 45min 1hour
Insulin Levels
Blood Glucose
HGH
Hormonal Response to Aerobic Exercise (cardiovascular)
Exercise begins
30min 45min 1hr.
Insulin levels drop. Glucagon levels rise. HGH levels rise.

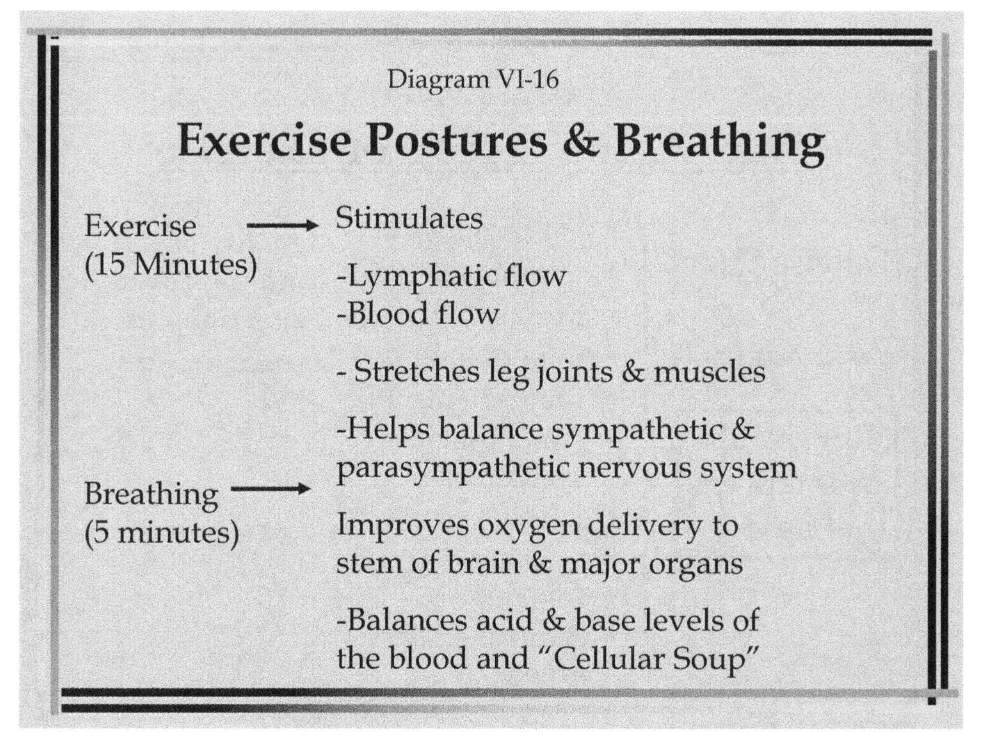

Diagram VI-16
Exercise Postures & Breathing
Exercise (15 Minutes)
Stimulates
-Lymphatic flow
-Blood flow
- Stretches leg joints & muscles
-Helps balance sympathetic & parasympathetic nervous system
Breathing (5 minutes)
Improves oxygen delivery to stem of brain & major organs
-Balances acid & base levels of the blood and "Cellular Soup"

also a "marked relaxation effect" on the central nervous system (CNS), leaving the performer in a calmer, more focused, mental state. Whether this breathing is performed in conjunction with yoga postures or on its own, the beneficial effects of a focused breathing session on many of the biomarkers of aging have been documented to be extremely effective.

Summary

Exercise has many positive effects on aging (Diagram VI-17). It should be an essential component of any anti-aging or age-management program designed for clinical or personal use. Exercise has many beneficial effects on the reversible biomarkers of aging, including:

1. Fat burning.
2. Optimizing hormonal levels.
3. Improving muscle repair and muscle mass.
4. Joint mobility.
5. Improving brain chemistry.
6. Balancing the autonomic nervous system.
7. Improving body balance.
8. Enhancing cardiovascular function.

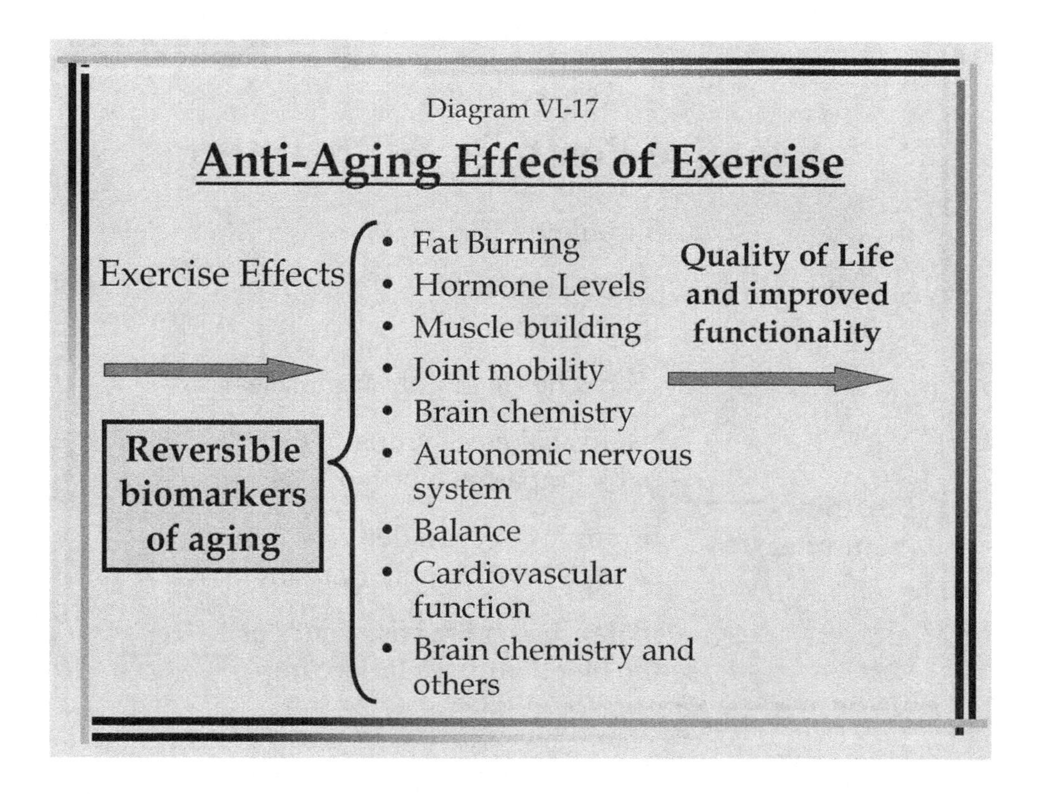

All of these key modifiable biomarkers of aging are directly related to an improvement in quality of life and functionality that we experience as we age. We need to make exercise an essential component or habit in our daily lives. It is an integral component of any age-management program.

References

[1] Yamanouchi KT, Shinozaki K, Childada T, et al. Daily walking combined with diet therapy is useful means for obese NIDDM patients not only to reduce body weight, but also to improve insulin sensitivity. *Diabetes Care.* 1995; 18: 775–778.

[2] Holloszy JO, Schultz J, et al. Effects of exercise on glucose tolerance and insulin resistance. *Acta Med Scand.* 1996; 711: 55–65.

[3] Mayor H, Davis EJ, D'Agostino R, et al. Intensity and amount of physical activity in relation to insulin sensitivity. *JAMA.* 1998; 279: 669–674.

[4] Thompson J., et al. Exercise and obesity: etiology, physiology, and intervention. *Psych Bull.* 1982; 91: 55–79.

[5] Gwinup G. Effect of exercise alone on the weight of obese women. *Arch in Med.* 1975; 135: 676–680.

[6] Leon A, et al. Effects of a vigorous walking program on body composition. *J Clin Nutr.* 1979; 32: 1776–1787.

[7] Folsom AR, Jacobs DR, Wagenknecht LE, et al. Increase in fasting insulin and glucose over seven years with increasing weight and inactivity of young adults. *Am J Epidemiol.* 1996; 144: 235–246.

[8] Blair SN, Kohl HW, Gordon NF, Paffenbarger RS. How much physical activity is good for health? *Ann Rev Pub Health.* 1992; 13: 99–126.

[9] Leon AS, Connett J, Jacobs DR, Rauramaa R. Leisure-time physical activity levels and risk of coronary heart disease and death: the Multiple Risk Factor Intervention Trial. *JAMA.* 1987; 258: 2388–2395.

[10] Alessio HM. Exercise-induced oxidative stress. *Med Sci Sports Exerc.* 1993; 25: 218–224.

[11] Viru A. *Hormones in Muscular Activity.* Vol. I. *Hormonal Ensemble in Exercise.* Boca Raton, FL: CRC Press; 1983.

[12] Viru A. *Hormones In Muscular Activity.* Vol. II. *Adaptive Effects of Hormones in Exercise.* Boca Raton, FL: CRC Press; 1983.

[13] Galbo H, Holst JJ, Christensen NJ. Glucagon and plasma catecholamine response to graded and prolonged exercise in man. *J Appl Physiol.* 1975; 38: 70–76.

[14] Galbo H, Holst JJ, Christensen NJ. The effects of different diets of insulin on the hormonal response to prolonged exercise. *Acta Physiol Scand.* 1979; 107: 19–32.

[15] Holloszy JO, Schultz J, Kusnierkiewicz J, et al. Effects of exercise on glucose tolerance and insulin resistance. *Acta Med Scand.* 1996; 711: 55–65.

[16] Van Helder W, et al. Growth hormone regulation in two types of aerobic exercise of equal oxygen uptake. *Eur J Appl Physiol.* 1986; 55: 236–239.

[17] Hagberg JM, Seals DR, Yerg JE, et al. Metabolic responses to exercise in young and older athletes and sedentary men. *Journal of Applied Physiology.* 1988; 65: 900–908.

[18] Pyka G, Wiswell RA, Marcus R, et al. Age-dependent effect of resistance exercise on growth hormone secretion in people. *J Clin Endocrinol Metab.* 1992; 75: 404–407.

[19] Hutchinson M. *Mega-Brain Power.* New York: Hyperion; 1994: 102–107.

7

Skin and Aging

The Cosmeceutical Approach

Nicholas V. Perricone, M.D.
Vincent C. Giampapa, M.D., F.A.C.S.

It is by logic that we prove, but by intuition that we discover.
Henri Poincaré

Introduction

We, as cosmetic surgeons and physicians, have become *acutely* aware of the signs and symptoms of aging skin. As the largest organ of the body, the skin *manifests* both the effects of the internal aging processes as well as the effects of the 21st century's toxic environment that surrounds us.

Recent scientific research has also supported the theme, developed in the preceding chapters, concerning key components of the **aging equation** on skin as we grow older. More specifically, the molecular effects of glycation and inflammation" have been directly linked to accelerated skin aging.

Also in accord with the scientific research presented in prior chapters on the effects of transcription factor kappa B (NF-kB) and deoxyribonucleic acid (DNA), we again can see immediate clinical relevance to controlling these essential reactions as a means of inhibiting the age-related changes found in the skin (Diagram VII-1).

The **membrane hypothesis of aging**, proposed in 1977 by Dr. Imre Zs-Nagy, briefly discussed in Chapter II, sheds light on the underlying cellular aging phenomenon, with direct clinical relevance to the physicians who now possess the significant treatment options to impede and alleviate many of the underlying causes of aging skin.

The Principles and Practice of Antiaging Medicine for the Clinical Physician, 109–128.

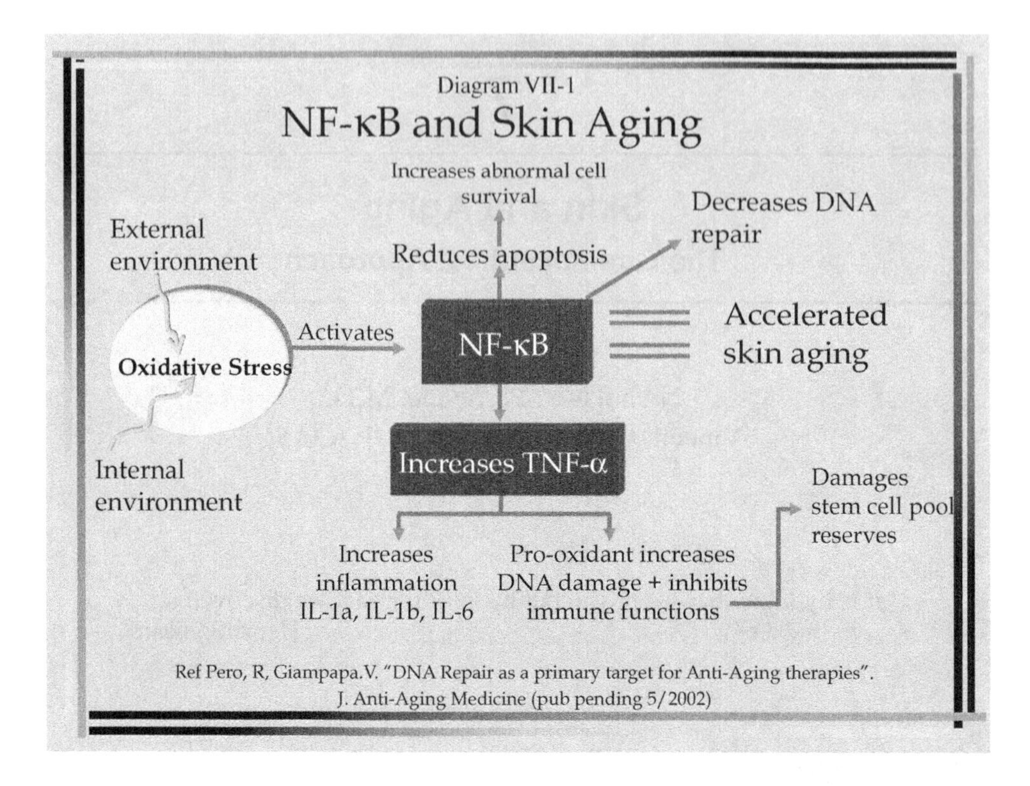

Although our research has evolved independently over the last decade, we have both come to the same conclusions concerning the key targets to attack in the war on skin aging.

It is with much gratitude and pleasure that I am able to include this chapter by Dr. Nicholas V. Perricone. It includes a wealth of clinically valuable research and treatments that can be immediately incorporated into the practice setting with minimal effort and maximal effects.

Skin and Aging

As practicing physicians, we must *understand the role of free radicals* and inflammation in health and aging. Most scientists agree that free radicals and the resultant inflammation are at the basis of such seemingly diverse illnesses as Alzheimer's disease, cardiovascular disease, Parkinson's disease, cancer and aging. My research over the past 15 years has indicated that free radicals, although they are a very short-lived species, do very little direct damage to cells under most physiological conditions; however, they trigger an inflammatory cascade on a cellular level that is perpetuated for hours and days. The *importance* of understanding the relationship between free radicals and inflammation *cannot* be overemphasized.

Over the past two decades, biomedical nutritional studies have emerged illustrating the importance of antioxidants in the prevention of acute and chronic disease as well as the aging process. My own research to date has shown that if we augment the antioxidant defenses, focusing on those agents that have the most potent anti-inflammatory activity, a significant clinical response can be obtained. This is especially evident in dermatological research, in which the target organ is visible and therefore clinical observations can be made and appreciated.

Free Radicals in Aging and Disease

As far back as the early 1950s, University of Nebraska researcher Denham Harman, M.D., Ph.D., began to observe connections between free radicals and cellular aging — links that prompted him to propose the now famous "free radical theory of aging."[1] Four decades later, most informed observers agree with his hypothesis that oxidative stress *promotes* premature aging and *causes* or exacerbates major degenerative disorders such as arteriosclerosis (clogged arteries), cataracts, Alzheimer's disease, macular degeneration, diabetes and certain cancers.[2,3]

The primary agents of excessive oxidative stress are free radicals generated by ultraviolet radiation, psychological stress, chemically unstable environmental pollutants, excessive dietary sugars, and chronic subclinical inflammation. The oxidative stress exerted by these factors — as well as the poor diets consumed by the public, which are inherently pro-inflammatory and antioxidant-deficient — results in the failure of the endogenous defense systems. Is it any wonder that we are witnessing a rise in diseases related to inflammation and an increased number of patients seeking solutions to the visible signs of premature skin aging?

The Membrane Hypothesis of Aging

Dr. Zs-Nagy's membrane hypothesis of aging (MHA)[3a] encompasses Harman's **free radical theory**; however, it goes on to the next step to elucidate the importance of the cell plasma membrane as a target of free radical activity.[4] Thus, building on the work of Harmon, Zs-Nagy explored mechanisms by which free radicals damage cells and promote premature aging.

At one time, most observers assumed that free radicals aged cells by damaging their DNA. By the late 1970s, however, Zs-Nagy demonstrated that free radicals weaken the oxygen-rich lipid bilayers of cellular membranes.[5,6] Being more compact, cell membranes are damaged at a higher rate than the components in the interior of the cell. Damage to their outer membranes renders cells less able to absorb nutrients or to expel toxic waste material, such as lipofuscin. (Human and animal studies show that high lipofuscin levels correlate with poor cellular health.) Waste material (e.g., lipofuscin) and salts (e.g., potassium) accumulate and osmotically exclude water. Once this happens, the dehydrated cell loses its ability to produce proteins and enzymes needed for repair and renewal. Within the cytoplasm, colloid density increases, resulting in a continued cycle of cell degeneration.

In line with the predictions of Zs-Nagy's hypothesis, the nootropic (memory enhancing) antioxidant drug centrophenoxine has been proven to slow age-related deterioration of the inner cell, largely by reducing lipofuscin levels. The chief active component of centrophenoxine is **dimethylaminoethanol (DMAE)**, a membrane stabilizer and metabolic precursor to the neurochemical **choline (trimethylaminoethanol)**.[7]

Zs-Nagy's well-supported hypothesis enables us to develop uniquely effective strategies to counter the ill effects of free radicals. For the most part, nutritional anti-aging strategies have focused too narrowly on water-soluble anti-oxidants, which are best suited to scavenging (inactivating) free radicals in intercellular spaces and the interior cytoplasm. But we now understand that while the watery cytoplasm needs antioxidant protection, the lipid bilayers of the cell membrane are an equally important focus of the struggle to control free radical activity. The two-front nature of the anti-aging battlefield dictates a defense that employs both lipid-soluble and water-soluble antioxidants as well as the uniquely versatile metabolic antioxidant **α-lipoic** acid (ALA) whose extraordinary capacities are described later in this chapter.

The Inflammation-Aging Theory of Aging

Free radicals are highly reactive species that damage biomolecules on contact. As mentioned, they exist only for a fraction of a second: however, they *trigger* inflammatory activity, which is long-lasting. This can be seen in the aging individual who has higher circulating levels of tumor necrosis factor-alpha (TNF-α), interleukin 6 (IL-6), cytokine antagonists and acute phase proteins. Dr. Claudio Franceschi of the University of Bologna has even coined a new term, "inflamm-aging,"[8] to describe the deleterious effects of evolutionary mechanisms that become counterproductive as we age and gradually lose endogenous antioxidant capacity.[9–17]

To understand the connection between Harman's free radical theory, Zs-Nagy's membrane hypothesis and my proposed **inflammation-aging connection**, we must take a closer look at oxidative stress and gene expression. The inflammatory chain follows this circular path[7,17–30] (Diagram VII-2):

1. Cell membranes, which are rich in polyunsaturated fatty acids (PUFAs), contain oxygen levels eight times higher than in the other portions of the cell. This results in greater generation of reactive oxygen species (ROS), which easily attack the multiple double bonds found in the polyunsaturated acids.[31,32]

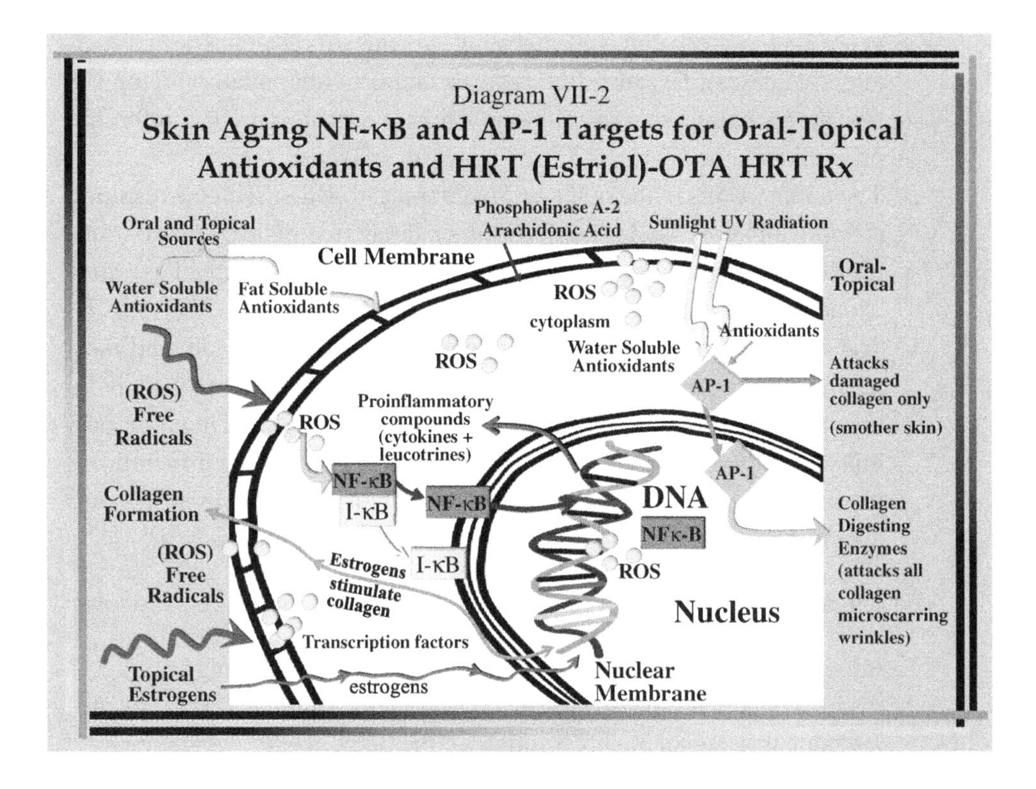

2. Free radical damage to cell membrane lipids results in the release of arachidonic acid (a precursor to pro-inflammatory prostaglandins) and some 200-plus PUFA fragments — some of which mimic the pro-inflammatory effects of platelet-activating factor (PAF).[31,32]
3. Release of pro-inflammatory arachidonic acid and PUFA fragments by free radical action on cell membranes initiates a cascade of events, including protein complex assembly and phosphorylation, that results in activation of oxidation-reduction (redox)–sensitive transcription factors, which in turn activate pro-inflammatory genes.[17,24–26] (Arachidonic acid also enters the mitochondria, where it generates free radicals and disrupts the process of oxidative phosphorylation by which adenosine triphosphate [ATP] produces cellular energy.[32])
4. The transcription factors NF-κB and activator protein 1 (AP-1) are activated by free radicals and pro-inflammatory cytokines generated by free radical activity. NF-κB and AP-1 also play critical roles in the regulation of pro-inflammatory cytokines, related proteins, and collagen-digesting enzymes. NF-κB stimulates pro-inflammatory cytokine genes such as IL-1, IL-2, IL-6, IL-8, and TNF-α along with genes that code for cell adhesion molecules and the Cox-2 enzyme. AP-1 mediates induction of

genes that express collagen-digesting enzymes (collagenases and metalloproteinases). Because free radicals and pro-inflammatory cytokines activate each of these transcription factors, the result is a self-reinforcing pro-inflammatory cycle[17–30] (see Diagram VII-1).
5. The initial insults to the outer cell membrane, together with the resulting pro-inflammatory cycle, produce intracellular free radicals that oxidize the PUFA-rich membranes surrounding the cytoplasm's organelles, mitochondria and nucleus.[27]
6. The mitochondria produce cellular energy via the ATP cycle, and most repair and reproduction activities occur via production of proteins by organelles, as directed by the cell's genetic material.[27] Thus, the pro-inflammatory cycle initiated by oxidative stress on the cell membrane gradually weakens the most basic functions of dermal cells while producing the enzymes that yield collagen damage and microscarring.[33]

The salient points with regard to reducing signs of skin aging are as follows:

1. Inflammation is a cornerstone of aging, including skin aging.
2. Antioxidants neutralize free radicals and thereby inhibit release of pro-inflammatory signal agents throughout the body.

Fortunately, it is not difficult to boost tissue levels of antioxidants — including endogenous antioxidants — via anti-inflammatory diets, supplements, and topical applications of antioxidants with powerful anti-inflammatory activity. Thus, my argument that the cell membrane is the site of inflammation and an etiologic factor in disease and aging is a reasonable hypothesis.

Excess Dietary Sugar: A Primary Pro-Inflammatory Influence

I would like to *stress* that together with ultraviolet radiation, psychic stress, and environmental toxin loads, "protein glycation" is one of the greatest forces driving subclinical inflammation. It is difficult to overemphasize the *impact* of a high-sugar diet on oxidative stress. Much attention has focused on increasing antioxidant intake to slow aging — which surely is wise. In our junk food culture, however, it is easy to overlook the other side of the equation. Every doughnut, candy bar and soda adds to a toxic excess of dietary glucose.

We need only look at the dramatic acceleration of aging parameters seen in diabetes to understand the insidiously pro-inflammatory nature of chronic hyperglycemia. We also know that chronic elevated serum insulin — an effect of chronic hyperglycemia and resulting diminished glucose sensitivity — is pro-inflammatory.[34] In addition, incubation of cells in sugar produces an inflammatory state, and advanced glycation and oxidation (glycoxidation) products, such as

N-ϵ-(carboxymethyl)lysine (CML) and pentosidine, are reduced in skin collagen by caloric restriction.[13,33]

Why is glycation so harmful? It generates highly active, persistent centers for catalyzing one-electron oxidation-reduction reactions.[35] In other words, glycation bonds are virtual free radical factories. I *urge* my patients to consume moderate amounts of low-glycemic carbohydrates rich in antioxidants and to shift their dietary balance in favor of lean protein and foods high in the anti-inflammatory omega-3 and omega-9 fatty acids (e.g., fish, nuts, seeds, dark green vegetables and olive, flax and hemp oils). The *effects* are both rapid and visible; within days, fine lines and puffiness diminish — and the effects grow over time. In truth, what we eat shows in our faces as we age.

Antioxidant Cosmeceuticals

Building on the work of Harman and Zs-Nagy and countless other scientists has resulted in the membrane-based inflammation-aging connection. This led to a therapeutic strategy designed to scavenge free radicals, stabilize lipid membranes, enhance dermal cell hydration and metabolic homeostasis, inhibit pro-inflammatory cytokines and transcription factors, and reduce release of collagen-digesting enzymes and subsequent microscarring. Substantial progress has been made in revealing the mechanisms of photoaging, which include inducement of matrix metalloproteinases by nuclear transcription factors AP-1 and the pro-inflammatory transcription factor NF-κB.[36–41] These events can be obviated by use of topical antioxidant, anti-inflammatory agents.[42]

The body's main endogenous defenses against free radicals are enzymatic antioxidants, such as glutathione and superoxide dismutase, and their ability to control free radicals is substantially enhanced by exogenous (dietary and topical) antioxidants. Supporting experimental data show that exogenous antioxidants can extend cell life and reduce two of the key agents of aging — glycation and lipid-protein peroxidation.[43,44]

Not surprisingly, epidemiological studies indicate that rates of degenerative, oxidation-related disease are reduced by diets high in antioxidant phytochemicals (e.g., vitamin C, vitamin E, carotenes and various polyphenols). Protein glycation generates free radicals and protein crosslinking closely associated with diabetes and prematurely aged skin,[45] whereas lipid-protein peroxidation produces the lipofuscin (dark brown lipochrome) closely associated with advanced aging and with degenerative diseases such as atherosclerosis and age-related cognitive decline.

Antioxidant nutrition offers exciting new prevention and treatment strategies for health conditions characterized by inflammation, including prematurely aging skin. A group of underappreciated antioxidants — some of which are not widely recognized as such — offer particular promise in the continuous struggle to control oxidative stress and consequent inflammation. In the sections that follow, we review relevant research into five leading skin anti-aging agents:

- Vitamin C ester (ascorbyl palmitate)
- Lipoic acid
- DMAE
- Tocotrienols
- Alpha and beta hydroxy acids

Vitamin C Ester: The Better Topical "C" Vitamin C (**ascorbic acid**), of course, is essential for the production and repair of collagen, but it is also capable of reducing inflammation in the skin. Indeed, the skin-protective and restorative powers of vitamin C are well established.[46–51] Thanks to its twin skin anti-aging attributes, adequate tissue levels of vitamin C are essential in helping to reduce wrinkles.

However, dietary vitamin C cannot maintain adequate skin levels in the face of oxidative stress.[50,51] Vitamin C is also unstable and lipid-insoluble — it neither penetrates skin nor prevents oxidative damage to cell membranes effectively, and it rapidly degrades in cosmetic products.[52–54] Being water-soluble, ascorbic acid can only work on the interior of a cell; it passes through the membrane, but it cannot heal free radical damage on the cell's surface. Topical ascorbic acid is even partially counterproductive, as it initiates inflammatory Fenton reactions that produce hydroxyl radicals and visibly inflame the skin.

Fortunately, a synthetic ester of vitamin C, called **ascorbyl palmitate**, realizes the nutrient's full potential as a topical anti-aging agent. Vitamin C ester consists of ascorbic acid molecules joined to a fatty acid from palm oil. Compared with topical ascorbic acid, the ascorbyl palmitate ester of vitamin C is a far preferable cosmetic antioxidant, as the following points demonstrate[55–58]:

1. Ascorbyl palmitate is physiologically active in human cells and at lower doses shows greater activity than ascorbic acid.
2. Ascorbyl palmitate is lipid-soluble and penetrates skin more effectively compared with ascorbic acid. Although results of vitamin C ester skin absorption tests have varied, most results suggest a substantial advantage over ascorbic acid. A comparative study by Procter and Gamble researchers showed that skin absorption of topical ascorbyl palmitate is six times higher than skin absorption of ascorbic acid when measured 2 and 24 hours after application.[59]
3. Because ascorbyl palmitate can reside in the cell membrane, it can regenerate the vitamin E radical on a continuous basis. Ascorbic acid interacts with vitamin E only at the interface of the water-soluble and lipid components.[46,52–54,56]

4. Unlike ascorbic acid, ascorbyl palmitate is stable, nonacidic (and nonirritating) and does not generate the potent hydroxyl radical when it comes into contact with the iron abundant in human skin.[52–54,56]
5. Ascorbyl palmitate possesses superior ability to stimulate growth of fibroblasts — cells that help produce collagen and elastin in human skin.[58,60]
6. Ascorbyl palmitate's superior chemical stability allows it to remain potent in creams and lotions for up to 2 years.[52]

The resulting cosmetic benefits of this specific vitamin C ester are not merely hypothetical, as shown by preliminary research testing its effects successfully against sunburn and psoriasis in placebo-controlled trials.[46–50,52–58] I am persuaded of vitamin C ester's benefits by the related experimental research, by my clinical evidence and by the striking results seen by my patients. It is clear that topical ascorbyl palmitate aids in reducing and preventing wrinkles and sagging skin in two ways: (1) by helping repair collagen and (2) by protecting skin from oxidative damage induced by ultraviolet radiation.

ALA: The Universal Antioxidant Without doubt, ALA is the most exciting skin anti-aging agent discovered to date. Its beneficial cosmetic properties center on its dual roles as a biphasic antioxidant and as a potent inhibitor of pro-inflammatory transcription factors.[61–65]

This disulfhydryl coenzyme was discovered in 1951, when it was recognized as an essential component of enzymes involved in mitochondrial electron transport (energy) reactions. ALA, which contains sulfur groups in a dithiol ring structure, is small and lipid-like in character, with a water-soluble portion that renders it both water-soluble and membrane-soluble. (This biphasic property, among others, has led to its being termed the "universal antioxidant.") In the late 1980s, research revealed ALA as a potent antioxidant that is readily transported through cellular membranes.[62,66]

ALA's Diverse Antioxidant Properties After reduction by enzymes in the body to its dithiol form (dihydrolipoic acid [DHLA]), lipoic acid gains markedly enhanced free radical–reducing power. ALA is a potent inhibitor of a key pro-inflammatory transcription factor. Together, ALA and DHLA form a uniquely versatile antioxidant, anti-inflammatory duo that vigorously "recycles" oxidized antioxidants:

1. ALA and DHLA have been shown to inactivate (scavenge) hydroxyl and peroxyl radicals as well as hypochlorous acid, singlet oxygen and nitric oxide. They can also chelate transition metals involved in oxidative reactions.[61,66]

2. ALA reduces (i.e., recycles to full potency) oxidized glutathione, vitamin E and ascorbate and boosts the cell's ability to synthesize glutathione by reducing cystine to cysteine.[67]
3. ALA possesses the ability to block protein glycation, a key factor in skin aging, owing to the protein crosslinking and free radicals it produces. ALA produces an indirect antioxidant effect in this context because the glycation bond continuously generates free radicals. In addition, ALA increases cellular glucose uptake through recruitment of the glucose transporter-4 to plasma membranes, thus reducing the amount of free glucose available for glycation reactions.[68–71]

ALA's Ability to Block Key Inflammation Signals As people age, cumulative damage by free radicals *lowers* the activation thresholds of redox-modulated transcription factors. The system that activates genes deteriorates, producing inflammation, immunosenescence, cancer and degenerative disease. ALA has special properties and functions within the antioxidant network that contribute to its impressive anti-aging properties.

ALA helps to *control* inflammation by regulating the redox-sensitive nuclear transcription factors known as activator protein 1 (AP-1) and NF-κB.[72,73] ALA exerts the following beneficial effects on these key transcription factors:

Activator Protein-1. AP-1 is activated bimodally. When the cell senses oxidative stress or inflammation, AP-1 is activated and results in the expression of genes that regulate production of matrix-degrading enzymes such as metalloproteinases, collagenase and gelatinase. At the other end of the spectrum, when AP-1 is activated by a powerful antioxidant like ALA, the result is a gene expression responsible for production of matrix metalloproteinases that work only on damaged collagen. Thus, when AP-1 is activated by oxidative stress it results in damage to intact collagen, which causes microscarring that eventually can be seen in skin as a wrinkle.

When AP-1 is activated by ALA, damaged collagen is degraded, resulting in resolution of scarring and wrinkling.[72,74] Thus, ALA interacts with transcription factor AP-1 in a positive mode, which repairs microscarring, restores skin elasticity and reduces wrinkles.

Nuclear Factor–κB. NF-κB can bind to DNA and can cause changes in the rate of expression of specific genes associated with the inflammatory response, as well as impede DNA repair. (The genes of older animals and humans have more NF-κB bound to them compared with the genes of younger ones.) ALA regulates NF-κB in the following ways[75–78]:

1. NF-κB is held in check by subunits called I-kB proteins. When an I-kB protein binds to NF-κB, the complex cannot pass from the cell cytoplasm through the porous lipid bilayer membrane of the nuclear envelope or

bind to nuclear genes. Free radicals, peroxides, and ultraviolet energy induce the inactive complex to dissociate, thus permitting NF-κB to penetrate into the nucleus and damage DNA.

2. Activation of NF-κB by free radical activity transforms it from a trimeric protein to a heterodimer — a process during which the inhibitory I-kB protein is cleaved and destroyed by a protease enzyme. The heterodimer then migrates through the nucleus, finds the DNA and produces an inflammatory gene expression, and/or inhibits DNA repair.
3. ALA inhibits both free radicals and activation of NF-κB, thereby halting NF-κB-mediated inflammatory damage more effectively and rapidly compared with any other antioxidant studied to date (e.g., coenzyme Q10 or *N*-acetylcysteine). For example, whereas dietary polyphenols (e.g., anthocyanins) are extremely potent antioxidants that are generally desirable as anti-aging factors, they are relatively large molecules that cannot readily migrate through cellular membranes. In contrast, ALA is very small and is readily transported through cellular membranes, including the nuclear membrane. It can terminate free radicals both in serum and on the cellular membrane while protecting nuclear DNA from dissociated NF-κB.

DMAE: Nature's Own "Facelift" Millions of people take dimethylaminoethanol (DMAE) capsules to increase cognitive function. DMAE occurs in high levels in fish — known as a traditional "brain food" — and for good reason, as it happens. DMAE is an endogenous nutritional complex that *enhances* nerve transmission to dramatically improve the appearance of sagging skin while stabilizing cell membranes and boosting the line-reducing, smoothing effects of antioxidants. Loss of firmness is a significant factor in visible skin aging, particularly in the face. We have all believed that surgery is the only way to address sagging; DMAE complex, however, offers another solution.

As we age, we begin to lose the baseline muscle tone that keeps our skin firm. This is because the chemicals and nutritional precursors that are responsible for production of the neurotransmitter acetylcholine (which controls baseline muscle tone) diminish as a result of free radical damage to the nerves that produce them. Whenever the proper amount of acetylcholine is delivered to muscle tissue, it responds with baseline muscle tone. The documented and observable tone-enhancing effects of DMAE appear related, in part, to its dual nature as a choline (trimethylaminoethanol) precursor and a choline oxidase inhibitor and, in part, to its inhibition of the enzyme acetylcholinesterase.[79]

The nervous system ages more rapidly, and production and efficacy of acetylcholine declines under chronic oxidative stress and suboptimal nutrition. One primary remedy is to boost acetylcholine levels with the help of DMAE, which is not traditionally thought of as an antioxidant. It acts as a de facto antioxidant because it stabilizes cell membranes, which protects them from free radical damage.[80,81] DMAE accomplishes membrane stabilization by helping cells discharge waste and retain nutrients and by stimulating a microsomal liver enzyme that synthesizes polyunsaturated phosphatidylcholines — key factors in regulating membrane functions.[82]

To a clinician, DMAE is remarkable for its rapidity of visible impact — firming tone within minutes of application. On the basis of my own clinical research, the effects are clearly visible to professional observers and patients alike. DMAE helped to increase circulation and tone to the lips, creating a fuller, firmer appearance for about 24 hours. I find that when applied together, ALA, vitamin C ester and DMAE can also improve thin or aging skin around the lips. These findings — and the long-term safety of topical DMAE — were confirmed in two clinical studies by researchers at Johnson & Johnson.[83,84]

To summarize, DMAE has several advantages as an anti-aging topical agent:

1. It enhances facial muscle tone via enhanced availability of the neurotransmitter acetylcholine.[83,84]
2. It stabilizes cell membranes, thus reducing damage to PUFAs and the resulting chain of pro-inflammatory, cell-degrading and collagen-degrading events described previously. In this sense, it is a significant, albeit indirect, antioxidant agent.[79,81,85,86]
3. It stabilizes lysosomal membranes, preventing the rupture of these and other organelles and the consequent release of hydrolytic and protein-degrading enzymes.[82]
4. It rapidly disperses into cell membranes, according to studies with radioactive tracers.[86]
5. It speeds the ejection of lipofuscin from cells.[87]

Tocotrienols: The Ideal Topical "E"? Vitamin E is made up of eight separate components: four tocopherol molecules and four tocotrienols. The antioxidant activity of each of these was originally tested decades ago in a laboratory using a hexane solvent system. It has become apparent that biological antioxidant activity cannot be predicted by performance in a test tube. The results of this laboratory study indicated that the components labeled **alpha-tocopherol** (α-tocopherol) were the most potent antioxidant component of vitamin E. When vitamin E was tested in a lipid bilayer system, which mimics the cell plasma membrane, the tocotrienol portion showed 40 times the antioxidant potential of α-tocopherol.[88,89]

Vitamin E tocopherols and tocotrienols are the *major endogenous antioxidants* protecting skin from the adverse effects of oxidative stress, including photoaging.

Tocotrienols, like tocopherols, can protect the skin from the damaging effects of exposure to ultraviolet radiation, pollution, cigarette smoke and other environmental and biological stress factors. Topically applied tocotrienols and tocopherols penetrate the entire skin to the subcutaneous fat layer within 30 minutes and significantly increase the concentration of these antioxidants in the deeper subcutaneous layers, and topical application of tocotrienols appears to preserve dermal vitamin E levels.[90,91]

Why might tocotrienols offer superior protection against ultraviolet light (photoprotection)? The antioxidant activity of tocotrienols is generally higher than that of tocopherols. Tocotrienols penetrate rapidly through skin to efficiently combat oxidative stress induced by ultraviolet light or ozone. (The higher antioxidant potency of D-α-tocotrienol is traceable to a higher recycling efficiency and a more uniform distribution in the cell membrane.[88])

In one study on the effects of topical application of a tocotrienol-enriched fraction of palm oil in hairless mice, the skin contained nearly 15% tocotrienols versus 1% gamma-tocopherol (γ-tocopherol). The authors concluded: "The unique distribution of tocotrienols in skin suggested that they might have superior protection against environmental stressors."[91]

The results of another rodent study showed that dietary tocotrienols accumulate in the adipose tissue and skin but not in plasma or other tissues. The same investigators found that dietary sesame seeds, naturally high in E-complex vitamins, elevated tocotrienol concentrations in the adipose tissue and skin of these experimental rodents but did not affect their concentration in other tissues or plasma.[92]

Glycolic Hydroxy Acid: An Overlooked Antioxidant/Anti-Inflammatory Agent

At one time, concerns had been expressed about the photosensitizing potential of alpha hydroxy acids. By 1985, however, I had noticed chemical similarities between vitamin C and glycolic acid. Over time, I found that glycolic acid was an effective treatment for patients with dermatitis.[93,94] In fact, alpha hydroxy acids — and glycolic acid in particular — have a photoprotective antioxidant, anti-inflammatory, collagen-synthesis stimulating action when applied after skin is exposed to ultraviolet B (UVB) light.[95–100]

In 1994, I published the results of clinical research to test the antioxidant and anti-inflammatory capabilities of glycolic acid.[94] In 1996, a second paper coauthored with Joseph DiNardo suggested the photoprotective effect of topical glycolic acid.[93]

DiNardo and I found that when a glycolic acid cream (8.0%; pH, 3.25) was applied to skin 4 hours after UVB exposure, skin erythema was markedly reduced at 48 hours in contrast to untreated skin areas exposed to UVB light. Even though the glycolic acid cream was applied after UVB exposure, the glycolic cream yielded results similar to those expected from a sun-blocking cream with a sun protection factor (SPF) of 24 applied before UVB exposure.

This second study confirmed my hypothesis that glycolic acid acts as an antioxidant as well as an exfoliating agent. In my own practice, I find that glycolic acid also seems to enhance the effects of antioxidant-based topical lotions.

The Next Step in Non-surgical Skin Rejuvenation — Vincent C. Giampapa, M.D., F.A.C.S.

Oral and Topical Antioxidants with Hormonal Replacement for Skin Aging

Although this chapter has stressed *topical* skin rejuvenation, focusing on restoring to optimal levels the key components of the aging equation — *glycation, inflammation* and *oxidation, oral supplements* can enhance skin quality to a remarkable level as well!

More important, in my own clinical experience at the Giampapa Institute over the last decade, the combined approach of both *oral* and *topical* anti-aging compounds has resulted in a synergistic clinical **skin rejuvenation effect**. In essence, feeding the skin from within and from without with these topical and oral compounds discussed earlier, is the most effective approach for a nonsurgical anti-aging program for the skin.

A topical cream has recently been designed as an anti-aging regimen. It is similar to that detailed in this chapter, except for the addition of the hormone estriol. The oral supplement component is taken twice a day (in the morning and evening with meals) and includes all the natural compounds to optimize the key categories of glycation, inflammation, oxidation and methylation as well as improve DNA repair.*

The oral supplement complex is used in combination with a topical cream preparation containing specific, clinically-tested concentrations and ratios of ALA, DMAE, copper, ester-C, carboxy-alkyl esters and estriol 0.3%. The addition of estrogen markedly enhances collagen production and intercellular matrix content.

In essence, this approach, called OTA–HRT Cream, is an anti-oxidant–hormonal replacement therapy (Diagram VII-3). Since no change in peripheral blood levels occurs with the estriol, this topical cream has been effective in both women and men with no undesirable side effects.

The combined use of both topical and oral antioxidant compounds, quite possibly, will set the stage for future nonsurgical treatments for skin aging, whether used alone or with the new array of non-ablative skin lasers or cultured fibroblast therapy.*

---

*Optigene-Time Machine™.

*Available through Isologen Inc.; see Resource section.

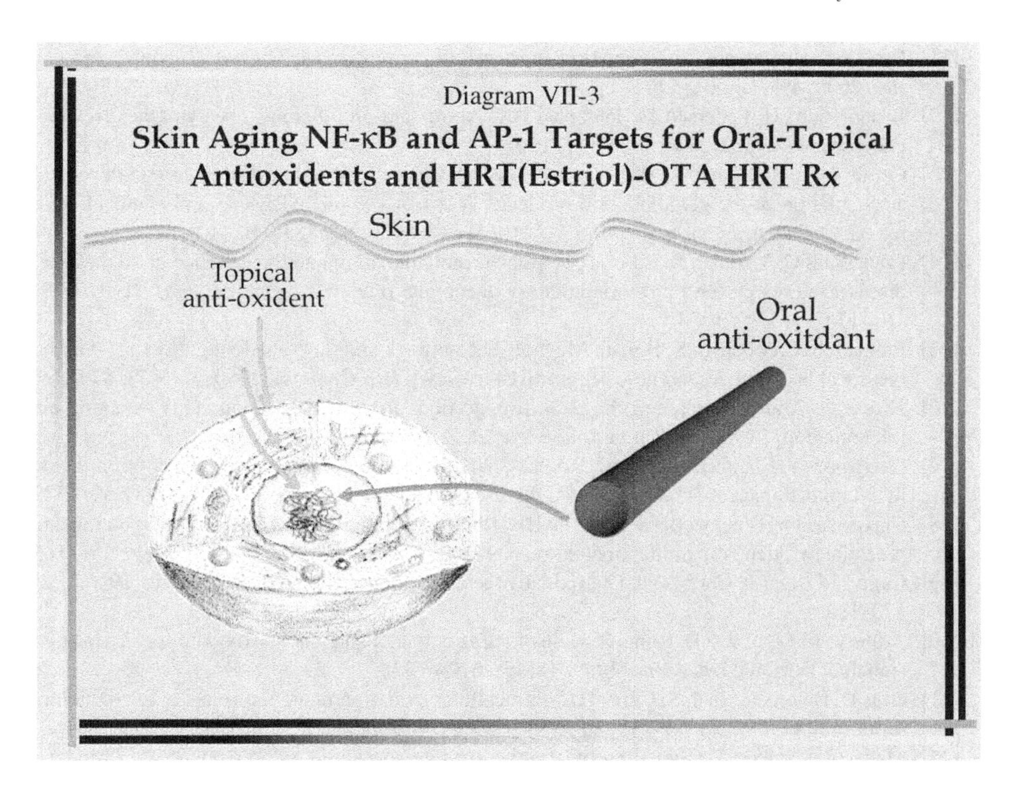

References

[1] Harman D. Nutritional implications of the free radical theory of aging. *J Am Coll Nutr.* 1982; 1(1): 27–34.

[2] Harman D. Extending functional life span [review]. *Exp Gerontol.* 1998; 33(1–2): 95–112.

[3] Harman D. Aging and oxidative stress [review]. *J Int Fed Clin Chem.* 1998; 10(1): 24–27.

[3a] Zs-Nagy I. The membrane hypothesis of aging: its relevance to recent progress in genetic research [review]. *J Mol Med.* 1997; 75(10): 703–714.

[4] Zs-Nagy I. The role of membrane structure and function in cellular aging: a review. *Mech Ageing Dev.* 1979; 9(3–4): 237–246.

[5] Zs-Nagy I. The membrane hypothesis of aging: its relevance to recent progress in genetic research [review]. *J Mol Med.* 1997; 75(10): 703–714.

[6] Zs-Nagy I, Cutler RG, Semsei I. Dysdifferentiation hypothesis of aging and cancer: a comparison with the membrane hypothesis of aging [review]. *Ann N Y Acad Sci.* 1988; 521: 215–225.

[7] Zs-Nagy I, Semsei I. Centrophenoxine increases the rates of total and mRNA synthesis in the brain cortex of old rats: an explanation of its action in terms of the membrane hypothesis of aging. *Exp Gerontol.* 1984; 19(3): 171–178.

[8] Franceschi C, Bonafe M, Valensin S, et al. Inflamm-aging: an evolutionary perspective on immunosenescence. *Ann N Y Acad Sci.* 2000; 908: 244–254.

[9] Bailey AJ. Molecular mechanisms of aging in connective tissues [review]. *Mech Aging Dev.* 2001; 122(7): 735–755.

[10] Black PH, Garbutt LD. Stress, inflammation and cardiovascular disease. *J Psychosom Res.* 2002; 52(1): 1–23.

[11] Brod SA. Unregulated inflammation shortens human functional longevity [review]. *Inflamm Res.* 2000; 49(11): 561–570.
[12] Bruunsgaard H, Pedersen M, Pedersen BK. Aging and proinflammatory cytokines [review]. *Curr Opin Hematol.* 2001; 8(3): 131–136.
[13] Cefalu WT, Bell-Farrow AD, Wang ZQ, et al. Caloric restriction decreases age-dependent accumulation of the glycoxidation products, *N*-ϵ-(carboxymethyl)lysine and pentosidine, in rat skin collagen. *J Gerontol A Biol Sci Med Sci.* 1995; 50(6): B337–B341.
[14] Franceschi C, Valensin S, Lescu F, et al. Neuroinflammation and the genetics of Alzheimer's disease: the search for a pro-inflammatory phenotype [review]. *Aging (Milano).* 2001; 13(3): 163–170.
[15] Franceschi C, Valensin S, Bonafe M, et al. The network and the remodeling theories of aging: historical background and new perspectives [review]. *Exp Gerontol.* 2000; 35(6–7): 879–896.
[16] Hirose N, Arai Y, Yamamura K, et al. [Suggestions from a centenarian study — aging and inflammation] [review]. *Nippon Ronen Igakkai Zasshi.* 2001; 38(2): 121–124.
[17] Lavrovsky Y, Chatterjee B, Clark RA, Roy AK. Role of redox-regulated transcription factors in inflammation, aging and age-related diseases [review]. *Exp Gerontol.* 2000; 35(5): 521–532.
[18] Christman JW, Blackwell TS, Juurlink BH. Redox regulation of nuclear factor κB: therapeutic potential for attenuating inflammatory responses [review]. *Brain Pathol.* 2000; 10(1): 153–162.
[19] Camhi SL, Lee P, Choi AM. The oxidative stress response [review]. *New Horiz.* 1995; 3(2): 170–182.
[20] Adler V, Yin Z, Tew KD, Ronai Z. Role of redox potential and reactive oxygen species in stress signaling [review]. *Oncogene.* 1999; 18(45): 6104–6111.
[21] Gius D, Botero A, Shah S, Curry HA. Intracellular oxidation/reduction status in the regulation of transcription factors NF-κB and AP-1 [review]. *Toxicol Lett.* 1999; 106(2–3): 93–106.
[22] Maher P, Schubert D. Signaling by reactive oxygen species in the nervous system [review]. *Cell Mol Life Sci.* 2000; 57(8–9): 1287–1305.
[23] Sen CK, Packer L. Antioxidant and redox regulation of gene transcription [review]. *FASEB J.* 1996; 10(7): 709–720.
[24] Allen RG, Tresini M. Oxidative stress and gene regulation [review]. *Free Radic Biol Med.* 2000; 28(3): 463–499.
[25] Thannickal VJ, Fanburg BL. Reactive oxygen species in cell signaling [review]. *Am J Physiol Lung Cell Mol Physiol.* 2000; 279(6): L1005–L1028.
[26] Roy AK. Transcription factors and aging [review]. *Mol Med.* 1997; 3(8): 496–504.
[27] Nose K. Role of reactive oxygen species in the regulation of physiological functions [review]. *Biol Pharm Bull.* 2000; 23(8): 897–903.
[28] Finkel T. Redox-dependent signal transduction [review]. *FEBS Lett.* 2000; 476(1–2): 52–54.
[29] Gamaley IA, Klyubin IV. Roles of reactive oxygen species: signaling and regulation of cellular functions [review]. *Int Rev Cytol.* 1999; 188: 203–255.
[30] Slater AF, Nobel CS, Orrenius S. The role of intracellular oxidants in apoptosis [review]. *Biochim Biophys Acta.* 1995; 1271(1): 59–62.
[31] Wilhelm J. Metabolic aspects of membrane lipid peroxidation [review]. *Acta Univ Carol Med Monogr.* 1990; 137: 1–53.
[32] Spiteller G. Peroxidation of linoleic acid and its relation to aging and age dependent diseases [review]. *Mech Aging Dev.* 2001; 122(7): 617–657.
[33] Fu MX, Wells-Knecht KJ, Blackledge JA, et al. Glycation, glycoxidation, and cross-linking of collagen by glucose: kinetics, mechanisms, and inhibition of late stages of the Maillard reaction. *Diabetes.* 1994; 43(5): 676–683.
[34] Lin Y, Rajala MW, Berger JP, et al. Hyperglycemia-induced production of acute phase reactants in adipose tissue. *J Biol Chem.* 2001; 276(45): 42077–42083.
[35] Yim MB, Yim HS, Lee C, et al. Protein glycation: creation of catalytic sites for free radical generation. *Ann N Y Acad Sci.* 2001; 928: 48–53.

[36] DeBuys HV, Levy SB, Murray JC, et al. Modern approaches to photoprotection [review]. *Dermatol Clin.* 2000; 18(4): 577–590.
[37] Emerit I. Free radicals and aging of the skin [review]. *EXS.* 1992; 62: 328–341.
[38] Fisher GJ, Wang ZQ, Datta SC, et al. Pathophysiology of premature skin aging induced by ultraviolet light. *N Engl J Med.* 1997; 337(20): 1419–1428.
[39] Giacomoni PU, Declercq L, Hellemans L, Maes D. Aging of human skin: review of a mechanistic model and first experimental data. *IUBMB Life.* 2000; 49(4): 259–263.
[40] Hanson KM, Simon JD. Epidermal trans-urocanic acid and the UV-A–induced photoaging of the skin. *Proc Natl Acad Sci U S A.* 1998; 95(18): 10576–10578.
[41] Miyachi Y. Photoaging from an oxidative standpoint [review]. *J Dermatol Sci.* 1995; 9(2): 79–86.
[42] Perricone NV. Aging: prevention and intervention. Part I: Antioxidants. *J Geriatr Dermatol.* 1997; 5(1): 1–2.
[43] Podda M, Grundmann-Kollmann M. Low molecular weight antioxidants and their role in skin ageing [review]. *Clin Exp Dermatol.* 2001; 26(7): 578–582.
[44] Yamamoto Y. Role of active oxygen species and antioxidants in photoaging. *J Dermatol Sci.* 2001; 27(Suppl 1): S1–S4.
[45] Yim MB, Yim HS, Lee C, et al. Protein glycation: creation of catalytic sites for free radical generation. *Ann N Y Acad Sci.* 2001; 928: 48–53.
[46] Smart RC, Crawford CL. Effect of ascorbic acid and its synthetic lipophilic derivative ascorbyl palmitate on phorbol ester–induced skin-tumor promotion in mice. *Am J Clin Nutr.* 1991; 54(6 Suppl): 1266S–1273S.
[47] Tebbe B, Wu S, Geilen CC, et al. l-Ascorbic acid inhibits UVA-induced lipid peroxidation and secretion of IL-1α and IL-6 in cultured human keratinocytes in vitro. *J Invest Dermatol.* 1997; 108(3): 302–306.
[48] Shindo Y, Witt E, Packer L. Antioxidant defense mechanisms in murine epidermis and dermis and their responses to ultraviolet light. *J Invest Dermatol.* 1993; 100(3): 260–265.
[49] Fuchs J, Kern H. Modulation of UV-light-induced skin inflammation by d-α-tocopherol and l-ascorbic acid: a clinical study using solar simulated radiation. *Free Radic Biol Med.* 1998; 25(9): 1006–1012.
[50] Eberlein-Konig B, Placzek M, Przybilla B. Protective effect against sunburn of combined systemic ascorbic acid (vitamin C) and d-α-tocopherol (vitamin E). *J Am Acad Dermatol.* 1998; 38(1): 45–48.
[51] Eberlein-Konig B, Placzek M, Przybilla B. Phototoxic lysis of erythrocytes from humans is reduced after oral intake of ascorbic acid and d-α-tocopherol. *Photodermatol Photoimmunol Photomed.* 1997; 13(5–6): 173–177.
[52] Perricone NV. Topical vitamin C ester (ascorbyl palmitate). *J Geriatr Dermatol.* 1997; 5(4): 162–170.
[53] Perricone NV. Photoprotective and anti-inflammatory effects of topical ascorbyl palmitate. *J Geriatr Dermatol.* 1993; 1(1): 5–10.
[54] Perricone NV. Topical vitamin C ester (ascorbyl palmitate). Adapted from First Annual Symposium on Aging Skin, San Diego, February 21–23, 1997. *J Geriatr Dermatol.* 1997; 5(4): 162–170.
[55] Perricone N, Nagy K, Horvath F, Dajko G, Uray I, Zs-Nagy I. The hydroxyl free radical reactions of ascorbyl palmitate as measured in various in vitro models. *Biochem Biophys Res Commun.* 1999; 262(3): 661–665.
[56] Perricone NV. Treatment of psoriasis with topical ascorbyl palmitate [abstract]. *Clin Res.* 1991; 39: 535A.
[57] Rosenblat G, Perelman N, Katzir E, et al. Acylated ascorbate stimulates collagen synthesis in cultured human foreskin fibroblasts at lower doses than does ascorbic acid. *Connect Tissue Res.* 1998; 37(3–4): 303–311.

[58] Ross D, Mendiratta S, Qu ZC, et al. Ascorbate 6-palmitate protects human erythrocytes from oxidative damage. *Free Radic Biol Med.* 1999; 26(1–2): 81–89.
[59] Bissett DL, Chatterjee R, Hannon DP. Photoprotective effect of superoxide-scavenging antioxidants against ultraviolet radiation–induced chronic skin damage in the hairless mouse. *Photodermatol Photoimmunol Photomed.* 1990; 7(2): 56–62.
[60] Geesin JC, Gordon JS, Berg RA. Regulation of collagen synthesis in human dermal fibroblasts by the sodium and magnesium salts of ascorbyl-2-phosphate. *Skin Pharmacol.* 1993; 6(1): 65–71.
[61] Fuchs J, Milbradt R. Antioxidant inhibition of skin inflammation induced by reactive oxidants: evaluation of the redox couple dihydrolipoate/lipoate. *Skin Pharmacol.* 1994; 7(5): 278–284.
[62] Packer L, Witt EH, Tritschler HJ. Alpha-lipoic acid as a biological antioxidant [review]. *Free Radic Biol Med.* 1995; 19(2): 227–250.
[63] Perricone N, Nagy K, Horvath F, Dajko G, Uray I, Zs-Nagy I. Alpha lipoic acid (ALA) protects proteins against the hydroxyl free radical–induced alterations: rationale for its geriatric application. *Arch Gerontol Geriatr.* 1999; 29: 45–56.
[64] Perricone NV. Topical 5% alpha lipoic acid cream in the treatment of cutaneous rhytids. *Clin Res.* 2000; 20(3).
[65] Podda M, Rallis M, Traber MG, et al. Kinetic study of cutaneous and subcutaneous distribution following topical application of [7,8-14C]rac-alpha-lipoic acid onto hairless mice. *Biochem Pharmacol.* 1996; 52(4): 627–633.
[66] Kagan VE, Shvedova A, Serbinova E, et al. Dihydrolipoic acid — a universal antioxidant both in the membrane and in the aqueous phase: reduction of peroxyl, ascorbyl and chromanoxyl radicals. *Biochem Pharmacol.* 1992; 44(8): 1637–1649.
[67] Han D, Handelman G, Marcocci L, et al. Lipoic acid increases de novo synthesis of cellular glutathione by improving cystine utilization. *Biofactors.* 1997; 6(3): 321–338.
[68] Bierhaus A, Chevion S, Chevion M, et al. Advanced glycation end product-induced activation of NF-κB is suppressed by alpha-lipoic acid in cultured endothelial cells. *Diabetes.* 1997; 46(9): 1481–1490.
[69] Kunt T, Forst T, Wilhelm A, et al. Alpha-lipoic acid reduces expression of vascular cell adhesion molecule-1 and endothelial adhesion of human monocytes after stimulation with advanced glycation end products. *Clin Sci (Lond).* 1999; 96(1): 75–82.
[70] Melhem MF, Craven PA, Liachenko J, DeRubertis FR. Alpha-lipoic acid attenuates hyperglycemia and prevents glomerular mesangial matrix expansion in diabetes. *J Am Soc Nephrol.* 2002; 13(1): 108–116.
[71] Podda M, Zollner TM, Grundmann-Kollmann M, et al. Activity of alpha-lipoic acid in the protection against oxidative stress in skin. *Curr Probl Dermatol.* 2001; 29: 43–51.
[72] Meyer M, Pahl HL, Baeuerle PA. Regulation of the transcription factors NF-κB and AP-1 by redox changes. *Chem Biol Interact.* 1994; 91(2–3): 91–100.
[73] Packer L, Roy S, Sen CK. α-Lipoic acid: a metabolic antioxidant and potential redox modulator of transcription. *Adv Pharmacol.* 1996; 38: 79–101.
[74] Meyer M, Schreck R, Baeuerle PA. H_2O_2 and antioxidants have opposite effects on activation of NF-κB and AP-1 in intact cells: AP-1 as secondary antioxidant-responsive factor. *EMBO J.* 1993; 12(5): 2005–2015.
[75] Saliou C, Kitazawa M, McLaughlin L, et al. Antioxidants modulate acute solar ultraviolet radiation–induced NF-κB activation in a human keratinocyte cell line. *Free Radic Biol Med.* 1999; 26(1–2): 174–183.
[76] Sen CK, Packer L. Antioxidant and redox regulation of gene transcription. *FASEB J.* 1996; 10: 709–720.
[77] Suzuki YJ, Aggarwal BB, Packer L. Alpha-lipoic acid is a potent inhibitor of NF-κB activation in human T cells. *Biochem Biophys Res Commun.* 1992; 189(3): 1709–1715.

[78] Suzuki YJ, Mizuno M, Tritschler HJ, Packer L. Redox regulation of NF-κB DNA binding activity by dihydrolipoate. *Biochem Mol Biol Int.* 1995; 36(2): 241–246.
[79] Alkadhi KA. End-plate channel actions of a hemicholinium-3 analog, DMAE. *Naunyn Schmiedebergs Arch Pharmacol.* 1986; 332(3): 230–235.
[80] Semsei I, Zs-Nagy I. Superoxide radical scavenging ability of centrophenoxine and its salt dependence in vitro. *J Free Radic Biol Med.* 1985; 1(5–6): 403–408.
[81] Zs-Nagy I, Semsei I. Centrophenoxine increases the rates of total and mRNA synthesis in the brain cortex of old rats: an explanation of its action in terms of the membrane hypothesis of aging. *Exp Gerontol.* 1984; 19(3): 171–178.
[82] Alvaro D, Cantafora A, Gandin C, et al. Selective hepatic enrichment of polyunsaturated phosphatidylcholines after intravenous administration of dimethyl-ethanolamine in the rat. *Biochim Biophys Acta.* 1989; 1006(1): 116–120.
[83] Cole AC, Gisoldi EM, Grossman RM. Clinical and consumer evaluations of improved facial appearance after 1 month use of topical dimethylaminoethanol [poster presentation]. American Academy of Dermatology, February 22–26, 2002, New Orleans.
[84] Grossman RM, Gisoldi EM, Cole AC. Long-term safety and efficacy evaluation of a new skin firming technology: dimethylaminoethanol [poster presentation]. American Academy of Dermatology, February 22–26, 2002, New Orleans.
[85] Yu MJ, McCowan JR, Thrasher KJ, et al. Phenothiazines as lipid peroxidation inhibitors and cytoprotective agents. *J Med Chem.* 1992; 35(4): 716–724.
[86] Zs-Nagy I. On the role of intracellular physicochemistry in quantitative gene expression during aging and the effect of centrophenoxine: a review. *Arch Gerontol Geriatr.* 1989; 9(3): 215–229.
[87] Nagy I, Floyd RA. Electron spin resonance spectroscopic demonstration of the hydroxyl free radical scavenger properties of dimethylaminoethanol in spin trapping experiments confirming the molecular basis for the biological effects of centrophenoxine. *Arch Gerontol Geriatr.* 1984; 3(4): 297–310.
[88] Serbinova E, Kagan V, Han D, Packer L. Free radical recycling and intramembrane mobility in the antioxidant properties of alpha-tocopherol and alpha-tocotrienol. *Free Radic Biol Med.* 1991; 10(5): 263–275.
[89] Serbinova EA, Packer L. Antioxidant properties of alpha-tocopherol and alpha-tocotrienol. *Methods Enzymol.* 1994; 234: 354–366.
[90] Thiele JJ, Traber MG, Packer L. Depletion of human stratum corneum vitamin E: an early and sensitive in vivo marker of UV induced photo-oxidation. *J Invest Dermatol.* 1998; 110(5): 756–761.
[91] Traber MG, et al. Diet derived topically applied tocotrienols accumulate in skin and protect the tissue against UV light–induced oxidative stress. *Asia Pac J Clin Nutr.* 1997; 6: 63–67.
[92] Ikeda S, Toyoshima K, Yamashita K. Dietary sesame seeds elevate alpha- and gamma-tocotrienol concentrations in skin and adipose tissue of rats fed the tocotrienol-rich fraction extracted from palm oil. *J Nutr.* 2001; 131(11): 2892–2897.
[93] Perricone NV, DiNardo JC. The photoprotective and antiinflammatory effects of topical glycolic acid. *Dermatol Surg.* 1996; 22(5): 435–437.
[94] Perricone NV. Treatment of pseudofolliculitis barbae with topical glycolic acid: a report of two studies. *Cutis.* 1993; 52(4): 232–235.
[95] Bernstein EF, Lee J, Brown DB, et al. Glycolic acid treatment increases type I collagen mRNA and hyaluronic acid content of human skin. *Dermatol Surg.* 2001; 27(5): 429–433.
[96] Kim SJ, Park JH, Kim DH, et al. Increased in vivo collagen synthesis and in vitro cell proliferative effect of glycolic acid. *Dermatol Surg.* 1998; 24(10): 1054–1058.
[97] Kim SJ, Won YH. The effect of glycolic acid on cultured human skin fibroblasts: cell proliferative effect and increased collagen synthesis. *J Dermatol.* 1998; 25(2): 85–89.
[98] Moon SE, Park SB, Ahn HT, Youn JI. The effect of glycolic acid on photoaged albino hairless mouse skin. *Dermatol Surg.* 1999; 25(3): 179–182.

[99] Moy LS, Howe K, Moy RL. Glycolic acid modulation of collagen production in human skin fibroblast cultures in vitro. *Dermatol Surg.* 1996; 22(5): 439–441.
[100] Van Scott EJ, Yu RJ. Hyperkeratinization, corneocyte cohesion, and alpha hydroxy acids. *J Am Acad Dermatol.* 1984; 11(5 Pt 1): 867–879.
[101] Boss W, Usal H, Chernoff G, Lask G, Fodor P. Autologous cultural fibroblasts as cellular therapy in plastic surgery. *Clin Plast Surg.* 2000; 27(4): 613–626.

8

Hormones

Basic Concepts and Cell Signaling

Vincent C. Giampapa, M.D., F.A.C.S.

> We are much more than molecules interacting with each other like falling domino pieces — our cells "chatter" with each other on the electromagnetic level.
>
> V. C. Giampapa,
> Fifth International Symposium on Anti-Aging Medicine

When we think of the term **hormones**, we should think in terms of two main concepts. This first concept is one of **communication** within cells. This intracellular communication is intimately dependent on the level of hormones that interact with the cell surface membrane and the interior of the cell itself.[1]

The second basic concept is **homeostasis**. Hormones are responsible for *integrating* information between different organ systems, as well as *maintaining* the optimal function of these key glands and organs (Diagram VIII-1). Hormones evolved over a million years as a means of communication from cell to cell. In this fashion, they are responsible for maintaining the balance between organ systems as diverse as the brain, immune system and gastrointestinal tract. In fact, it has recently been discovered that the same hormones that are found within the brain are found within the immune system.[2] This has further supported the link between body wellness and mental health and has spawned the new field of psychoneuroimmunology.[3,4] A series of other new studies directly *link* the elevation of the hormone cortisol with suppressed immune function.[5–12]

The diverse organ systems found within the body actually communicate with each other through a weblike model, and the "words" that make up this "language" are what we call "hormones." The degree of balance, or **homeostasis**, has much to

The Principles and Practice of Antiaging Medicine for the Clinical Physician, 129–137.

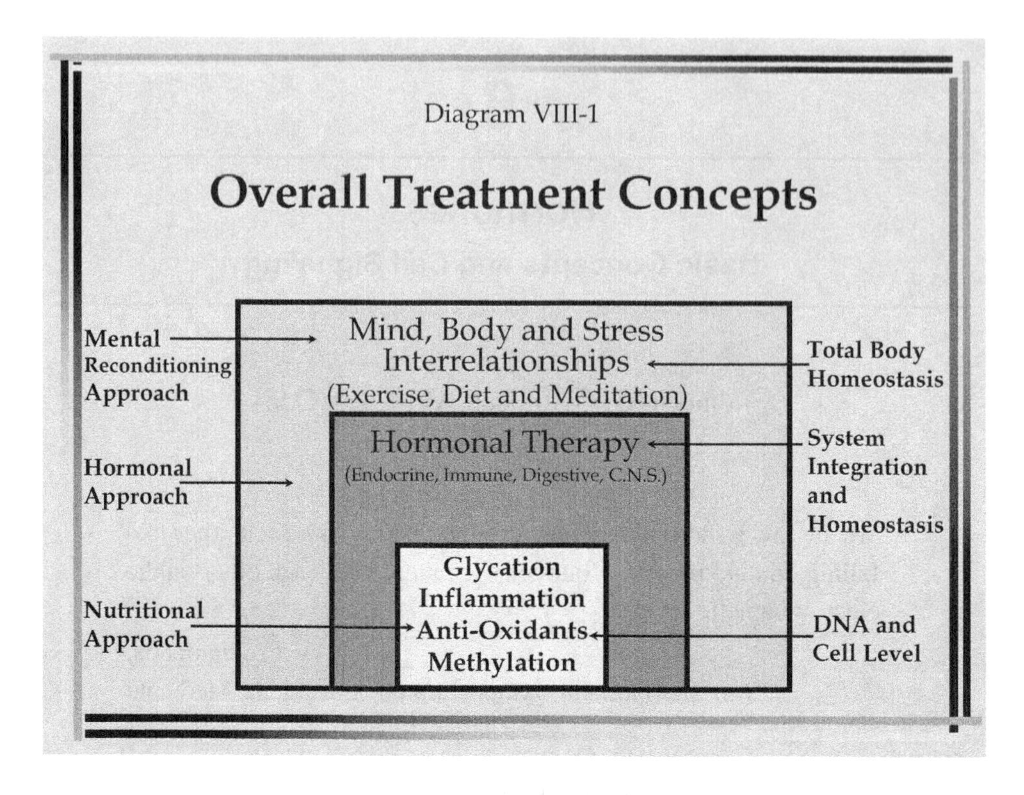

do with our level of health and how we age as well as the physical appearance we present.

New research has also documented that hormones *affect* a number of areas within and on the surface of the cell itself. It has been documented that, along with hormones, *nutrients* can directly interact and affect the deoxyribonucleic acid (DNA) within the nucleus.[13] Although the hormonal response is significantly stronger than nutrients, in general both hormones and nutrients act at the same region of our DNA, the promoter region, with similar effects — signaling **genes** to do their work.

Hormones also directly interact with cell receptors on the surface of the cells, as well as with other key complexes within the cell. When activated, these receptors initiate secondary changes at the level of DNA as well (Diagram VIII-2). This concept of "cell signaling" is also referred to as **signal transduction**.

The origin of the hormone's signal can be traced originally to the brain, or central nervous system, where primary messages are sent in the form of pre-hormones, or releasing factors from the hypothalamus to central brain receptors, and, eventually, to every one of the 100 trillion cells of the body. These chemical, electromagnetic messages[14,15] *initiate* changes within key organs, which then feed back to the source again — the brain — through a series of additional hormonal messengers. This series of events usually occurs in seconds, or less!

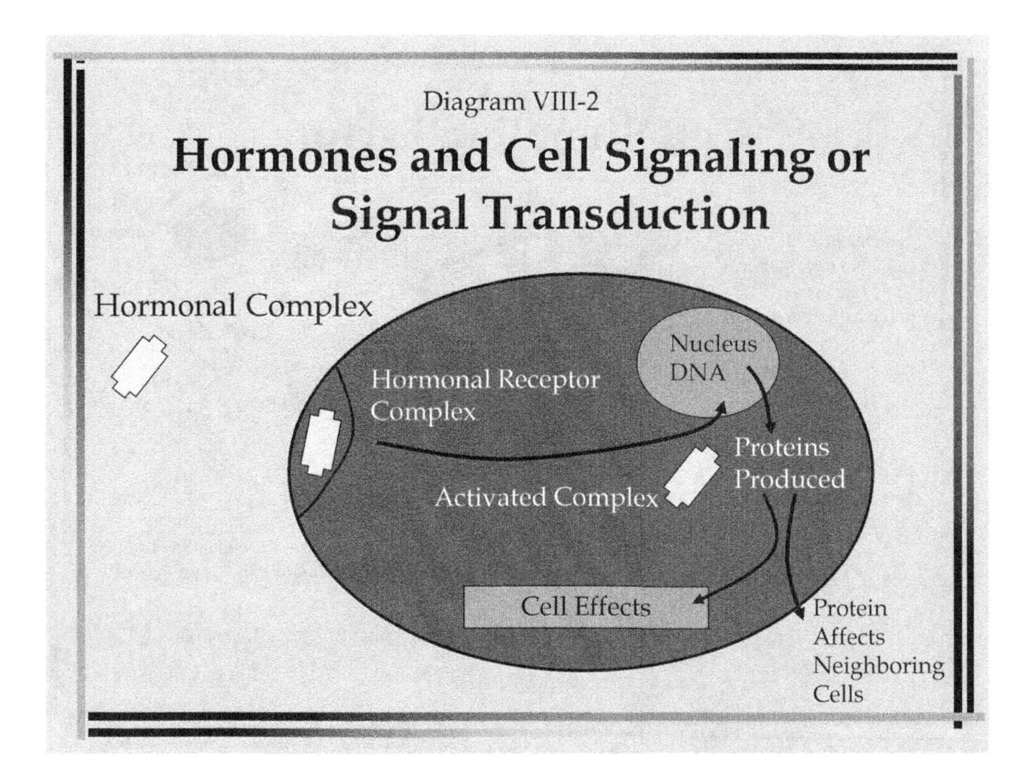

The Basic Signaling Process

It is important to understand the basic signaling process involved in hormones (Diagram VIII-3). Usually, a hormone released from the brain travels through the circulatory system to a cell via a carrier protein. At the surface of the cell, a cell membrane receptor allows the interlocking of this hormonal complex, without the carrier, via a complicated three-dimensional (3D), electrostatic and electromagnetic interaction to the specific cell membrane site. This membrane site then releases a message (through both electrical and magnetic changes) within the cell through a series of **second messenger complexes**. These complexes can do one of two things:

1. *Directly enter the nucleus* and activate or repress genes and/or their subsequent genetic information.
2. *Activate another complex already within the cell itself*, which then interacts with the nucleus, initiating the same changes as described in the first example. The net effect is that the final changes initiated by the hormone result in production of changes in key proteins and enzymes that occur by "turning on" specific genes in the cell. These proteins affect the cell that has originally been stimulated. These newly manufactured proteins, or enzymes, may also affect their neighboring cells, initiating intercellular "chatter," which is the means of communication used throughout the

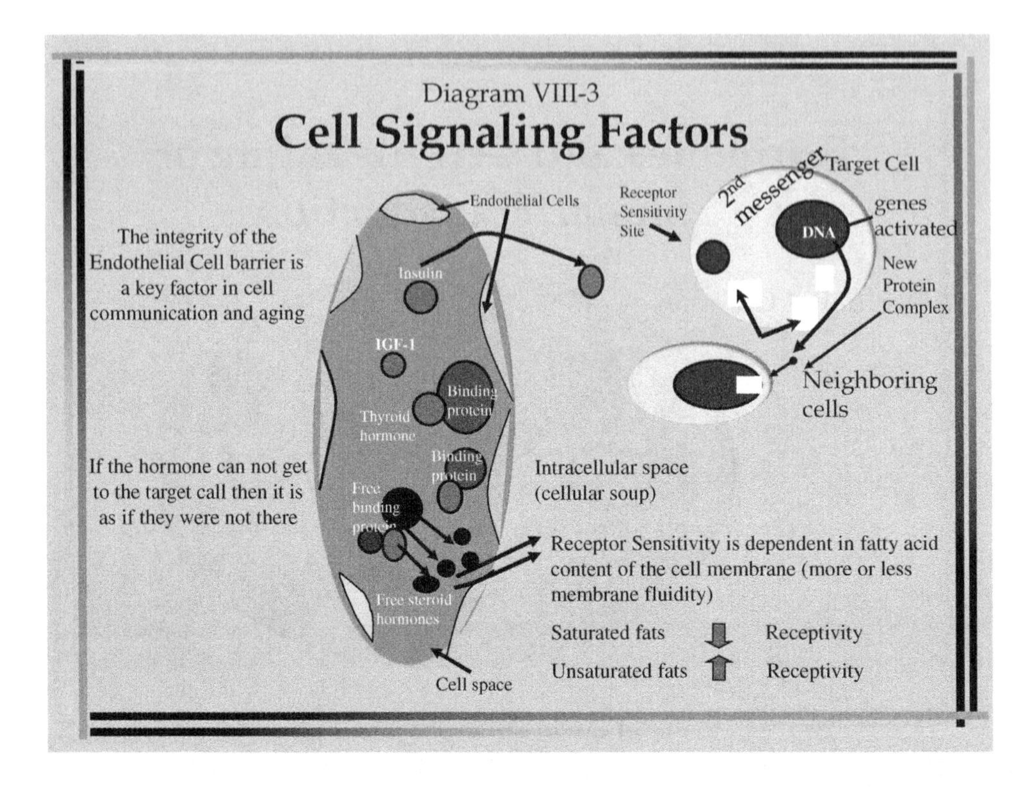

body. In this fashion, cell communication at both the *intracellular* and *extracellular* levels occurs and homeostasis is restored and maintained. This process evolved early in the evolution of life on earth as a key means of cellular survival.[16–18a]

The Second Messenger Complex

A closer look at the second messenger complex is also essential. When an activated hormonal complex interlocks "three-dimensionally" and fits into the cell receptor site, it interacts with the fatty acids within the cell membrane's bilipid layer complex, which makes up the cell receptor.

Virtually all hormones interact with receptor sites on the cell membrane surface. The cell membrane is composed of fatty acid complexes of both the omega-3 and omega-6 fatty acid varieties. If the fatty acid content in the diet is poor in omega-3 and omega-6 fatty acids, the receptor sites, instead of being ideally formed as a specific 3D structure, may be distorted. This distortion is due to the fact that the cell membrane structure itself may not be fluid or *mobile* enough to interact with its docking hormone. Think of it as a large boat trying to dock into a small slip. If this happens, it can become stiff from the lack of fatty acids, and the inappropriate balance or ratio of omega-3 and omega-6 fatty acids. This receptor molecule interaction occurs in three ways[19–21]:

1. Three-dimensionally (as a result of the 3D structure of both the receptor and hormone).
2. Electrically.
3. Electromagnetically.

What, then, happens in this large, overall picture? When a message is sent from the brain, it is transmitted, biochemically, in the form of a 3D hormonal molecule, with a specific electrical charge to each one of the one hundred trillion cells in the body. That message is then amplified like a strong "drumbeat" from the surface of the cell membrane to the DNA within the nucleus. The message is again amplified and copied from the DNA and changed into ribonucleic acid (RNA) and proteins, which eventually affects the cell itself as well as neighboring cells. In this fashion, the original message from the brain, which was sent to one cell, is then sent to all other cells in contact with it. This is a biological process that evolved over the millennia.[16,17]

Although this is a complicated process, the message originating from the brain is communicated all the way to each one of the trillions of cells within the body. These cells, in turn, send messages to neighboring cells. It is in this fashion that the resulting original message is again received by "feedback" to the brain in the form of these new proteins that are produced by the initial message. This process occurs in a few seconds to microseconds. It is, obviously, not the result of one molecule hitting another in a "domino effect," but, rather, the outcome of complicated processes on the electromagnetic level.

If we look at this process in a number of steps, it can be analyzed in a simple fashion. The first step is a message sent from the brain and communicated to the cell membrane. The second step occurs from the activated receptor site on the cell membrane to the DNA itself within the nucleus. It is here that the second message, which can be thought of as a drumbeat, is produced. If for some reason that receptor site is poor and the receptor site reacts as an inexpensive, or loose, drum skin might, we can get a muffled sound along with a poor transmission of the original energy and message to the end target — in this case the nucleus. The events of step 2 are directly related to a process called **phosphorylation**. Although it is a multistep process that takes place, it is this step that amplifies the signal from the cell membrane to the nucleus.

The detail of this biological response — the process of amplifying the message from the cell membrane surface to the DNA — is actually dependent on a series of second messengers. These second messengers are most commonly called cyclic adenosine monophosphate (cAMP), inositol triphosphate (IP_3), or diacylglycerol (DAG). Depending on which of the second messengers is predominant at a given time, the quality (strength) of this second hormonal messenger signal can be modified.

The actual *threshold level* (quantity of molecules) of these second messengers within the cell control the amplitude, or loudness of the signal from a specific hormone. The greater the quantity or baseline level of the appropriate second messenger maintained (the higher the level of cAMP), the smaller the quantity of a specific hormone is required at the cell membrane site to send its biological message to the nucleus.

Simply stated, if the level of cAMP in the cell is high, just a small amount of a given hormone is necessary that will result in a strong hormonal response. This is the situation when we are young and healthy. If, however, the level of second messengers is very low, then even high levels of hormones will give a very poor response or perhaps no message at all to the nucleus. This is the condition when we age or are in a state of disease. This is frequently the cause of a patient's inability to respond to administered hormones and inability to see or feel a difference in health.

Aging affects all steps in this process. It also affects the quantity and the balance, or ratio, of these second messengers, receptor sensitivity and the quantity of the initial hormones. Let us take a look at the second messenger complex in a little more detail (Diagram VIII-4).

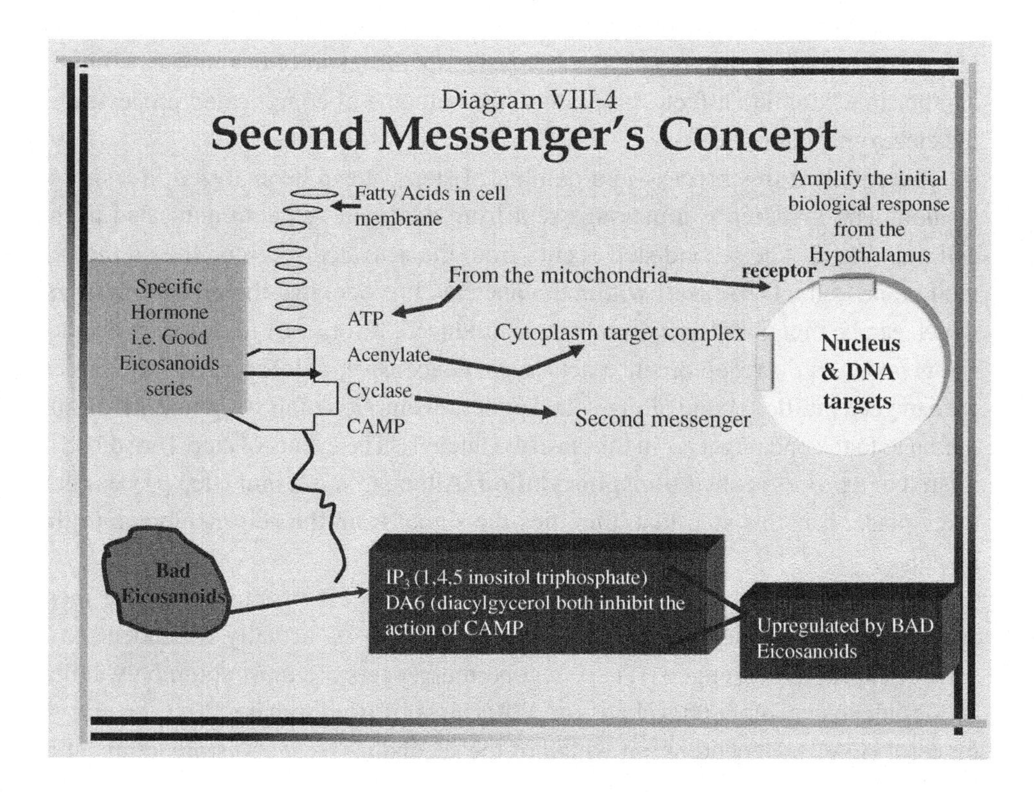

In this instance, a growth hormone (HGH) molecule is used. As the HGH molecule "docks" into its 3D receptor site and creates a change in the membrane receptor, a message is initiated. It is once again important to understand the importance of the fatty acids in the cell membrane at this initial point and that they are essentially responsible for receptor site sensitivity.

Next, the adenosine triphosphate (ATP) that is present in the cell (intracellular matrix, or "cellular soup") is activated by adenylate cyclase and is transformed into cAMP. This is the most important second messenger. It is through high levels of cAMP that we amplify this second message to the nucleus, the location of genes to be acted upon. The effect of cAMP is analogous to tightening the drum skin and creating a strong drumbeat.

cAMP is produced from the mitochondria, lying within the cellular soup, which converts food into energy originally stored as ATP. We have already discussed mitochondria as "quantum energy" devices, or organelles, that convert food "matter" into ATP "energy." There are many ways in which this quantum transformation process can be interrupted.

If the mitochondria have been damaged by excess free radicals or if the level of the mitochondria here do not have the appropriate quantity of key nutrients (coenzyme Q10, lipoic acid (ALA), L-carnitine and others), ATP is not available in sufficient amounts. This second messenger signal is then muffled. The drumbeat is weak. If there are adequate quantities of ATP, the signal is sent loudly to DNA at the center of the cell and specific genes are *turned on* to make specific proteins.

Although cAMP[22] is the key essential messenger, not only are there good (amplifying) messengers; bad *(dampening)* second messengers exist as well. IP_3 and DAG both inhibit the action of cAMP. Thus, the net response of the second messenger has to do with which one of these second messengers predominate. If there are large quantities of cAMP, we are still in good shape to send the appropriate message to the nucleus. If we have too much of the bad second messengers, the message is sure to get muffled, and regardless of the quantity of hormones, a poor message is sent and little gene activation occurs.

It is now easier to see that the message that reaches the nucleus of the cell through this second messenger step is actually an amplification process of the initial message, or **biological response**, that originated in the hypothalamus of the brain. The origin of the brain signal may have been a thought, an emotion or a change in the environment.

We must keep in mind that what occurs in the cellular level *originates* in the brain. This vital comment underscores why it is important to maintain a healthy central nervous system, functioning at an optimal level at all times. If the brain is *not* sensing what is going on, it cannot even initiate the first step in the cell communication process. If the membrane status in the brain (composed mainly of

Diagram VIII-5

Second Messenger Baseline Concept

The final biological response of a cell depends on which second messenger (CAMP or IP_3/DAG) predominates at a point in time.

The threshold level of these second messengers within the cell controls the amplitude of hormonal communication.

The more the **baseline level** of the appropriate second messenger is maintained, the less specific hormone is required to exert its **biological action.**

Aging affects the quantity and balance of these second messengers, the receptor **sensitivity**, and the quantity of the initial hormones secreted.

fatty acids) is poor, if the brain is dehydrated, or if the level of neurotransmitters is low, few of the initial steps in this process can get started.

This is a key example of why, in regard to hormone replacement in anti-aging therapies, **single-hormone therapy** is not the best choice.

To summarize this concept of the second messenger, we should recall a few salient points (Diagram VIII-5):

1. The final biological response of a cell depends on which second messenger group, that is, cAMP, as the *upregulating* messengers, or IP_3 and DAG, as the *downregulating* messengers, predominates at a point in time.
2. The threshold level of these second messengers within the cell controls the amplitude or power of hormonal responses and hormonal communication.
3. The more the baseline level of the appropriate second messenger (cAMP) is maintained, less of the specific hormone is required to observe and feel its biological action on the body.
4. Single-hormone aging replacement therapies are usually ineffective or even detrimental to the aging process when they are administered over any significant period of time.

5. Aging affects the quantity and balance of the second messengers, the receptor sensitivity (the 3D structure, electrical and magnetic properties) and the quantity of the initial hormone secreted. Therefore, hormonal responses should be viewed as the result of a multifactorial environment. It is also for this reason that single-hormone therapies are usually ineffective or even detrimental to the aging process when administered over any significant period of time.

References

[1] Felig P, Baxter JD, Frohman LA. *Endocrinology and Metabolism*, ed 3. New York: McGraw-Hill; 1995.

[2] De Groot LJ, Besser M, Burger HG, et al, eds. *Endocrinology*, ed 3. Philadelphia: WB Saunders Co.; 1995.

[3] Timiras PS, Quy WB, Bernakdakis A, eds. *Hormones and Aging*. Boca Raton, FL: CRC Press; 1995.

[4] Timiras PS, ed. *Physiological Basis of Aging and Geriatrics*, ed 2. Boca Raton, FL: CRC Press; 1994.

[5] Martin P. *The Sickening Mind: Brain, Behavior, Immunity and Disease*. London: HarperCollins; 1997.

[6] Becker JB, Breedlove MS, Crews D. *Behavioral Endocrinology*. Cambridge, MA: MIT Press; 1992.

[7] Sapolsky RM, Krey L, McEwen BS. Stress down-regulates corticosterone receptors in a site-specific manner in the brain. *Endocrinology*. 1984; 114: 287–292.

[8] Sapolsky RM, Krey L, McEwen BS. Glucocorticoid-sensitive hippocampal neurons are involved in terminating the adrenocortical stress response. *Proc Natl Acad Sci USA*. 1984; 81: 6174–6177.

[9] Sapolsky RM, Krey L, McEwen BS. Prolonged glucocorticoid exposure reduces hippocampal neuron number: implications for aging. *J Neurosci*. 1985; 5: 1222–1227.

[10] Sapolsky RM, Packan DR, Vale WW. Glucocorticoid toxicity in the hippocampus: *in vitro* demonstration. *Brain Res*. 1988; 453: 367–371.

[11] Sapolsky RM, Uno H, Rebert CS, Finch CE. Hippocampal damage associated with prolonged glucocorticoid exposure in primates. *J Neurosci*. 1990; 10: 2897–2902.

[12] Sapolsky RM. *Stress, the Aging Brain and the Mechanisms of Neuron Death*. Cambridge, MA: MIT Press; 1992.

[13] Cousins RJ. Nutritional Regulation of Gene Expression. In: Shils ME, Olson JA, Shike M, Ross AC. *Modern Nutrition in Health and Disease*, ed 9. Baltimore: Lippincott Williams & Wilkins; 1999: 573–603.

[14] Popp FA. *Electromagnetic Bio-Information*. Urban and Schwartzenberg; 1989.

[15] Becker R. *Cross Currents*. Los Angeles: Jeremy Tarcher, Inc.; 1990.

[16] Gesteland RF, Atkins JF, eds. *The RNA World*. Cold Spring Harbor, New York: Laboratory Press; 1993.

[17] Woese C. The universal ancestor. *Proc Natl Acad Sci USA*. 1998; 95: 6845–6859.

[18] Poole AM, Jeffares DC, Penny D. The path from the RNA world. *J Mol Evol*. 1998; 46: 1–17.

[18a.] Jeffares DC, Poole AM, Penny D. Relics from the RNA world. *J Mol Evol*. 1998; 46: 18–36.

[19] Lahkovsky G. *Radiations and Waves: Sources of Our Life*. New York: E.L. Cabella; 1941.

[20] McClare CWF. Resonance in bioenergetics. *Ann N Y Acad Sci*. 1974; 227: 74–91.

[21] Popp FA. Electromagnetic Bio-Information. Urban and Schwartzenburg, 1989.

[22] Norman AW, Litwack G. *Hormones*, ed 2. New York: Academic Press; 1992.

9

Hormones

Balancing and Restoring Hormone Levels

Vincent C. Giampapa, M.D., F.A.C.S.

A man's character is his fate.
Heraclitus

From Chapter VIII, "Hormones: Basic Concepts and Cell Signaling," we recall that the key concepts are that hormones evolved as a means of communications from *cell to cell* and from *organ to organ*.[1,2,2a] Again, the essential function of hormones, in general, is to coordinate with the interorgan system communication network (i.e., the nervous system, through a weblike interaction with the gastrointestinal system, and the immune function with the endocrine function (Diagram IX-1). Hormones are responsible for homeostasis between organ systems as well as integration of the key primary messages sent from the brain to the intimately connected cells throughout the body.

Remember, the brain is the *essential* monitoring center that acts like a rheostat and is responsible for initiating the message and receiving the feedback to control both organ-to-organ and brain-to-organ communication.[3] The final result is maintenance of health and optimal quality of life. This is why maintenance of appropriate neurotransmitter levels is an essential part of an anti-aging program.

In this chapter, we will begin to look at the key effects of specific hormones and their functions from alterations in personality traits to physical appearance. Obviously, as cosmetic surgeons and physicians, we are concerned with focusing on the effects of hormonal declines as they are directly related to changes in body shape and composition as well as skin texture as we age (Diagram IX-2).

The Principles and Practice of Antiaging Medicine for the Clinical Physician, 139–174.

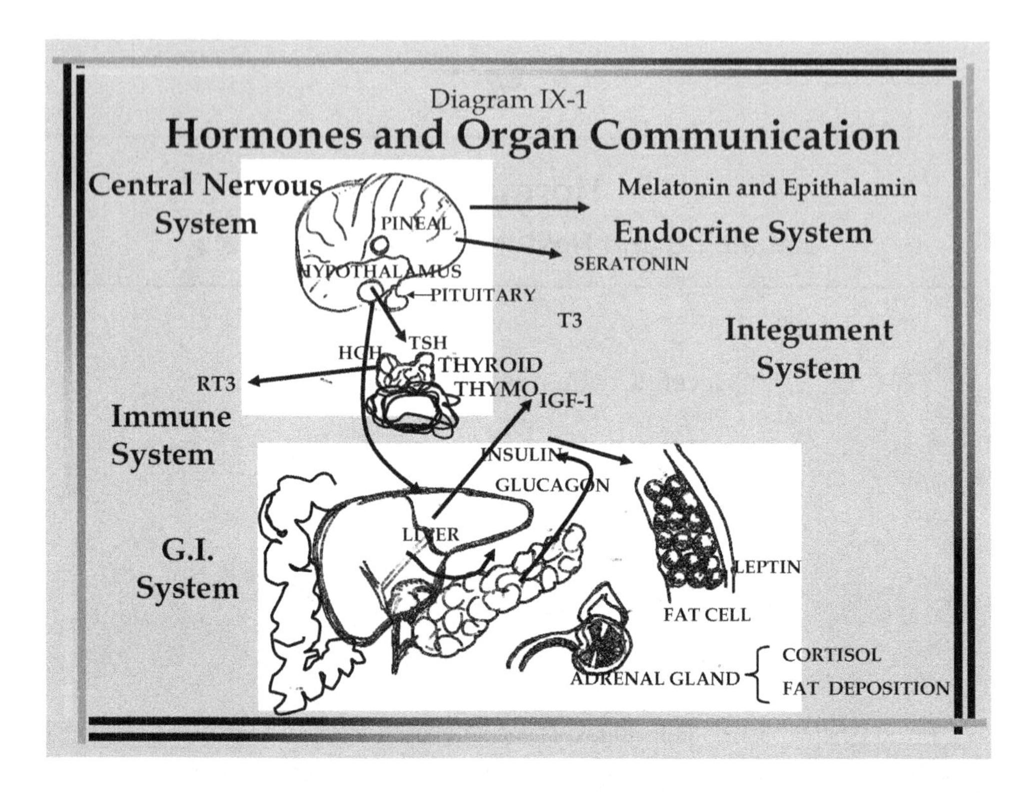

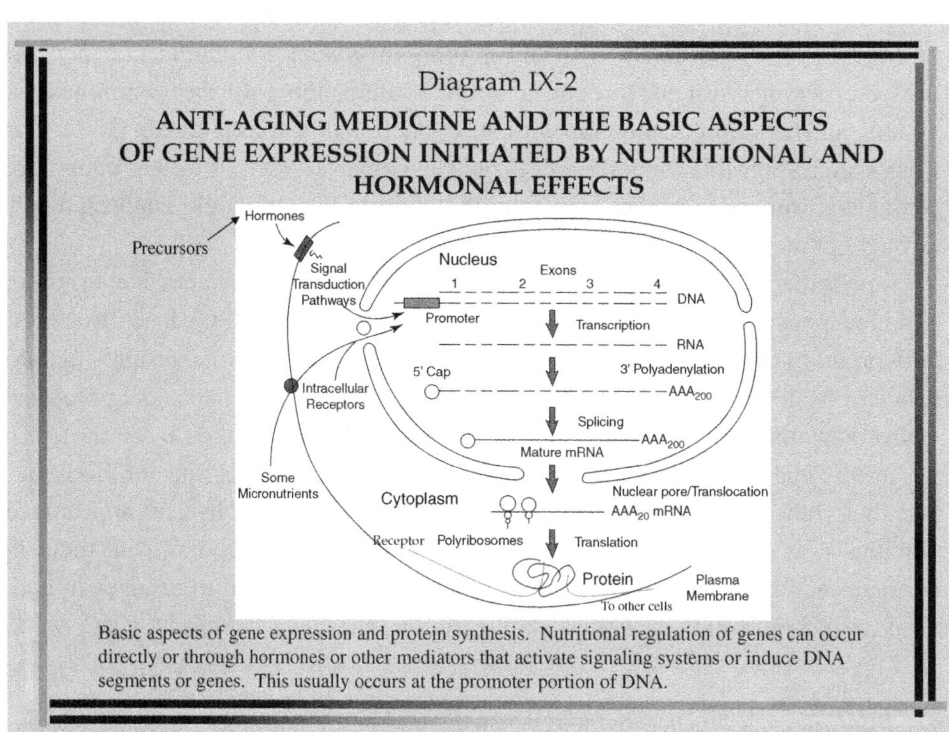

Basic aspects of gene expression and protein synthesis. Nutritional regulation of genes can occur directly or through hormones or other mediators that activate signaling systems or induce DNA segments or genes. This usually occurs at the promoter portion of DNA.

Before we review the key hormones of aging, it is important to present some essential rules in regard to **hormonal replacement**, or **balancing** of the hormonal system in general (Diagrams IX-3 and IX-4).

1. Hormone therapy should be accomplished in an overall general restorative program that focuses on *multiple* hormonal replacements rather than a single-hormone modality approach. The present trend of stimulating or replacing growth hormone alone, without consideration of its effects on other hormones, is medically inappropriate. In the long run, this practice can create problems in health and, possibly, in our stem cell pools.
2. Restoring hormonal values to the range of the mid-30s for both men and women, initially with a secretagogue trial, is the best first-step approach.
3. Certain hormones have been shown to be more effective in restoring and improving life quality, or functionality, for example, growth hormone (HGH), dehydroepiandrosterone (DHEA), testosterone, the estrogens and progesterone.
4. All hormone replacement–balancing programs should be initiated and followed up with appropriate laboratory monitoring on an ongoing basis.
5. Hormonal declines are directly related to changes in body shape and body composition and skin quality as we age (Diagrams IX-5 to IX-7).[4]

Diagram IX-3

B. Hormonal Balancing Criteria and Goals:

1. **Restore levels of hormones in the reference range of mid 30's.** (Optimal range is typically considered the upper part of the reference range for a healthy population of age mid 30's)
2. **Maintain optimal physiological function:**
   a. Mental, sexual, physical performance
   b. Bone density, body composition
   c. Skin thickness and elasticity
   d. Hair
3. **Reduce risk of degenerative disease:**
   a. Cardiovascular
   b. Diabetes
   c. Alzheimer's
   d. Cancer

Diagram IX-4

A. Hormonal Imbalances Causes (cont.)

4. **Impaired availability of hormones** due to changes in the regulation of binding globulins; Sex Hormone Binding Globulin (SHBG), Thyroid Binding Globulin (TBG), Albumin, Transthyretin (TTR), CBG (Corticotrophin Binding Globulin), IGFBP (IGF Binding Protein).
5. **Impaired circulation** of hormones due to circulatory problems.
6. **Inadequate nutrition** (total calories, macronutrients rations, micronutrients as cofactors).
7. **Excessive stress** (mental, physical).

Diagram IX-5

HORMONES AND BODY COMPOSITION

Fat Decrease Caused By:

↑ HGH, IGF = < FAT

↑ GLUCAGON = < FAT

↑ SERATONIN = < FAT

↑ T_3 = < FAT

↑ GOOD ECOSANOIDS = < FAT

Diagram IX-6

HORMONES AND BODY COMPOSITION

Fat Deposition Caused By:

↑ WEIGHT GAIN = FAT ↓ LEPTIN

↑ MAO = ↓ SERATONIN

↑ INSULIN = > FAT

↑ CORTISOL = > FAT

↓ MELATONIN = > FAT

↑ RT_3 = > FAT

↑ BAD EICOSANOIDS = > FAT

Diagram IX-7

Effects of Testosterone Therapy on Body Fat & Bone Density

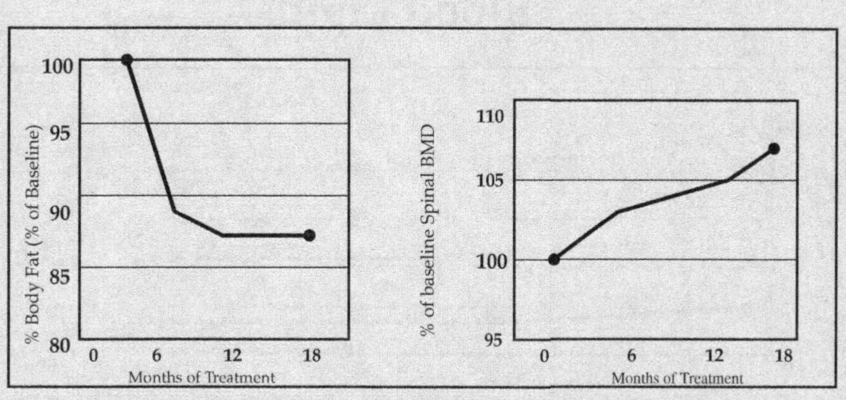

Reduction in body fat and increased bone density (BMD) in hypogonadal men receiving treatment with "testosterone".

Some other key goals in regard to hormonal replacement should focus on the following:

1. Maintaining optimal physiological function is essential. With optimal hormonal levels, mental, sexual and physiological performance is optimized; bone density and body composition are improved. The aspects of aging that we as cosmetic surgeons and physicians encounter (e.g., skin thickness, elasticity and body composition) are also improved.[5]
2. Optimal hormonal levels are also directly related to *reducing* the risk of key degenerative diseases of aging,[6] such as cardiovascular disease,[7] diabetes, Alzheimer's disease and cancer (Diagram IX-8).
3. Hormonal imbalances also cause *impaired* bioavailability of hormones owing to changes in the regulation of key binding globulins that are responsible for transporting these essential hormones. Although it is beyond the scope of this chapter to discuss binding globulins, the key categories include albumin, sex hormone-binding globulin (SHBG), thyroid hormone-binding globulin (TBG), transthyretin (TTR), corticotropin-binding globulin (CBG), insulin-like growth factor binding protein-III (IGF-BP-III) and growth hormone-binding globulin.
4. The other key factors causing poor hormonal utilization (see Chapter VIII) concern *impaired* circulation of hormones resulting from

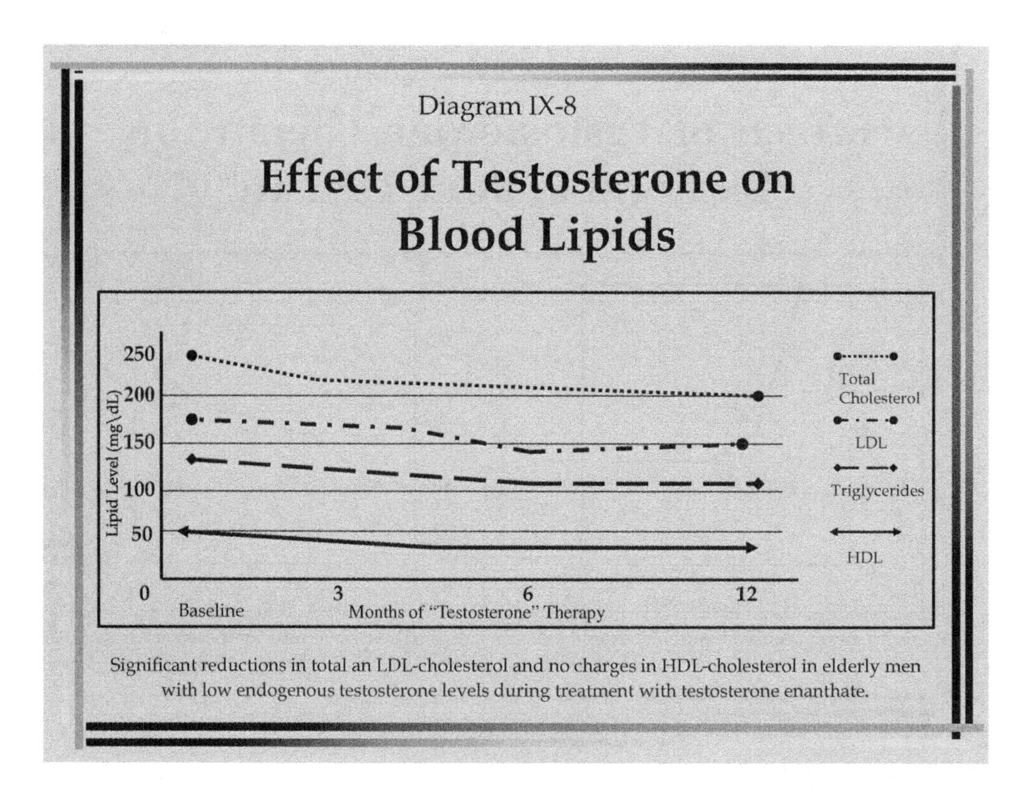

Significant reductions in total an LDL-cholesterol and no charges in HDL-cholesterol in elderly men with low endogenous testosterone levels during treatment with testosterone enanthate.

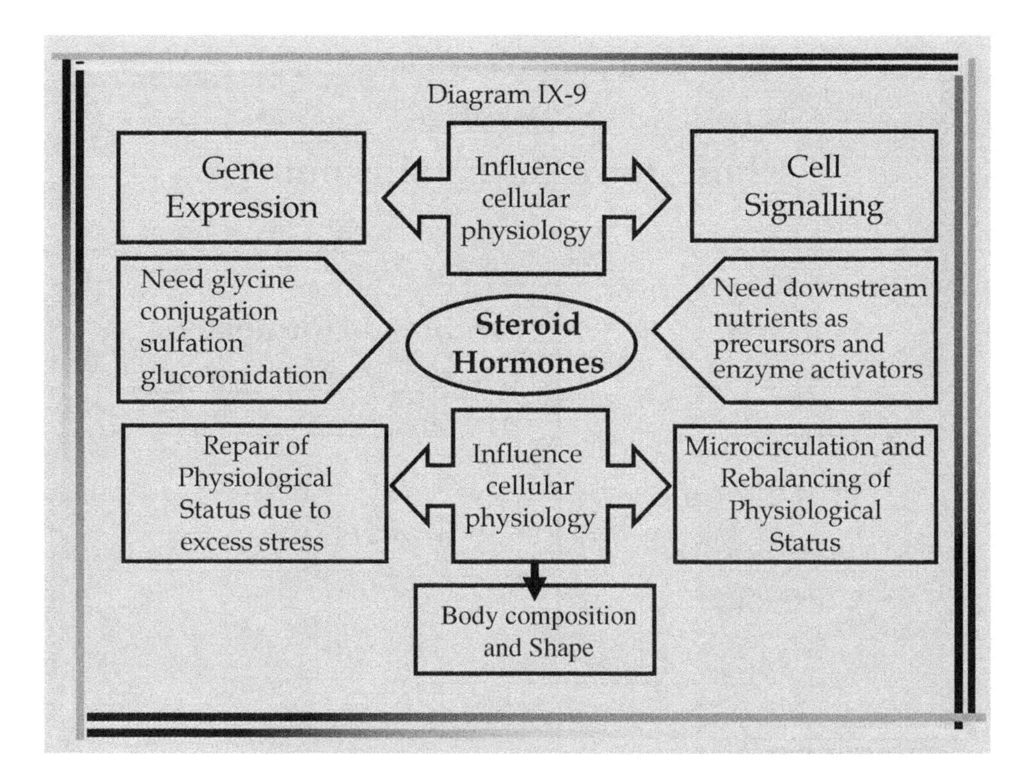

decreases in the microcirculatory capacity of the vascular system, most commonly caused by arteriosclerosis, as we age.

5. Inadequate nutrition (i.e., an inappropriate quantity and quality of total calories and macronutrients and micronutrients, which are essential as cofactors in optimizing the function of hormones) is also a key cause of hormonal imbalance.
6. The concept of *excessive* stress, whether it originates in mental time urgency or is secondary to physiological or dietary stress, is directly related to elevated cortisol levels, which have a *major negative* impact on all hormonal levels in general. These overall considerations are summarized in Diagram IX-9).[3,8]

General Classification of Hormones

Hormones can be divided into a few main categories, whether they are cholesterol (steroid)-based (Diagram IX-10), amino acid-based (Diagram IX-11), or fat-based hormones or require binding proteins (Diagram IX-12). The size of the hormone molecule is also one of the key factors that affect a hormone's effectiveness and responsiveness as well.

To begin, it is important to understand the concept of **functional organ reserve**. Although we have touched on this topic briefly, it is essential here to redefine this

Diagram IX-10

Cholesterol Based Hormones

Steroid Endocrine Hormones
- DHEA
- Estrogen
- Progesterone
- Testosterone
- Cortisol

- Small Hormones
- Bound to Binding Proteins
- Long Acting
- Travel Long Distance

Diagram IX-11

Types of Hormones

Amino Acid Based

Polypeptide Endocrine Hormones —— Most Large Hormones Protein Derived (No Binding Proteins-Short Action)
- Insulin
- Glucagon

- IGF * —— Binding Protein *
- HGH

Amino Acid- endocrine hormones
- Thyroid * —— Binding Protein*

Amino Acid- paracrine hormones
- Seratonin
- Melatonin

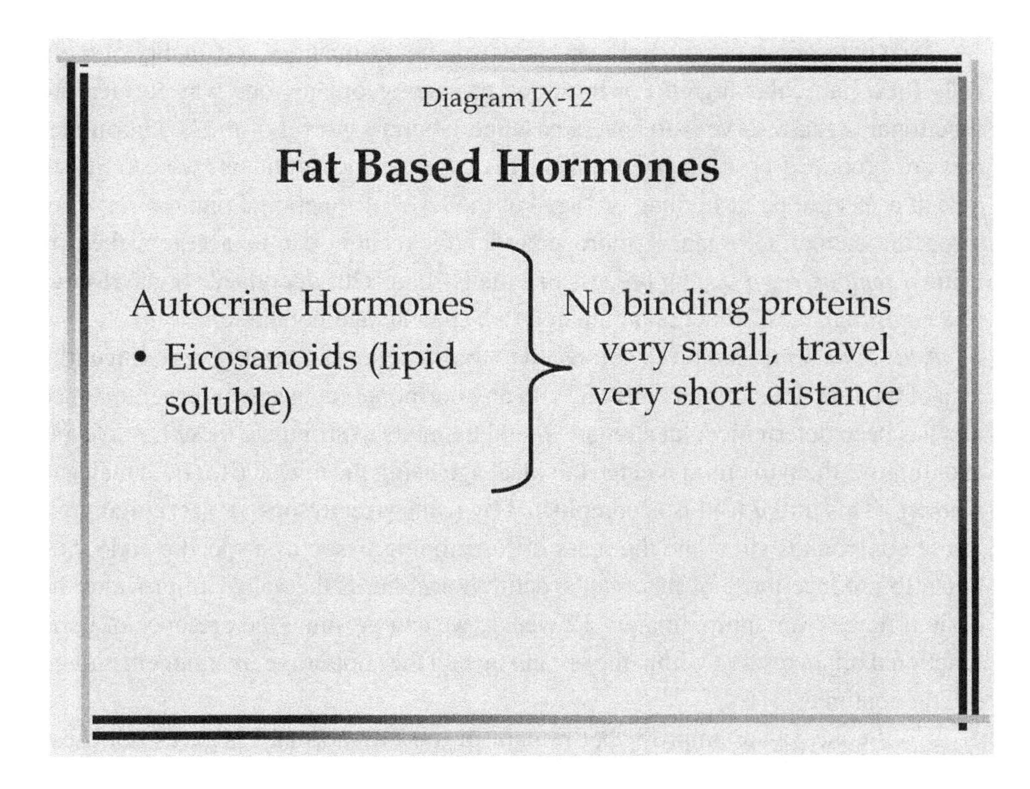

concept. **Functional organ reserve** has been defined as "the capacity of an organism to mobilize a margin of defense or functional ability when it is exposed to various stress factors, which allows the organism to return to its normal physiological function and maintenance."

In simpler terms, when we are young, within each key organ system we not only have enough functional ability within the quantity of cells of each organ to handle an insult or damage that may arise from infection, environmental stress or abuse; at this healthy early age, we also can overcome these stresses and still maintain optimal health and function within the specific organ involved.

For example, when we are in our 20s, we have about 300% to 400% of the amount of organ function and reserve necessary to maintain that key organ functioning at a level that will maintain its health and integrity. When we are in our 40s or 50s, we probably have 150% of reserve left within the same specific organ system. When we arrive in our 60s and 70s, we may have perhaps 85%, 95%, or 98% of reserve; thus, the *slightest* amount of stress may put us into what is considered *mild organ failure*. In essence, we lose the amount of functional organ reserve as we age.

This *concept* of functional organ reserve is extremely important. The more we can do to *maintain organ reserve*, the more we will be able to handle the aging process as well as the stresses of the increasing growing toxic environment we now find ourselves in in the 21st century.

What determines our organ reserve seems to be the quantity and quality of **stem cells** for a particular organ. For hormone-producing organs, one way to measure functional organ reserve is to have a baseline laboratory level of the key hormones that are produced by each organ. With this baseline measurement, we can get an idea at a specific point in time, or "age" of the level of functional organ reserve for a specific endocrine organ. A more superficial screening can be accomplished by using a *computerized* testing process like the H-scan. This documents organ reserve as a *measurement* of function compared to a specific age population.

A *baseline* hormonal level, compared to the *ideal* normal level, gives us an initial idea of the functional organ reserve. With any hormonal replacement program, once this has been determined, an attempt should be made to stimulate these basic levels and improve them to optimal range (an ideal age being the mid-30s for both men and women.) This initial trial is accomplished by using **precursors**, or **secretagogues**. These compounds stimulate the mass of functioning tissue of a specific endocrine organ to produce more of its organ-specific hormone. If the values improve to the desired range over approximately 12 weeks, we can presume the presence of good functional organ reserve within the organism, and the supportive treatment instituted should continue.

If hormone values improve but remain in the low-normal range, use of these organ-specific precursors should continue and a small quantity of actual hormonal replacement compounds should be added. In this condition, we are looking at *marginal* functional organ reserve.

If hormone values after the initial secretagogue trial (12 weeks of stimulation) are below the normal range, we can deduce that the *functional organ reserve* within the specific endocrine gland is poor (Diagram IX-13).[9] Specific secretagogues are available* for each endocrine organ that needs to be tested.

Hormonal Protocols

A rational way of looking at hormonal effects and replacement has to do with evaluating key hormones whose effects are male-specific, female-specific, and those that are generally common to both males and females (Diagram IX-14).

Male-Specific Hormones

Testosterone Male-specific hormones can mainly be viewed as those that focus on **testosterone**. In essence, *testosterone is really three hormones.*[10] First, it is converted to *estradiol* and, subsequently, to *estrone* by the effects of aromatase enzyme (estrogen synthetase) systems found mainly in the fat cells of the body.

*Prepared secretagogaue combinations are available through Optigene-X Inc; see Resource section.

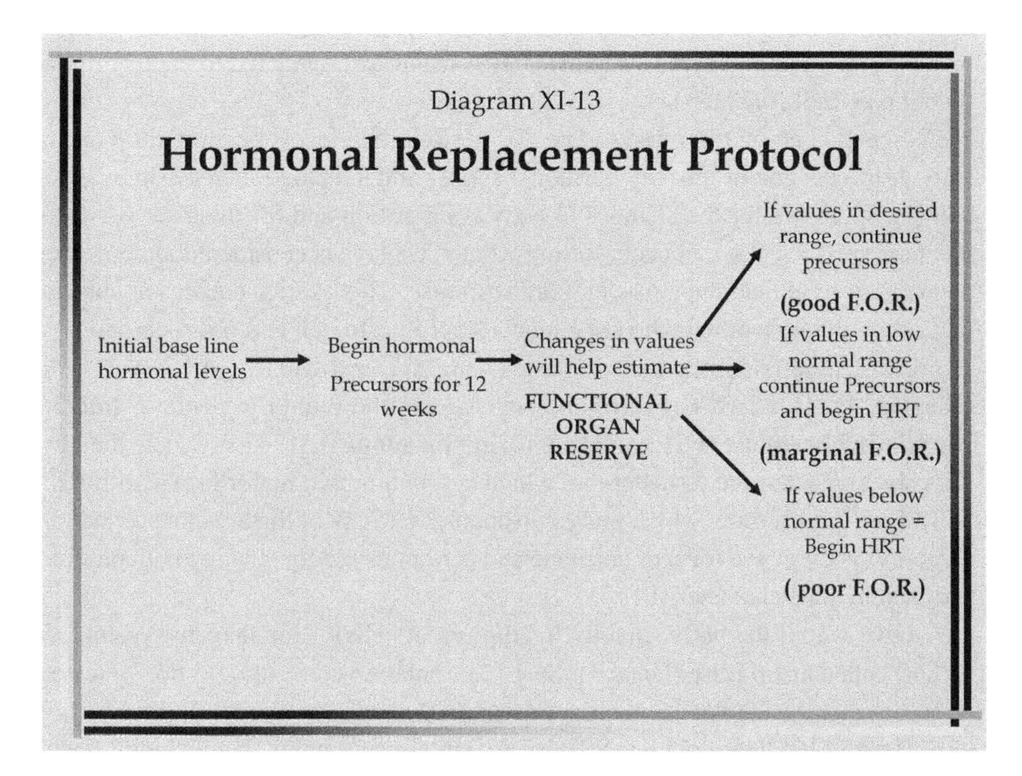

Diagram IX-14

C. HORMONAL PROTOCOLS WHOSE EFFECTS ARE:

1. **Common to male / female:**
   a. Thyroid
   b. Growth Hormone
   c. Melatonin
   d. Cortisol
2. **Male Specific:**
   a. Testosterone, Dihydrotestosterone
   b. Estradiol, Estrone
   c. Progesterone
   d. DHEA
3. **Female Specific**:
   a. Estradiol, Estrone, Estriol
   b. Progesterone
   c. Testosterone, Dihydrotestosterone
   d. DHEA

Second, it is converted in the opposite direction into 5-*dihydrotestosterone* (DHT) by the 5α-reductase enzyme systems. DHEA is intricately related to male hormone function as well (see later).

When men are in their mid to late 30s, testosterone levels begin to drop drastically. It is a decline of this key hormonal supply and a relative increase in estrogen that is responsible for the changes in body composition and for the male personality characteristics that can occur during what nowadays is considered a recognized counterpart to female menopause — **andropause**. This change equates to changes that occur among women who experience drops in estrogen and progesterone.

Testosterone is secreted by the testes and, more particularly, the Leydig cells (Diagram IX-15). Two key hormones released in the pituitary gland — **follicle-stimulating hormone** (FSH) and **luteinizing hormone** (LH) — stimulate the Leydig cells to make more testosterone, which is found in two major forms within the body. Free testosterone, which makes up about 2% to 3% of the amount secreted by the testes, is the active form of hormone and is responsible for shaping both physical and mental male characteristics.

Active within the body (mainly in adipose fat cells) is an enzyme system (see earlier) called **aromatase** (Diagram IX-16). Aromatase converts a portion of testosterone into estrogen. In males, estrogen acts on other cells to stimulate the production of sex steroid-binding globulin (SSBG), an inhibiting complex that, in binding with testosterone, makes less of the free testosterone available to do its metabolic job

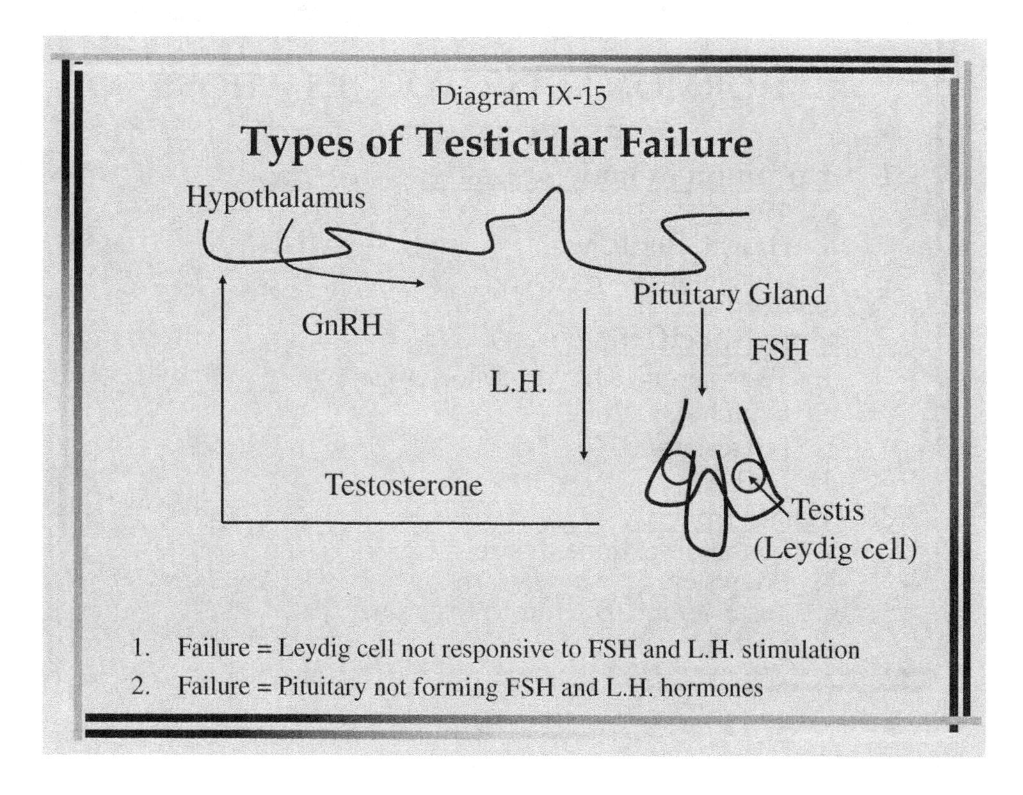

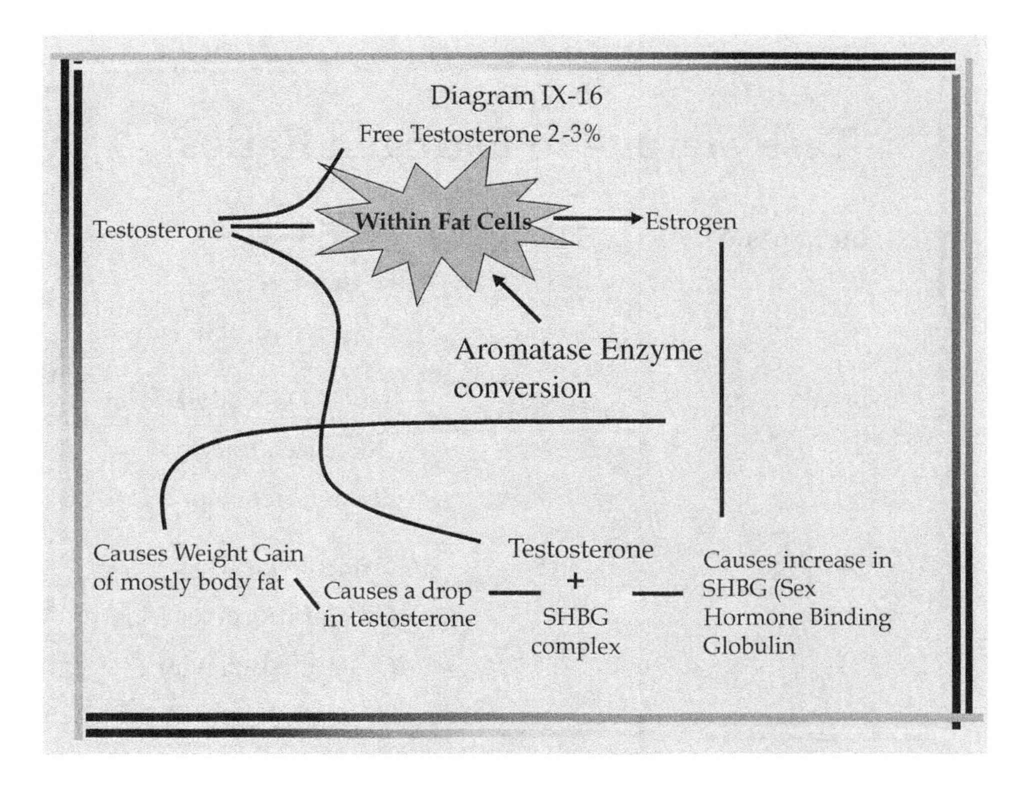

within the brain, muscles and fat cells. The greater the amount of sex hormone-binding hormone (SHBH) produced, the more testosterone that is removed from the circulatory system, and as time goes on, the greater the weight gain sustained in the form of body fat. This *results* in the large abdominal girth and waistline associated with many middle-aged adult males.

As the aging process continues, the *ratio* of testosterone to estrogen declines (Diagram IX-17). How fast it drops is usually influenced by genetic factors as well. Among lean young adults, the testosterone/estrogen ratio is approximately 50:1 and is accompanied by high energy levels and assertive behavior that for most of us in middle age exists in memory only.

In the middle-aged male, the testosterone/estrogen ratio drops to approximately 20:1. This ratio begins to manifest itself in truncal obesity, or what we refer to as "love handles," as well as an increase in fatty tissue around the chest, which we as cosmetic surgeons and physicians designate as *gynecomastia*.[11]

The dominant male personality that may have prevailed during high school and early college days begins turning more passive and tolerant. All in all, however, these are not such terrible changes. In the 20:1 ratio, the hormone levels are still sufficient to maintain an excellent state of health and quality of life.

Unfortunately, as the aging process progresses and our bodies begin to deteriorate, the testosterone/estrogen ratios drop to approximately 8:1. At this point, there

Diagram IX-17

Testosterone to Estrogen Ratios

Young Male	50:1	-Lean Body -Higher Energy -Aggressive Behavior
Middle Age	20:1	-Truncal Obesity Begins -Decreased Energy -Passive Behavior
Old Age	8:1	-Truncal Obesity -Lose of Muscular Mass -Lethargic Behavior

is a more marked increase in fatty deposition in the abdomen, chest and waist as well as a loss of muscle mass and bone mass. In many men, a lethargic state of mind may occur.

It is important to keep in mind, too, that certain lifestyle routines and health habits, specifically *alcoholic ingestion*, are also directly related to elevations in estrogen (Diagram IX-18).[12] Alcohol has been shown to *decrease* key enzymes found in the liver (the cytochrome P450 enzyme detoxification system). When this happens, even with moderate amounts of alcohol intake, estrogen cannot be metabolized out of the body; it builds up in the bloodstream, causing a more rapid deterioration of body composition described earlier.

Zinc deficiency has also been directly related to an increase in activity of the aromatase enzyme system, which also forces more testosterone to be converted to estrogen. This again causes testosterone/estrogen ratios to drop. Another consideration related to the decrease of testosterone is that as men age, more testosterone is converted to the secondary nonactive form, 5-DHT. This compound is much less effective than pure testosterone in keeping the body young and strong and occupies key receptor sites, further decreasing the effectiveness of whatever free testosterone is available. DHT is also directly related to hair loss among men as well as prostatic hypertrophy and the symptoms of difficulty with urination as men age.

Diagram IX-18

Seven Reasons for Estrogen Elevations in Males

The most common causes of midlife estrogen increases in males are:

- Age-related increases in aromatase activity
- Alteration in liver function
- Zinc deficiency
- Obesity
- Overuse of alcohol
- Drug-induced estrogen imbalance
- Ingestion of estrogen-enhancing food or environmental substances

Testosterone's *main* function is to support the development and maintenance of the *sexual organs* and secondary sex characteristics in the male. It contributes to libido and supports the ability to maintain and build muscles and burn fat. It is essential in supporting optimal skin tone, bone strength, immunity and mental performance.

Testosterone deficiencies are directly related to *low libido* and *mental depression* (Diagram IX-19).[13,14] A sparse mustache, diminished body hair and very slow beard growth are also commonly related to deficiencies in this hormone as well. The inability to build muscle, maintain bone mass and burn fat is another key sign of testosterone inefficiency. Osteoporosis, anemia and labile emotions, along with indecisiveness, are again related to deficiencies in testosterone as men age.[15]

Excess testosterone results in an overactive libido, excess body hair, reduced scalp hair and excessively aggressive behavior. Greasy, oily skin and acne are other key signs of excess testosterone levels.[12]

Dietary Interventions and Supplements So, what are some of the things we can do for our male patients to keep their testosterone/estrogen ratios in balance? Prescribing 20 to 50 mg of zinc twice a day helps to inhibit aromatase enzyme activity, thus making more testosterone available to do its key effects in optimizing aging. Reducing body fat through exercise and a healthy diet also serves as a buffer, limiting testosterone to estrogen conversion. Also important is *avoiding*

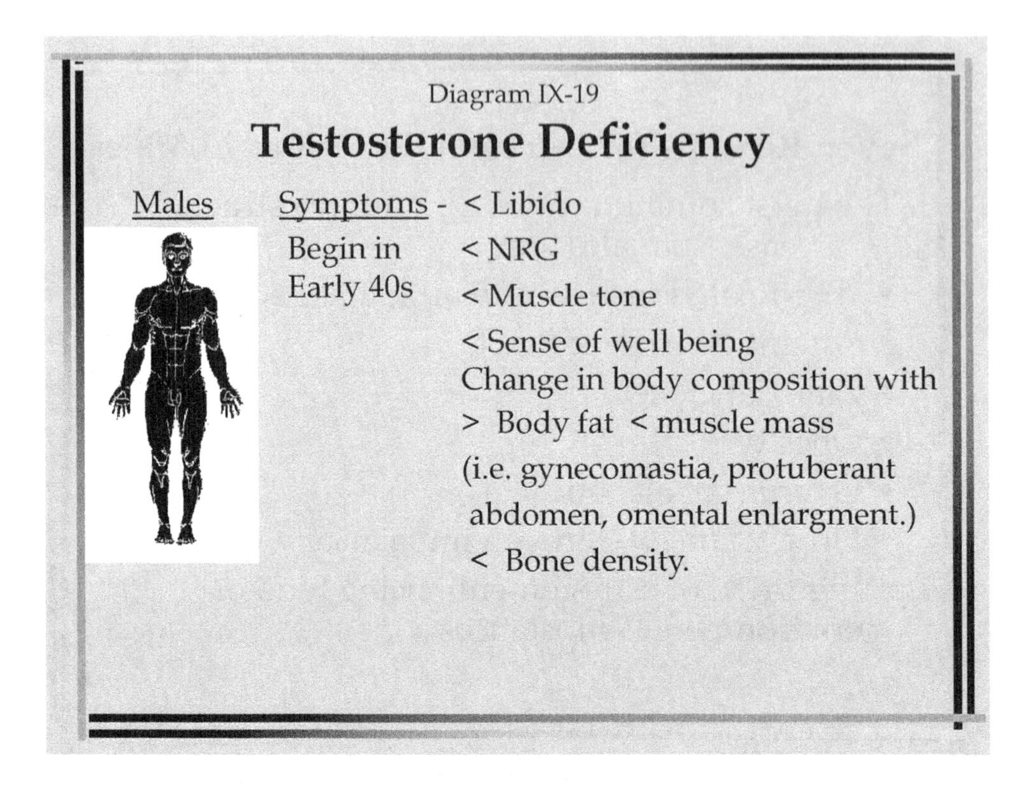

excess alcohol consumption, which improves cytochrome P450 function of the liver and works to maintain better testosterone/estrogen ratios by allowing the liver to excrete and metabolize unneeded estrogen. Eating more soy products containing genistein and isoflavones, which are both plant estrogens that help to displace estrogen from its receptors. Cruciferous vegetables like broccoli and cauliflower stimulate excretion of estrogen, also from the liver.

A worthwhile additional supplement is a twice-daily dose, 500 mg, of chrysin, an herbal extract, that helps block the aromatase enzyme activity very effectively as well. For this reason, patients presenting to the office with gynecomastia should be placed on a zinc and chrysin regimen to help limit the activity of key enzymes so that gynecomastia will not recur after it has been alleviated surgically.

Finally, if DHEA levels are also being augmented for a man on an anti-aging program, it is important to note that the form of DHEA called **7-keto DHEA** is more effective as a precursor, or secretagogue, to testosterone production in many men, because this 7-keto form cannot be converted into estrogen (Diagram IX-20).

With an age-management or anti-aging program that seeks to *optimize functional organ reserve*, it is important and convenient to note a number of key concepts according to the age group of the male patient (Diagram IX-21). In the 40- to 50-year-old age group, the physician should emphasize supplementing the diet with

Diagram IX-20

Balancing Testosterone: Estrogen in the Male Patient

- 50 mg b.i.d. of zinc until balanced
  - inhibits aromatase activity
- Reduced body fat to decrease testosterone to estrogen conversion
- Avoid alcohol consumption to improve cytochrome P450 function
- Increase consumption of:
  - soy:genistein inhibits tyrosine kinase receptors
  - soy:isoflavones displace estrogen from its receptors
  - crucifers: promote estrogen Phase II detoxification
- 500 mg b.i.d. of chrysin to block aromatase activity
- 7 – keto form of DHEA: not converted into estrogens

Diagram IX-21

Male Hormone Replacement Therapy Overview

Ages	Treatment
• 40 - 50 years	• Balance Estrogen to Testosterone ratio (i.e. zinc, soy, precursors, weight loss.)
• 50 - 60 years	• H.C.G. and precursors (increase production of own Testosterone)
• 60 - 70 years	• Testosterone replacement (creams, pellets, patches, gels, pills.)

zinc and soy products, coupled with a moderate dietary program and exercise for weight loss and body composition improvement.

In 50- to 60-year-olds, the use of hormonal precursors (e.g., DHEA, pregnenolone, and androstenedione) can help to stimulate production of testosterone from the testes.

In still later years, the use of testosterone replacement, in the form of transdermal creams, pellets, patches, gels and pills, is something that might seriously be considered under the *guidance* of a physician familiar with more advanced anti-aging therapeutic programs and hormonal replacement regimens. In all men, the appropriate screening tests should be completed and should include prostate-specific antigen (PSA) levels and a prostate gland examination *before* hormone replacement therapy (HRT) is begun.

It is now well recognized that andropause, as previously established, is an eventuality in all men. The power to manipulate these biochemical and physiological changes is easily within our grasp with a thorough understanding of the aforementioned concepts.

DHEA DHEA is produced in the adrenal glands from the precursor molecule pregnenolone, which is synthesized from cholesterol. Its main function is to serve as a precursor to testosterone, mainly in the form of androstenediol.

DHEA has been well established to *positively* impact the immune system by augmenting its efficiency. It is also directly related to body fat composition and helps to maintain optimally lean and fat free body composition. DHEA has been directly related as well to maintaining *optimal* function within the brain.[13,16]

In males, deficiencies of DHEA have been directly related to poor immunity, psychological depression, low metabolic rates, extremely low estrogen levels and low HDL cholesterol levels, which are also related to cardiovascular disease. Normal replacement levels for men are 20 to 50 mg/day orally or transdermally. (In women, the daily dose is 5 to 20 mg orally, or transdermally.)

Excess DHEA levels have been directly related to high estrogen levels, especially estradiol. As mentioned earlier, this can cause gynecomastia as well as truncal obesity in the waist and abdominal area and an inability to lose weight.[17,18] These negative effects of DHEA in men can be counteracted by the application of the steps previously mentioned while the positive aspects of its conversion of testosterone are maintained.[19–22]

Female-Specific Hormones When one is considering hormone replacement, only "natural" bioidentical hormones should be used. There is *no* justification for horse-derived hormones for chemical use in humans. This is the basis of the recent controversy with hormone replacement in women and should *not* be associated with the natural hormone replacement that is discussed in this text.

The ovaries produce the majority of female hormones. Small amounts are derived from the adrenal secretion of DHEA and androstenedione and from the aromatization of testosterone.

Estrogens The female hormones estradiol, estrone and estriol contribute mainly to the development of sex organs, ovulation and the menstrual cycle. They are also essential for collagen synthesis, brain function and memory, bone density, muscle and fat cell metabolism and blood lipid levels.

Deficiencies in the three estrogens in general can frequently result in wrinkled skin secondary to collagen production deficiency.[23] In our female cosmetic surgery patients, this is frequently noted in the perioral and periocular areas by the presence of fine rhytid ("crow's feet").[24] Bone loss is also a major problem that affects women, mainly in the lumbar and other spinal regions as well as in the long bones and hips. Age-related mental impairment or loss of recent memory is also a direct sequela to estrogen deficiencies in many women as they age.[25,26]

Low HDL and high LDL cholesterol fragments are directly related to diminished estrogen levels and are one of the key reasons that postmenopausal women have an *accelerated* rate of arteriosclerosis and heart attacks.[27]

The obvious symptoms of menopause (hot flashes and vasomotor irregularities along with mood disorders) are a direct result of estrogen insufficiency. In general, potential detrimental effects, which can also occur with estrogen imbalances, include difficulty in losing weight, water retention or bloating and proliferative effects on breast tissues, uterine tissues and ovarian tissues. There is a potential of a carcinogenic effect of metabolites with estradiol and estrone, but not estriol.[28] **Estriol** has been dubbed the "protective estrogen" because it seems to inhibit the proliferative effects of the other estrogen components.[29]

The symptoms of *excess* estrogen production manifest as general water retention, painful breasts, cysts and cystic masses, anxiety, endometrial fibroids, difficulty in losing weight, and hot flashes, which can be a symptom of deficiency as well.

Testing for the three key estrogen levels can be accomplished through salivary tests,* or blood tests.† In this fashion, we can establish the general functional reserve of the ovaries at a given point in time.

Estrogen replacement can be accomplished by using transdermal estrogen creams, which can be compounded* to contain a personally tailored balance of the three main estrogens. The standard norm values have been established for each decade, but for *optimal* aging purposes a range in the early to *mid-30s* is ideal, along with subjective improvement in estrogen deficiency symptoms. The use of transdermal creams is preferable for a number of reasons over oral replacement. The appropriate baseline screening tests should also be completed (Diagram IX-22).

---

*Aeron Labs; see Resource section.

†Immunosciences Lab; see Resource section.

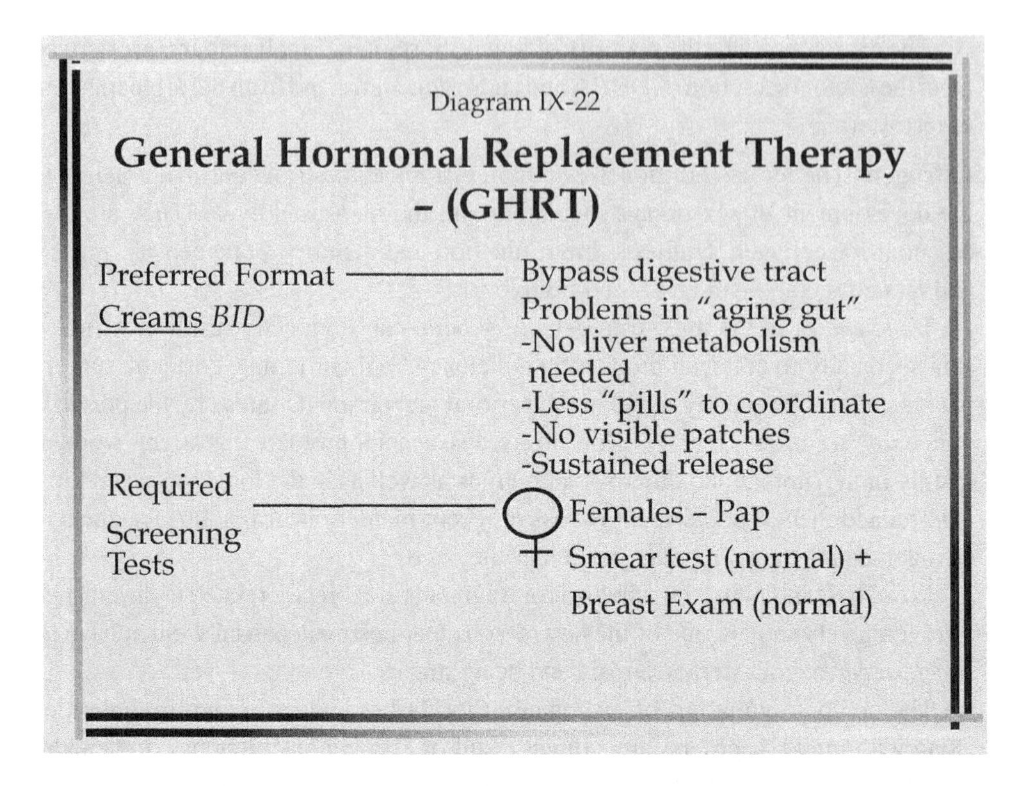

One of the best ways to supply estrogen replacement is in the form of transdermal creams or gels twice daily, once in the morning and once in the evening. This results in a more even blood level of the three key essential estrogens throughout the day.

An optimal replacement level during menopause can often eliminate the symptoms of estrogen deficiency, most commonly hot flashes, irritability and mood swings. If a patient has a family history of breast, ovarian or endometrial cancer or has been a cancer patient herself, estriol alone or with progesterone as a protective estrogen can be used in the form of a transdermal cream or gel.

One of the easiest ways to incorporate estrogen is to apply it to the face, neck and hands. This method not only elevates intravascular levels but also improves the quality of the skin of the face, neck, and hands.[5]

Progesterone Progesterone in women is produced by the corpus luteum within the ovary following ovulation only. It is also produced in small amounts in the adrenals by the conversion from pregnenolone. Levels of progesterone decline starting at age 35.

Progesterone *deficiency* is due to anovulatory cycles. This is considered the main cause of premenstrual syndrome (PMS). Progesterone's main function is a supportive role of the midluteal phase in order to aid the uterine lining and prepare it for egg implantation.

Progesterone has an important antiproliferative effect on the breast and uterus, which counteracts the potentially carcinogenic effects of the other two estrogens, estradiol and estrone.[30–31] It has a γ-aminobutyric acid (GABA)-like effect on the brain, calming, enhancing sleep and counteracting the anxiety-producing effects of estrogens. Progesterone also has a diuretic effect. It supports bone buildup and reverses bone loss. It has an anti-inflammatory effect and supports the myelination of nerve fibers in the peripheral nervous system.

The effects of progesterone deficiency and excess in women are summarized in Diagram IX-23.

Deficiencies in progesterone frequently result in water retention, weight gain, bone loss, anxiety, sleep problems, painful breasts and shorter menstrual cycles. Heavy bleeding is also commonly present. Episodes of arthritis or a general inflammatory state are also frequently present with a deficiency of progesterone.

Excess progesterone can result in dizziness and sleepiness; nausea may be associated with high levels. Excessive conversion to cortisol, the "age-accelerating hormone" as well as *excessive* conversions to testosterone can also create problems.

Estrogen or progesterone replacement in females can be used in many different forms, including pills, injections and transdermal creams or gels.

During perimenopause, progesterone is usually started 2 days before ovulation and is stopped when menstrual bleeding occurs. This represents the *luteal phase*,

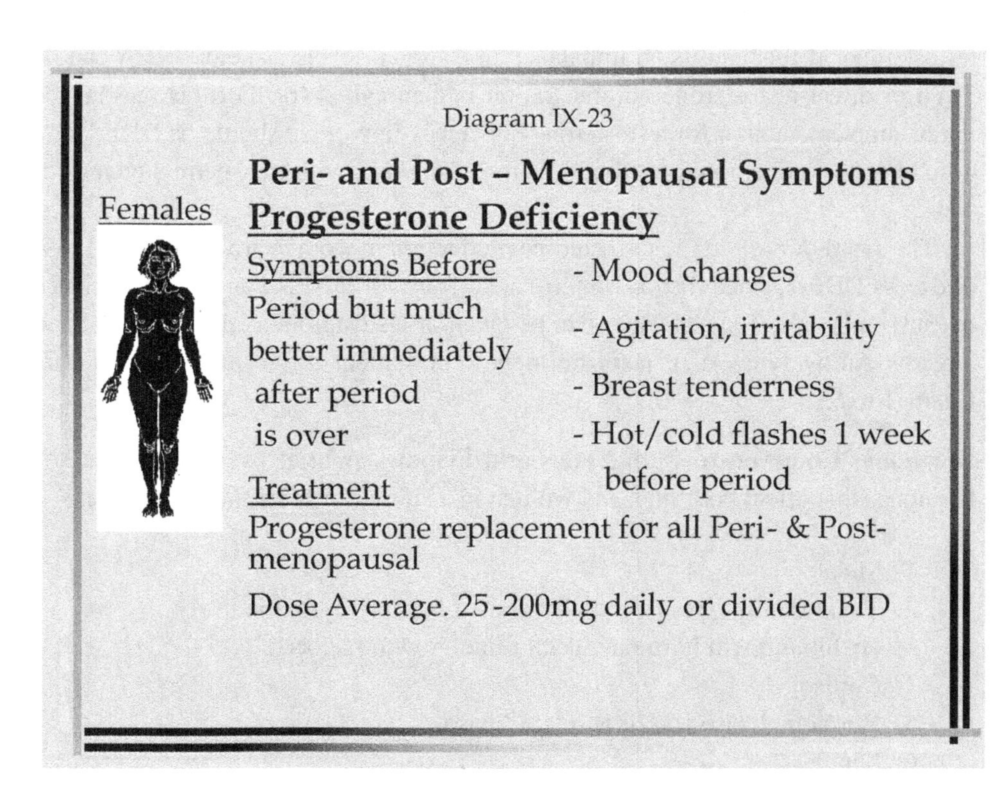

during which estrogen levels tend to peak. During menopause, progesterone is used on a 25-day continuous schedule with 5 to 6 days off at the end of the month. Progesterone is started on the first calendar day of each month. The dose of progesterone is titrated to alleviate deficiency symptoms, as well as maintain an adequately high laboratory value. (See Chart IX-5 in this chapter.)

DHEA DHEA is produced in the adrenal glands from pregnenolone, which is in turn synthesized from cholesterol. Its main functions are to serve as a precursor to the female hormones (estrogen, estradiol and estriol) and to testosterone. These are important roles in maintaining optimal immune function, brain function, memory and body metabolism, which help to restore and maintain lean body mass and body fat deposition.[32,33]

DHEA *deficiency* in women can result in low libido, poor immunity, depression and low metabolic rate. In excess, symptoms of masculinization occur, including excess acne, loss of scalp hair, greasy skin, excessive libido and facial hair growth. If taken late in the day as a supplement, DHEA can interfere with sleep.

Replacement of DHEA in women can be accomplished via sublingual sprays, capsules or topical creams.[20]

Testosterone When faced with testosterone *deficiency* in women, the easiest approach to restoring *optimal* levels is through the administration of DHEA. With DHEA supplementation, a large amount of the compound is converted directly to testosterone. If this causes an imbalance in estrogen levels, 7-keto DHEA can be given or direct testosterone administration is then called for. For this reason, the initial supplementation for testosterone deficiency in women should be DHEA or a dose of testosterone, ranging from 0.25 to 2.0 mg/mL, with transdermal creams as well.

The symptoms of *excess* testosterone production in women are identical to those of excess DHEA, since they are interrelated. If use of DHEA causes an imbalance in estrogen levels, 7-keto DHEA can be given or testosterone can be administered directly. An overview of general hormone replacement for women is detailed in Charts 1 to 6.

Hormones Common to Both Males and Females A brief overview of the key hormones that affect both men and women in a similar fashion should include:

1. Melatonin.
2. Thyroid hormones (triiodothyronine [T_3] and thyroxine [T_4]).
3. Grehlin/growth hormone/insulin-like growth factor-1.
4. Cortisol.
5. Hormone lipase and lipoprotein lipase.
6. Leptin.

Melatonin Melatonin (Greek for "night worker") is a hormone produced by the pineal gland (whose function is relatively unknown) in response to many unknown factors. It is suppressed by light during the day and is noted to be enhanced by darkness.

Along with melatonin, the hormone epithalamin is also produced by the pineal gland but may have an inhibitory effect, much the same as somatostatin does with human growth hormone (HGH).

The functions of melatonin appear to be multiple![34] It enhances sleep and helps the brain reach specific metabolic phases, which are favorable to optimal function (e.g., increased HGH production). It has a role in controlling the circadian (daily) rhythms, which may directly affect and balance the parasympathetic and sympathetic nervous function. Adequate levels of melatonin production are also thought to keep cortisol levels low and influence production of many sex hormones as well as the thyroid hormones.[35,36]

Melatonin is also considered an extremely potent antioxidant and a DNA protector.[37] In fact, melatonin is the only neurohormone produced that does *not* need a binding protein to deliver itself to the cells throughout the body. It is also the only hormone known to have a direct receptor on the chromatin of DNA. Many studies have documented that melatonin is an *extremely* potent antioxidant and has an obvious function in protecting DNA from free radical damage.[38]

Melatonin *deficiency* is usually manifested by symptoms that include trouble falling asleep and restless sleep in general.[39] *Excess* melatonin symptoms may be exhibited by early morning awakening or difficulty awakening and vivid or hallucinogenic-like dreams.

In breast cancer patients, melatonin has been suspiciously lacking in adequate levels. In patients who have undergone chemotherapy or radiation therapy as their initial treatment, those who took melatonin supplements had a better overall long-term prognosis.[40] Recent studies by Lissani have documented remarkable effects of melatonin when used with other primary oncologic treatment of a number of tumor types.[40–49]

Melatonin supplementation is usually begun with 0.5 to 3 mg orally anywhere from 1/2 hour to 1 hour before sleep.

Thyroid Hormones The thyroid hormones consist of thyroid-stimulating hormone (thyrotropin, or TSH), which is produced by the pituitary and stimulates the thyroid gland to produce T_4 (thyroxine). The liver and other peripheral tissues convert T_4 to T_3 (triiodothyronine), the more active of the two thyroid hormones.[50]

It is important to note that **selenium** helps to convert T_4 to T_3 in both the liver and peripheral tissues. **Tyrosine**, a precursor to T_4 and T_3, is required for hormonal production. It is part of the daily dietary requirement. An excess or deficiency of iodine directly controls T_4 production as well.

Thyroid hormones, specifically T_4 and T_3, have many major functions. They *regulate* the entry of oxygen into the cells, thus controlling most physiological processes, including metabolic rate, body temperature, ovulation, heart rate, brain activity, skin turnover and hair growth.[51]

Thyroid hormone deficiency usually results in symptoms that include a lower metabolic rate, lower basal temperature, cold extremities, morning grogginess, constipation and dry brittle skin and hair. In women, anovulatory cycles and menstrual irregularities are also common. Inability to perspire and a goiter (enlarged thyroid gland) are also common signs of deficiency.

Thyroid hormone excesses, especially T_3, can produce heart palpitations; sleep disturbances; a sense of nervousness; rapid bone growth, specifically around the orbit; and protruding eyes.

Normal thyroid supplementation is most easily accomplished with Armour Thyroid, which contains natural porcine glandular T_3 and T_4. It is normally given in the dose of 0.25 to 1 grain daily.

Growth Hormone Growth hormone (HGH) is one of the most commonly mentioned hormones and has maintained the major focus in most longevity journals. Because of its many beneficial effects, growth hormone has been touted as the "fountain of youth," or the key component of an anti-aging program. *Each of these statements could not be further from the truth.* Although growth hormone is essential in high levels during adolescence, its levels drop drastically in the early 20s and then continue to decline (Diagram IX-24).

First, let's have a brief but important review of growth hormone production and feedback. **Somatotropin** is produced by the pituitary gland in response to growth hormone-releasing hormone (GHRH) which is manufactured within the hypothalamus. Growth hormone production is also stimulated by **grehlin**, which is produced by specific gastrointestinal cells. Its production is inhibited by somatostatin, which is produced within the hypothalamus. In this fashion, somatostatin and GHRH form a feedback loop that *regulates* growth hormone levels.

Also now considered part of this regulation of the overall pathway is grehlin, since it appears to work in synergy with GHRH to augment growth hormone levels.

Growth hormone can have a direct effect by itself on tissues and organs, including muscle, brain, skin and bone and on immune function, or it can then be converted into a more potent and active form, called **insulin growth factor-1** (IGF-1) within the liver and other selected areas of the body as well.[33]

Growth hormone's main functions focus on increasing anabolic activity and in building lean tissue (i.e., more muscle, denser bones and thicker skin).[52] Growth hormone is directly associated with increased fat burning. It *improves* brain function as well as overall *immune function.*

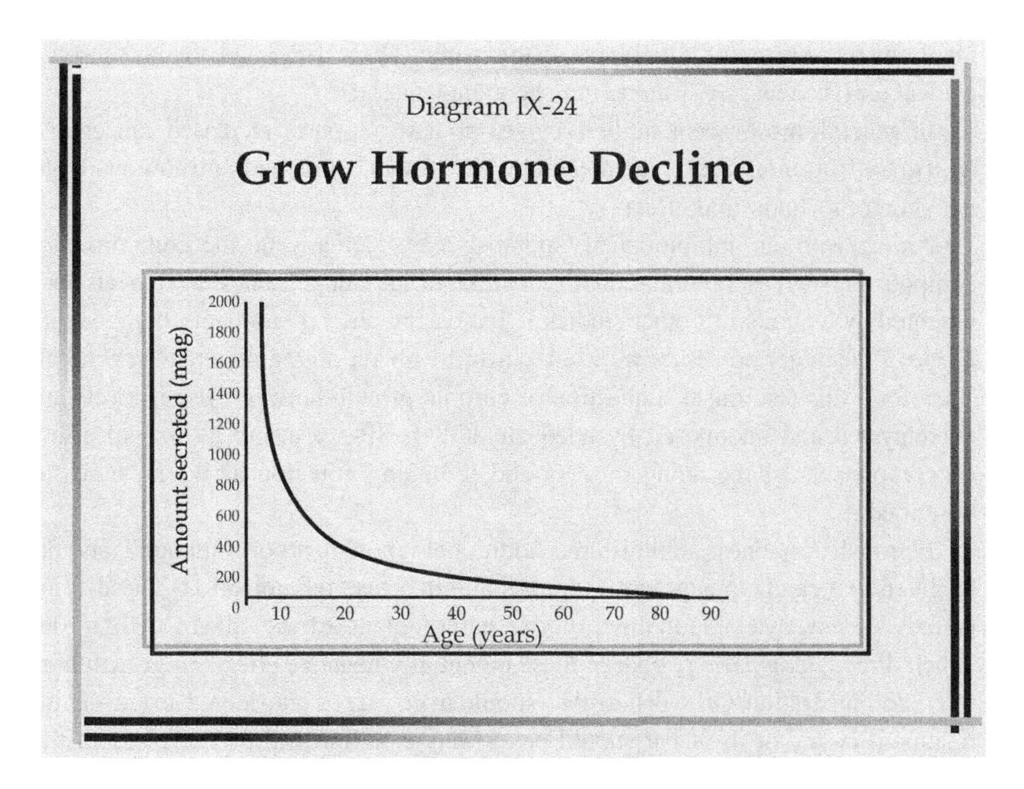

Symptoms of growth hormone *deficiency* include muscle and joint weakness, thin skin, increased body fat, poor bone density or osteoporosis, poor memory and impaired immune function. *Excess* growth hormone secretion is directly associated with acromegalia, edema, joint pain, dizziness and an increase in soft tissues, including tongue size.[53]

Growth Hormone and Potential Side Effects Abnormally high growth hormone levels have been associated with a *decrease* in apoptosis and an *increase* in inflammatory compounds.[54] **Apoptosis** is the cellular protective program in humans that helps us search out and find abnormally growing cells or early tumors. It is through an adequately functioning apoptosis pathway that these abnormal cells are identified and destroyed.

Chronic elevation of growth hormone has been shown to inhibit apoptosis! Therefore, the important question arises whether long-term use of growth hormone will create a problem with increased tumor growth or tumors in general.[55] This question appears to have been answered!

The key concept is that the cellular effects of IGF-1 are directly related to its interaction with IGF-1–BP-3 (binding proteins). IGF-1–BP-3 appears to have an inhibitory effect on cell growth[56] in vivo. In vitro experiments on cell cultures involving the sole use of IGF-1 have no opposing IGF-1–IBP-3 and, therefore, some evidence exists that IGF-1 *alone* may be mutagenic.

In general, most recent studies suggest *no* association of increased cancer risk with growth hormone replacement therapy.[57–59] Other compounds are now available to counter this undesired effect.

Along with an inhibition of apoptosis, an increase in the inflammatory compounds, such as key inflammatory cytokines and leukotrienes, have been documented by a number of other studies.[54] Increasing any inflammatory components is one of the major processes we want to avoid with anti-aging therapies in general. Therefore, this also raises a question of chronic growth hormone use if levels are overelevated and unopposed by adequate IGF-1–BP-3 use and its overall effect on components of the aging process and the aging equation as we have so far described it.[57,58]

If growth hormone is administered, additional supplements or compounds should be given that can help augment apoptosis and decrease inflammation. These compounds are now available in the form of a natural plant extract called C-MED-100, which directly *counteracts* both of these potentially negative effects of growth hormone administration. Growth hormone should in *no* way be considered an anti-aging treatment in and of itself but should be viewed as an integral part or piece of the puzzle in treating the aging process.[60] It should be used with moderation and appropriate laboratory monitoring of IGF-1 values, and IGF-1–BP-3.

Modest endogenous growth hormone augmentation can be accomplished in most patients by the use of an oral secretagogue or by combinations of key amino acid blends such as arginine, ornithine, and lysine, that can stimulate the anterior pituitary to make more growth hormone itself, especially when used with heavy exercise sessions. Recently available are transdermal secretagogues (e.g., Trans-D Tropin*) that have been very effective in some patients in raising HGH levels as well.

Actual growth hormone administration is accomplished by injection and is recommended to commence with 0.5 to 1.0 unit/day for men or women 5 days a week with weekends usually injection-free, as a rest period.

Recently available are GHRH and IGF-1 (Serrono Pharmaceuticals, Inc.). Both of these compounds are relatively new as clinical therapeutic anti-aging medicine and should be used only under the direction of a knowledgeable anti-aging physician who strongly believes in close laboratory monitoring. Caution: Both of these more potent means of elevating HGH, and IGF-1 are also associated with a *decrease* in apoptosis and an *increase* in inflammatory compounds as well.

Cortisol Cortisol has been dubbed the *age-accelerating hormone*. In Chapter V, "Diet and Aging," we have seen its relation to elevated insulin levels. It is produced

---

*Trans-D Tropin Plus is available through Optigene-X Inc.; see Resource section.

by the adrenal glands in response to adrenocorticotropic hormone (ACTH), which is released by the hypothalamus. ACTH, in turn, is produced by the pituitary gland as a result of corticotropin-releasing hormone (CRH) from the hypothalamus.

The main function of cortisol is to produce an anti-inflammatory effect, and at first glance, it would seem to be beneficial to the aging equation! First, let us look at its positive effects. Cortisol helps break down glycogen, fat and muscle for energy in order to make sure that this energy in the form of glucose is available in the bloodstream. It is one of the key backup systems utilized by the brain to ensure that it has appropriate energy in the form of glucose to keep it functioning optimally.[61]

In chronic elevations, which are usually the case as we age, cortisol is *directly* related to a breakdown in skin collagen and elastin as well as joint, bone and muscle tissue. It has a major inhibitory effect on sex hormones, thyroid hormones, growth hormone, and DHEA as well.[62]

Elevated levels of cortisol have been shown to cause damage to the central nervous system neurons, causing memory loss.[63] Relatively new research has also documented damage to cortisol receptors in the hypothalamus, which then results in further elevations of cortisol due to poor feedback receptivity.[64,65]

Chronic cortisol elevations also cause damage in both thymic and lymphatic tissues decreasing the immune function.[66,67] Clinical depression also occurs as a result of a decrease in neurotransmitter levels, which are directly related to elevated cortisol.

Cortisol also directly increases formation of bad eicosanoids (pro-inflammatory hormones). It decreases good eicosanoids (anti-inflammatory hormones) and decreases second messenger levels that have been discussed previously, such as cyclic adenosine monophosphate (cAMP), therefore interfering with hormone communication and efficiency.[68–70] For these reasons, cortisol has been called the age-accelerating hormone, a name that could not be more appropriate.

Cortisol *deficiency*, clinically, can result in hypoglycemia and sugar cravings. Dizziness due to hypotension and fatigue and accelerated inflammatory conditions are all quite common. These inflammatory conditions include arthritis, asthma and, in general, allergies.

Excess cortisol **(Cushing's syndrome)** clinically can result in a puffy moon face; tiny, thin wrinkles, especially around the face; and obesity, which is more truncal in nature and located in the neck and shoulders.[64,70]

High insulin levels, which result in glucose intolerance, are another key problem (see Chapter V). Lipid disorders that include elevated triglycerides are frequently related to high cortisol levels. Hypertension due to fluid retention is also a frequent complication of clinically elevated cortisol levels.

Simple ways to decrease cortisol include age-management programs and stress relaxation methods. Moderate exercise, frequent meal spacing with adequate macronutrient content, and adequate sleep as well as melatonin supplementation are all effective components that combat high cortisol production. Nutrients that can counteract some of the effects of cortisol in the central nervous system are phosphatidyl serine, phosphatidyl choline and phosphatidyl ethanolamine.

If cortisol levels are significantly lower than normal, possible beneficial ways to increase them include an easy regimen of oral pregnenolone, 50 to 100 mg/day in men and women. In women, progesterone, 25 to 200 mg/day has also helped to improve cortisol levels.

When cortisol levels are being evaluated, it is *important* to measure them at the same time each day, usually first thing in the morning. This is one of the most accurate times to obtain levels of adrenal function.

The importance of controlling elevated cortisol levels and keeping them well within the range of normal *cannot* be overstated. This is a main goal of any age-management or anti-aging program.

Hormone-Sensitive Lipase and Lipoprotein Lipase Both of these hormones are directly related to fat storage and fat deposition. Both reside on the surface of fat cells. Each is regulated by insulin and glucagon and is responsible for regulating the flow of fat into or out of fat cells.[71]

Lipoprotein lipase transports fatty acids into the fat cells and keeps them there. **Hormone-sensitive lipase** does the opposite. It releases fat from the fat cells into the blood, where it is then transported to other cells so that it burns as fuel, usually within the mitochondria. Insulin stimulates the activity of lipoprotein lipase, and glucagon inhibits its activity. Hormone-sensitive lipase is stimulated by glucagon to release fat from adipocytes and is inhibited from releasing fat by insulin.

Also produced within the fat cells is another hormone called **leptin**, which is directly related to body fat and body composition (see later). It becomes obvious that controlling insulin and glucagon levels are the key ways of maintaining optimal levels and functioning of hormone-sensitive lipase and lipoprotein lipase.

In other words, the chief *activator* of body fat deposition, which promotes lipoprotein lipase, is insulin; the chief *inhibitor* of lipoprotein lipase is glucagon. The importance of regulating insulin and glucagon by our diet should by now be obvious.

Leptin Leptin is produced by adipocytes (fat cells). This hormone directly inhibits somatostatin, thus helping to stimulate release of growth hormone from the hypothalamus. *Decreased* leptin values are usually found in thin, or *underweight*, people. This leptin decrease sends a signal to the hypothalamus to increase food consumption and, at the same time, to decrease energy expenditure so that energy can be stored and body mass will increase. It also creates an increase in parasympathetic tone.

Elevated leptin levels are present in fat, or *obese*, patients. This leptin increase causes a decrease in food consumption, an increase in energy expenditure and an increase in sympathetic tone.

Eicosanoids Eicosanoid hormones are a biologically powerful group of compounds termed **autocrine hormones**. They are derived from a group of polyunsaturated fatty acids, usually containing 20 carbon atoms. They consist of the compounds that include *prostaglandins, thromboxanes, leukotrienes, lipoxins and hydroxylated fatty acids.*[72–75]

The autocrine eicosanoids are being produced continuously in minute quantities at the cellular level and persist for mere nanoseconds. They are powerful local response modifiers, or feedback hormones, which help to coordinate and fine-tune cellular reactions. They are the control factors of the "inflammation" levels in the aging equation.

Prostaglandins, or the **PG-1 series**, are derived from fatty acids, mainly γ-**linoleic acid** (GLA) and are generally considered "good" eicosanoids.[72] Prostaglandins of the **PG-2 series** are considered "bad" eicosanoids, at least when they are present beyond bare minimum amounts. The **PG-2s**[73] are derived from the fatty acid **arachidonic acid**, which in turn can either be made from GLAs or obtained from the diet.[74]

One of the key properties of the prostaglandins is their ability to alter intracellular cAMP levels. cAMP is the same second messenger discussed earlier in this chapter and is used by numerous endocrine hormones that transmit their biological information to the appropriate component within the cell and the neighboring target cell. By *maintaining* adequate levels of these key eicosanoids, we are guaranteed that a certain baseline level of cAMP is always present in the cell to amplify the message sent by key hormones.

The good eicosanoids (PG-1) act as vasodilators and immune enhancers. They decrease inflammation, decrease pain, increase oxygen flow, increase endurance, prevent platelet aggregation, dilate airways and decrease cellular proliferation.

The bad eicosanoids (PG-2) act as vasoconstrictors and immunosuppressors. They increase inflammation, increase pain, decrease oxygen flow, decrease endurance, cause platelet aggregation, constrict airways and increase cellular proliferation.

It becomes obvious that there is a pivotal importance of prostaglandin E (PGE_1 and PGE_2) eicosanoids for controlling insulin and cAMP levels and for modulating the effects of the age-related decrease in levels of cAMP.

The easiest way to gain control over both PG-1 and PG-2 levels is through dietary and nutrient influence at **three key control points** in their production pathways.

The *first control point* involves increasing the effectiveness of the conversion of key fatty acids into GLA. Increasing dietary intake of borage oil, evening primrose oil and black currant oil help with this activity. GLA is then rapidly converted into dihomo-γ-linoleic acid (DGLA).

The *second control point* rests upon the fate of DGLA, since it can end up either as a good or bad eicosanoid, depending on whether or not it is converted to the appropriate fatty acid by delta-5-desaturase. This enzyme controls the balance of good and bad eicosanoids. *If delta-5 desaturase is active, more arachidonic acid is produced to make more bad eicosanoids.* If its activity is blocked, DGLA produces more good eicosanoids. One of the easiest ways to inhibit delta-5 desaturase is to increase our intake of eicosapentaenoic acid (EPA), most easily obtained from cold water fish oils, mainly cod liver oil.

The *third control point* comes from restricting the dietary intake of arachidonic acid. This is the compound most commonly found in the yellow egg yolks. Another way to handle excess levels of bad PGE_2 eicosanoid hormones involves eliminating as much red meat as possible from our diet.

It is the balance between DGLA and arachidonic acid that controls the inflammatory response in each one of the 100 trillion cells of our bodies. Increased arachidonic acid equals increased inflammation, which results in the production of more free radicals, accelerated aging and a poor health span. A test has recently become available that determines the arachidonic acid/EPA ratio. It is an excellent indication of the eicosanoid status within our cells, and, therefore, the level of cellular inflammation.

Dosing Protocols Dosing protocols for men and women are summarized in Charts IX-1 to IX-6. They include the use of natural hormone replacement (i.e., hormones that are molecularly identical to the human biological form). The use of equine (horse)-derived hormone replacement is not recommended.*

As stated at the beginning of this chapter, hormones not only are responsible for the physical changes we note as physicians and surgeons but also are intimately connected to the psychological and personality changes we observe in our patients. These personality characteristics are the *result* of changing combinations and ratios of myriad hormones that occur throughout life and are *responsible* for the joy or depression we may experience at various times.

Truly, our character directly affects not only our fate but also the quality of our life.

---

*Hopewell Pharmacy; see Resource section.

Chart I

DOSING PROTOCOLS

PMS (Pre-Menstrual Syndrome)

Progesterone	25 mg to 1200 mg daily (usually 100 to 400 mg)
	Dose at least twice a day
	Give cyclically days 14-25

Chart II

PERI-MENOPAUSAL

Progesterone	25 mg to 200 mg daily
	Dose once or twice daily
	Give cyclically days 14-25
Tri-estrogen	0.625 mg to 0.5 mg daily, dosed once or twice a day
	Cyclically days 1-25
	If progesterone alone doesn't alleviate symptoms
	Continue progesterone as above

Chart III

MENOPAUSE – NATURAL or SURGICAL

Tri-estrogen	1.25 to 5.0 mg daily, dosed once or twice a day May use continuously or stop 5 days a month
PLUS	
Progesterone	25 to 200 mg daily, dosed once or twice a day May use continuously or stop 5 days a month
Testosterone	0.25 to 2.0 mg once daily
DHEA(Optimal)	5 to 25 mg daily

with Natural menopause

Protocol same as surgical menopause, *if symptoms indicate*

Chart IV

POST- MENOPAUSE

Hormone Therapy needed to maintain vital functions (heart protection, bone mass, lipid profile, etc.). Even when patient asymptomatic, use lower doses if asymptomatic

Tri-estrogen (or equivalent)	0.625-5.0 mg daily, dosed once or twice a day
PLUS	
Progesterone	25-200mg daily, dosed once or twice a day
Testosterone	0.25-2.0 mg daily
DHEA (optional)	5-25 mg daily

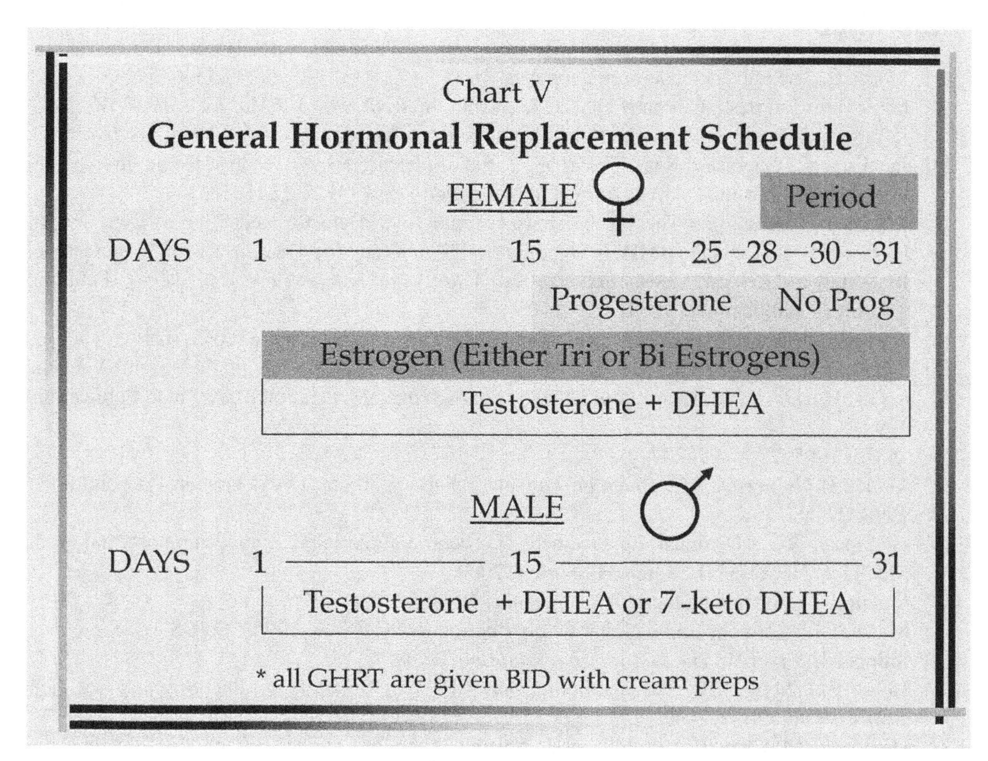

Chart VI

Dosing Protocols for Both Men and Women: daily doses (adults)

GH 0.05 –1.0 I.U. @ day Mon.- Fri.
growth hormone

THYROID-> Armour thyroid (T_3 and T_4)
0.25 grams to 1 gram daily

MELATONIN -> (0.5 – 3mg) at bedtime

*Basic line Blood or Salivary valves must be established Before replacement Rx and again in 6-12 weeks.

References

[1] Woese C. The universal ancestor. *Proc Natl Acad Sci U S A.* 1998; 95: 6845–6859.

[2] Poole AM, Jeffares DC, Penny D. The path from the RNA world. *J Mol Evol.* 1998; 46: 1–17.

[2a.] Jeffares DC, Poole A, Penny D. Relics from the RNA world. *J Mol Evol.* 1998; 46: 18–36.

[3] Jacobson L, Sapolsky RM. The role of the hippocampus in feedback regulation of the hypothalamic-pituitary-adrenocortical axis. *Endocr Rev.* 1991; 12: 118–134.

[4] Effects of Testosterone Therapy on Body Fat and Bone Density: Reduction in Body Fat and Increased Bone Density (BMD) in Hypogonadal Men Receiving Treatment with "Testosterone." In: Wright JV, Lenard L, eds. *Maximize Your Vitality and Potency: For Men over 40.* Petaluma, CA: Smart Publications, 1999.

[5] Topical estrogen and its effects on skin aging. *Int J Dermatol.* 1996; 35: 669–674.

[6] Effect on Testosterone and Blood Lipids. Natural Hormone Replacement. In: Wright JV, Lenard L, eds. *Maximize Your Vitality and Potency: For Men over 40.* Petaluma, CA: Smart Publications; 1999.

[7] *JAMA.* 1995; 293: 199–207.

[8] Yanick P, Giampapa VC. *Quantum Longevity.* Santa Barbara, CA: Christian Publishing Co.; 1998: 27–38.

[9] Giampapa VC. Hormonal Replacement Protocol. International Symposium on Anti-Aging Medicine, Newark, NJ. March 31–April 1, 2000.

[10] Mazur A. Aging and endocrinology. *Science.* 1998; 279: 305–306.

[11] Marin P. Testosterone and regional fat distribution. *Obesity Res.* 1995; 3: 609S–612S.

[12] Shippen E, Fryer W. *The Testosterone Syndrome.* 1998.

[13] Heller CG, Myers GB. The male climacteric: Its symptomatology, diagnosis and treatment. *JAMA.* 1994; 126: 472–479.

[14] Davidson JM, Chen JJ, Crapo L, et al. Hormonal changes and sexual function in aging men. *J Clin Endocrinol Metab.* 1983; 57: 71–77.

[15] Stearns EL, MacDonnell JA, Kaufman BA, et al. Declining testicular function with age. *Am J Med.* 1974; 57: 761–766.

[16] Morales AJ, Haubrich RH, Hwang JY, et al. The effects of six months treatment with a 100-mg daily dose of dehydroepiandrosterone (DHEA) on circulating sex steroids, body composition, and muscle strength in age-advanced men and women. *Clin Endocrinol (Oxf).* 1998; 49: 421–432.

[17] Haffner SM, Valdez RA, Mykkanen L, et al. Decreased testosterone and dehydroepiandrosterone sulfate concentrations are associated with increased insulin and glucose concentration in nondiabetic men. *Metabolism.* 1994; 43: 599–603.

[18] Kalimi M, Shafagoi Y, Loria R, et al. Anti-glucocorticoid effects of dehydroepiandrosterone (DHEA). *Mol Cell Biochem.* 1994; 131: 99–104.

[19] Labriel F, Belanger A, Cusan L, Candas B. Psychological changes in dehydro-"epiandrosterone are not reflected by serum levels of active androgens and estrogens, but of their metabolites: Intracrinology. *J Clin Endocrinol Metab.* 1997; 82: 2403–2409.

[20] Morales AJ, Nolan JJ, Nelson C, Yen SS. Effects of replacement dose of dehydro-"epiandrosterone in men and women of advancing age. *J Clin Endocrinol Metab.* 1994; 78: 1360–1367.

[21] Paoletti R, ed. *Drugs Affecting Lipid Metabolism.* New York: Elsevier Publishing Co.; 1980.

[22] Regelson W, Kalimi M. Dehydroepiandrosterone–the multifunctional steroid. *Ann NY Acad Sci.* 1994; 719: 564–575.

[23] Topical estrogen and its effects on skin aging. *Int J Dermatol.* 1996; 35: 669–674.

[24] Sherwin BB. Sex hormones and psychological functioning in post-menopausal women. *Exp Gerontol.* 1994; 29: 423–430.

[25] Phillips SM, Sherwin BB. Effects of estrogen on memory function in surgically menopausal women. *Psychoneuroendocrinology.* 1992; 17: 485–495.

[26] Simpkin JW, Singh M, Bishop J. The potential role for estrogen replacement therapy in the treatment of cognitive declines and neurodegeneration associated with Alzheimer's disease. *Neurobiol Aging.* 1994; 5: S195–S197.
[27] *JAMA.* 1995; 293: 199–207.
[28] Colditz MJ, Stampfer MJ, Hennekens C, et al. The use of estrogen and progestins and the risk of breast cancer in post-menopausal women. *N Engl J Med.* 1995; 332: 1589–1593.
[29] Lemon HM, Wotiz HH, Parsons L, Mozden PJ. Reduced estriol excretion in patients with breast cancer prior to endocrine therapy. *JAMA.* 1996; 196: 112–120.
[30] Stanford JI, Weiss NS, Voigt LF, et al. Combined estrogen and progestin hormone replacement therapy in relation to breast cancer in middle-aged women. *JAMA.* 1995; 274: 178–179.
[31] Ziel HK, Finkle WD. Increased risk of endometrial carcinoma among users of conjugated estrogens. *N Engl J Med.* 1975; 293: 1167–1170.
[32] Regelson W, Kalami M. Dehydroepiandrosterone — the multifunctional steroid. *Ann N Y Acad Sci.* 1994; 719: 564–575.
[33] Regelson W, Colman C. *The Super-Hormone Promise.* New York: Simon and Schuster; 1996.
[34] Pierpaoli W, Regelson W. *The Melatonin Miracle.* New York: Simon and Schuster, 1995.
[35] Reiter RJ, Robinson J. *Melatonin.* New York: Bantam Books; 1995.
[36] Sewerynek ED, Melchiorri D, Ortiz GG, et al. Melatonin reduces H_2O_2-induced lipid peroxidation in homogenates of different rat brain regions. *J Pineal Res.* 1995; 19: 51–56.
[37] Reiter RJ, Melchiorri D, Sewerynek E, et al. A review of the evidence supporting melatonin's role as an antioxidant. *J Pineal Res.* 1995; 18: 1–11.
[38] Pierpaoli W, Dall'Ara A, Pedrinis F, Regelson W. The pineal control of aging: the effects of melatonin and pineal grafting on the survival of older mice. *Ann N Y Acad Sci.* 1991; 621: 291–313.
[39] Dawson D, Encel N. Melatonin and sleep in humans. *J Pineal Res.* 1993; 15: 1–12.
[40] Lissoni P, Barni S, Meregalli S, et al. Modulation of cancer endocrine therapy by melatonin: a phase II study of tamoxifen plus melatonin in metastatic breast cancer patients progressing under tamoxifen alone. *Br J Cancer.* 1995; 71: 854–856.
[41] Lissoni P, Barni S, Ardizzoia A, et al. Randomized study with the pineal hormone melatonin versus supportive care alone in advanced non-small cell lung cancer resistant to a first-line chemotherapy containing cisplatin. *Oncology.* 1992; 49: 336–339.
[42] Lissoni P, Barni S, Ardizzoia A, et al. A randomized study with the pineal hormone metalonin versus supportive care alone in patients with brain metastases due to solid neoplasms. *Cancer.* 1994; 73: 699–701.
[43] Lissoni P, Pittalis S, Rovelli F, et al. Interleukin-2, melatonin, and interleukin-12 as a possible neuroimmune combination in the biotherapy of cancer. *J Biol Regul Homeost Agents.* 1995; 9: 63–69.
[44] Lissoni P, Barni S, Tancini G, et al. A randomized study with subcutaneous low-dose interleukin-2 alone vs. interleukin-2 plus the pineal neurohormone melatonin in advanced solid neoplasms other than renal cancer and melanoma. *Br J Cancer.* 1994; 69: 196–199.
[45] Lissoni P, Eregalli S, Nosetto L, et al. Increased survival time in brain glioblastomas by a radioneuroendocrine strategy with radiotherapy plus melatonin compared to radiotherapy alone. *Oncology.* 1996; 53: 43–46.
[46] Lissoni P, Paolorossi F, Ardizzoia A, et al. A randomized study of chemotherapy with cisplatin plus etoposide versus chemoendocrine therapy with cisplatin, etoposide, and the pineal hormone melatonin as a first-line treatment of advanced non-small cell lung cancer patients in a poor clinical state. *J Pineal Res.* 1997; 23: 15–19.
[47] Lissoni P, Paolorossi F, Tancini G, et al. Is there a role for melatonin in the treatment of neoplastic cachexia? *Eur J Cancer.* 1996; 32A: 1340–1343.
[48] Lissoni P. Melatonin and cancer treatment. In: Watson RR, ed. *Melatonin in the Promotion of Health.* Boca Raton, FL: CRC Press; 1998.

[49] Lissoni P, Barni S, Madala M, et al. Decreased toxicity and increased efficacy of cancer chemotherapy using the pineal hormone melatonin in metastatic solid tumor patients with poor clinical status. *Eur J Cancer.* 1999; 12: 1688–1692.

[50] Brent GA. The molecular basis of thyroid action. *N Engl J Med.* 1994; 331: 847–853.

[51] Rosenthal MS. *The Thyroid Sourcebook.* Los Angeles: Lowell House; 1996.

[52] Rudman D, Feller AG, Nagraj HS, et al. Effects of human growth hormone in men over 60 years of age. *N Engl J Med.* 1990; 323: 1–6.

[53] Klatz R. *Grow Young with hGH.* New York: HarperCollins, 1997.

[54] *J Natl Cancer Inst.* 2000; 92(18).

[55] Chan JM, Stampfer MJ, Giovannucci E, et al. Plasma insulin-like growth factor-I and prostate cancer risk: a prospective study. *Science.* 1998; 279: 563–566.

[56] Cohen P, et al. *Endocr Metab Clin.* 1996; 25(3).

[57] *J Clin Endocrinol Metab.* May 2001.

[58] Giovannucci E. Insulin-like growth factor-I and binding protein-3 and risk of cancer. *Horm Res.* 1999(51, Suppl.); S3: 34–41.

[59] Hankison SE, Willet WC, Colditz GA, et al. Circulating concentration of insulin-like growth factor-I and risk of breast cancer. *Lancet.* 1998; 351(9113): 1393–1396.

[60] Morley JE, Kaiser F, Raum WJ, et al. Potentially predictive and manipulable blood serum correlates of aging in the healthy male: progressive decreases in bioavailable testosterone, dehydroepiandrosterone sulfate, and the ratio of insulin-like growth factor-I to growth hormone. *Proc Natl Acad Sci U S A.* 1997; 94: 7537–7542.

[61] Norman AW, Litwack G. *Hormones*, ed 2. New York: Academic Press; 1997.

[62] Orth DN. Cushing's syndrome. *N Engl J Med.* 1995; 332: 791–803.

[63] Sapolsky RM, Packan DR, Vale WW. Glucocorticoid toxicity in the hippocampus: in vitro demonstration. *Brain Res.* 1988; 453: 367–371.

[64] Sapolsky RM, Krey L, McEwen BS. Prolonged glucocorticoid exposure reduces hippocampal neuron number: implications for aging. *J Neurosci.* 1985; 5: 1222–1227.

[65] Zimmet P, Baba S. Central obesity, glucose intolerance, and other cardiovascular risk factors. *Diabetes Res Clin Proc.* 1990; 6: S167–S171.

[66] Burr GO, Burr MR. On the nature and role of the fatty acids essential in nutrition. *J Biol Chem.* 1930; 86: 587–621.

[67] Munch A, Crabtree GR. Glucocorticoid-Induced Lymphocyte Death. In: Bower ID, Locksin RA, eds. *Cell Death in Biology and Pathology.* New York: Chapman and Hall; 1981: 329–357.

[68] Coleman RA, Humphrey PPA. Prostanoid Receptors. In: Vane JR, O'Grady J, eds. *Therapeutic Applications of Prostaglandins.* London: Edward Arnold; 1993: 15–25.

[69] Brenner RR. Nutrition and hormonal factors influencing desaturation of essential fatty acids. *Prog Lipid Res.* 1982; 20: 41–48.

[70] Schapira DV, Kumar NB, Lyman GH, Cox CE. Abdominal obesity and breast cancer risk. *Ann Intern Med.* 1990; 12: 182–186.

[71] Kern PA, Ong JM, Soffan B, Carty J. The effects of weight loss on the activity and expression of adipose tissue lipoprotein lipase in very obese individuals. *N Engl J Med.* 1990; 22: 1053–1059.

[72] Kunkel SL, Fantone JC, Ward PA, Zurier RB. Modulation of inflammatory reaction by prostaglandins. *Prog Lipid Res.* 1982; 20: 633–640.

[73] Bergstrom S, Rhyhage R, Samuelson B, Sorval J. The structure of prostaglandins E_1, $E_{1\alpha}$, and $F_{1\beta}$. *J Biol Chem.* 1963; 238: 3555–3565.

[74] Chatzpanteli K, Rudolph S, Axelrod L. Coordinate control of lipolysis by prosta-"glandin E_2 and prostacyclin in rat adipose tissue. *Diabetes.* 1992; 41: 927–935.

[75] Sears B. *The Anti-Aging Zone.* New York: HarperCollins; 1999.

10

Observing the Aging Body Through the Eyes of the Anti-Aging Clinician and the Cosmetic Surgeon

Vincent C. Giampapa, M.D., F.A.C.S.
Oscar M. Ramirez, M.D., F.A.C.S.

Sometimes the obvious can stare you in the face.

Matt Ridley, *Genome*

Introduction

For the first time within the field of cosmetic surgery, we have been prompted to utilize a different perspective of our function as cosmetic surgeons. We have begun to examine the process we call "aging," not just from the point of view of effect and surgical procedure but from a perspective of "cause and effect." The effects are represented by age-related changes that we have so often encountered.

We must realize that the physical changes we address with our surgical procedures are due to alterations previously described in the form of an **aging equation** caused by the **internal aging process** (Diagram X-1). These changes, ultimately, are due to alterations in "gene expression patterns"[1–4] as we age. These noticeable physical changes in different areas of the body can be halted and, in some cases, reversed with both anti-aging medical therapy and external cosmetic procedures for longer-lasting and more optimal results. A convenient way to understand the fundamental basis interrelationships of the aging process, as well as surgical procedures we choose, is to *examine* the effects of aging and their underlying conditions on a regional basis.

The Principles and Practice of Antiaging Medicine for the Clinical Physician, 175–188.

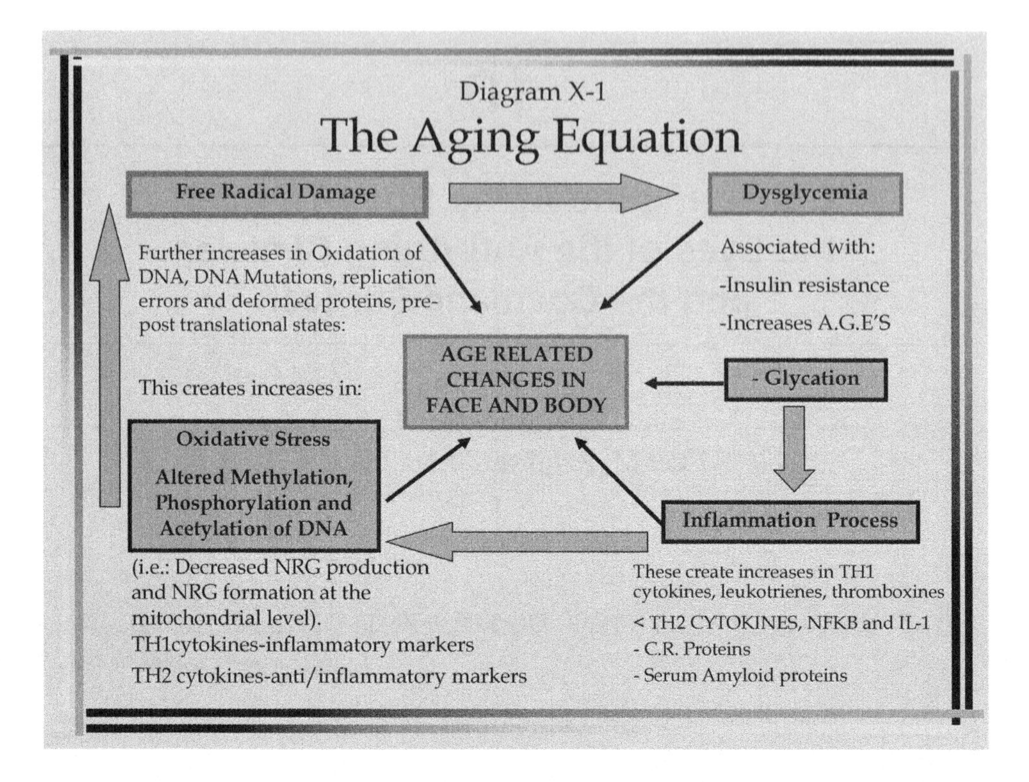

Regional Aging Concepts and Conditions

The Face On the face, we are accustomed to observing loss of skin tone resulting from loss in skin elasticity, which is due to altered ratios of collagen and elastin. One biochemical cause of these conditions is a *constant elevation of glucose levels*, resulting in glycation[5] and crosslinking of these skin proteins. This is actually the origin, or birth, of a wrinkle (see Chapter VII, "Skin and Aging."[6–19] On the other hand, a drop in glucose levels creates a demand for energy, and the body therefore releases cortisol as its backup mechanism to utilize proteins for sources of glucose, or fuel. This hormonal irregularity is the most common finding in today's fast-paced modern-day diet, which reinforces irregular meals throughout the day with large quantities of carbohydrates. For the same reasons, trendy "starvation diets" also contribute to accelerated skin aging. Also noted on the face are marked changes in the overall shape and thickness of the skin owing to loss of the underlying facial fat, which is also negatively affected by elevated cortisol levels.

The loss of muscle tone and, to some degree, the loss of bone and cartilaginous prominence and support also occur on the face at the areas of the nose and ears. Within the neck, the decrease in collagen and muscle tone is shown by a loss of the cervical mental angle, a loss of overall skin elasticity and the presence of wrinkling (Diagram X-2).

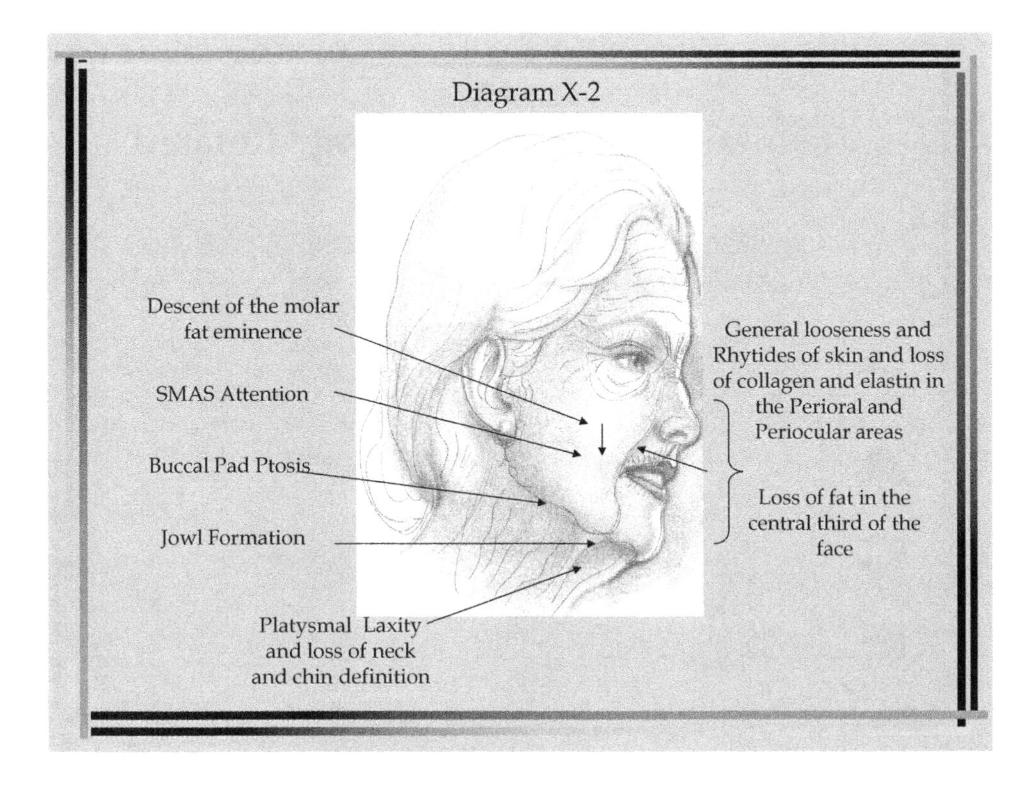

The actual causes of these changes are a result of (1) damage to the skin[20] secondary to elevated free radical levels,[21,22] (2) a decrease in collagen synthesis and (3) wrinkling of skin due to the crosslinking of glycated and oxidized proteins occurring in the dermis. Inflammation at the subclinical level is an essential component to collagen loss as well.[23]

Decreases in key hormones that occur with the normal aging process (e.g., lower levels of total and free testosterone and declining levels of estrogen, growth hormone, progesterone and thyroid hormones) also significantly affect the facial tissues and neck.

The plastic surgical treatments for these changes are focused on skin redraping, blepharoplasty, laser resurfacing, chemical peels, fat grafting and muscle-suspension and muscle-tightening procedures (Diagram X-3).

Clinical Anti-Aging Treatment Concepts The anti-aging treatment for the listed facial and neck age-related changes should also focus on a different dimension, the *cellular* level. This nonsurgical approach should emphasize:

1. Restoring antioxidant levels thus decreasing free radical damage to the muscles and the skin. In the skin, this is paramount to preserving the integrity of DNA in stem cells, particularly in the fibroblasts, which make collagen.

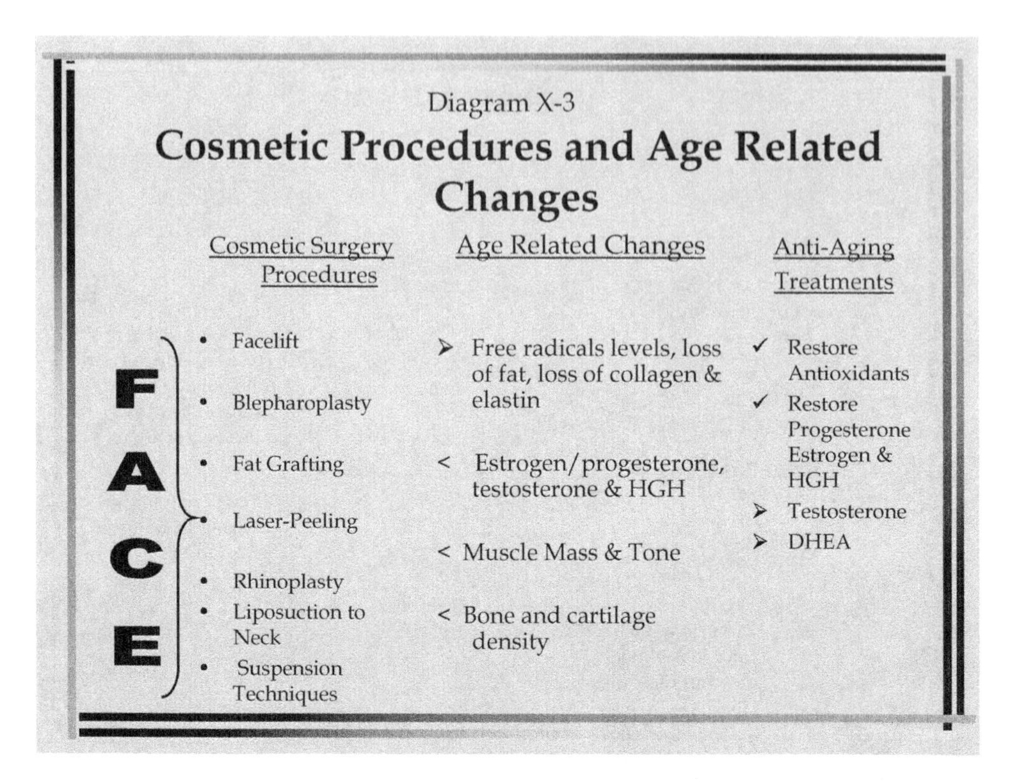

2. Decreasing glycated proteins by controlling serum glucose levels and by limiting protein crosslinking is another essential process to decrease rhytide formation and progression.[24,25]
3. Maintaining an optimal level of plasma amino acids (through meal composition and frequency) in order to provide the building blocks for tissue regeneration.
4. Restoring the ratio and quantity of fatty acids as building blocks of cell membranes and hormone precursors (eicosanoids) that control blood flow, inflammation and hydration.[26]
5. Optimizing the supply of cofactors (catalysts) for tissue synthesis (vitamin C for collagen synthesis) and all other vitamins and minerals as cell energy cofactors and repair compounds (coenzyme Q-10, alpha-lipoic acid [ALA], all B vitamins, magnesium, calcium, and selenium).
6. Restoring steroid hormone levels by initially giving a secretagogue, the first natural step in improving deteriorating levels of these key hormones. A natural hormonal replacement treatment, such as improving steroid hormone levels of testosterone, dihydroepiandrosterone (DHEA), estrogen and progesterone with transdermal creams, pills or gels, is also essential for counteracting the effects of aging at this level.

7. Reducing the catabolic effect of cortisol on collagen by proper timing of meals, lifestyle changes (relaxation techniques), supplements and the inhibitory effect of other hormones (DHEA).
8. Restoring growth hormone levels by use of prescription medications only after an attempted secretagogue therapy trial is found to be unsuccessful. Such restoration with simultaneous improvement of overall steroid hormones, along with a decrease in cortisol, can create a marked improvement in skin tone and elasticity as well as improved body fat and muscle mass ratios.[27]

The General Truncal Regions Within the truncal region, aging changes include loss of muscle tone and the abnormal appearance of excess body fat. Lipodystrophy usually presents itself within the abdominal area, legs and arms. The actual cause of excess body fat deposition is due to the *loss* of insulin receptor sensitivity[28] and *elevated* insulin and cortisol levels, which shift the deposition of body fat to the central truncal areas primarily and to the extremities secondarily (Diagrams X-4 and X-5).[29] The secondary cause of loss of a youthful body composition and a youthful appearance is a decline in key muscle mass and muscle maintenance hormones, such as testosterone and growth hormone.

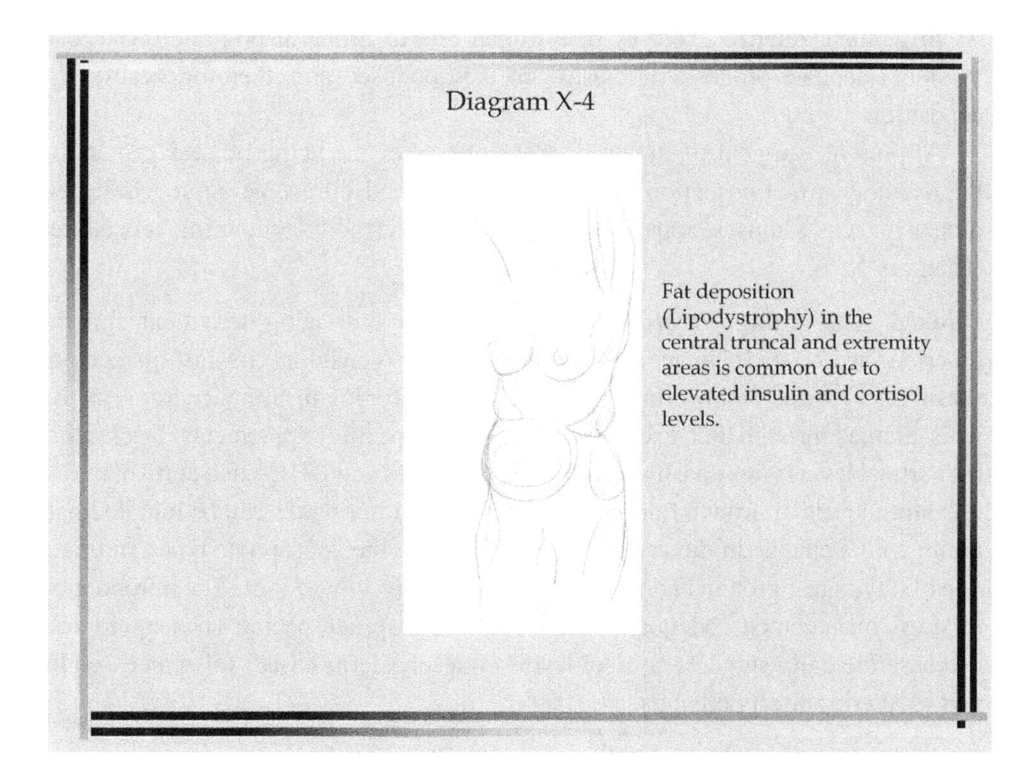
Diagram X-4

Fat deposition (Lipodystrophy) in the central truncal and extremity areas is common due to elevated insulin and cortisol levels.

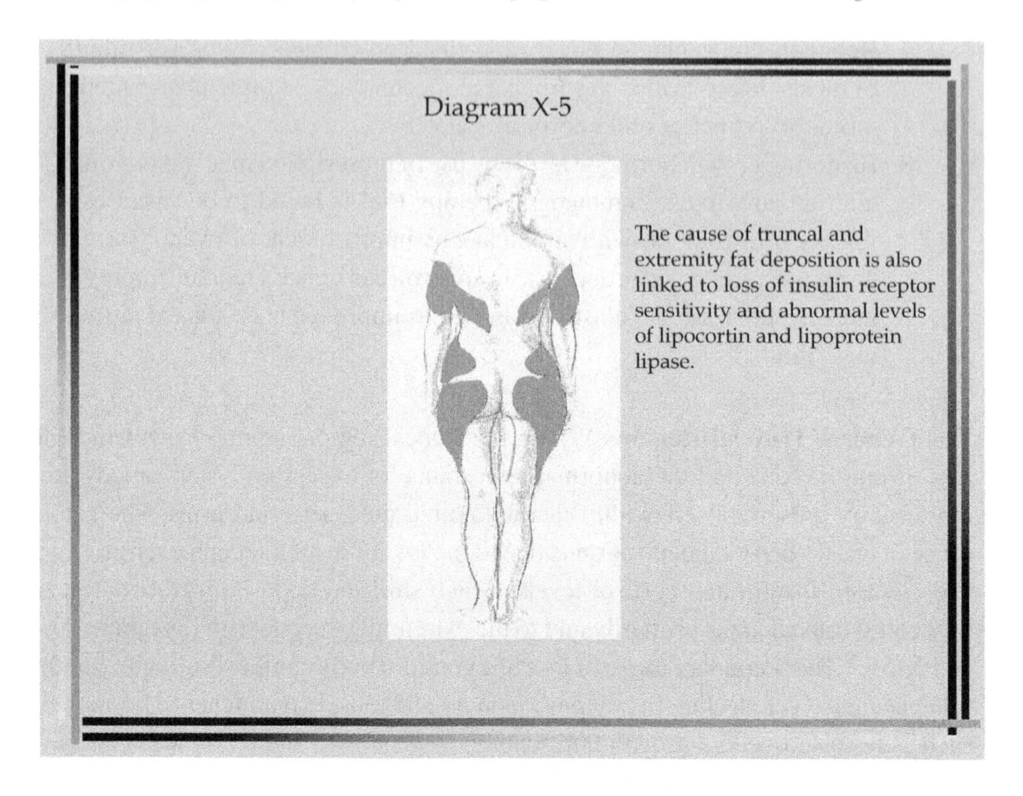

In women, relative excesses of estrogen due to drops in progesterone occur. Without adequate progesterone, estrogen is unopposed and, therefore, causes fat deposition.

At present, surgical treatment of these effects of the aging process centers on the techniques of liposuction — both tumescent and ultrasonic approaches; dermoplasty; rectus muscle repair; thigh lifts and other body-contouring procedures (Diagram X-6).

Clinical Anti-Aging Treatment Concepts The anti-aging treatment for the described age-related changes in the body should *focus* on improving insulin receptor sensitivity[30, 31] and on lowering cortisol levels. Improving insulin receptor sensitivity is attained through diet, exercise and the use of specific supplements. Decreasing the cortisol level is most easily accomplished by restoring DHEA and cortisol ratios. Elevating levels of growth hormone and key steroid hormones can be initiated naturally with a change in dietary programs to include the appropriate types and ratio of carbohydrates, protein and fats along with low-glycemic foods. The introduction of stress management, adequate exercise and appropriate pacing of meals to help decrease the daily surge of cortisol levels on a long-term basis is also an essential part of altering lifestyle habits.

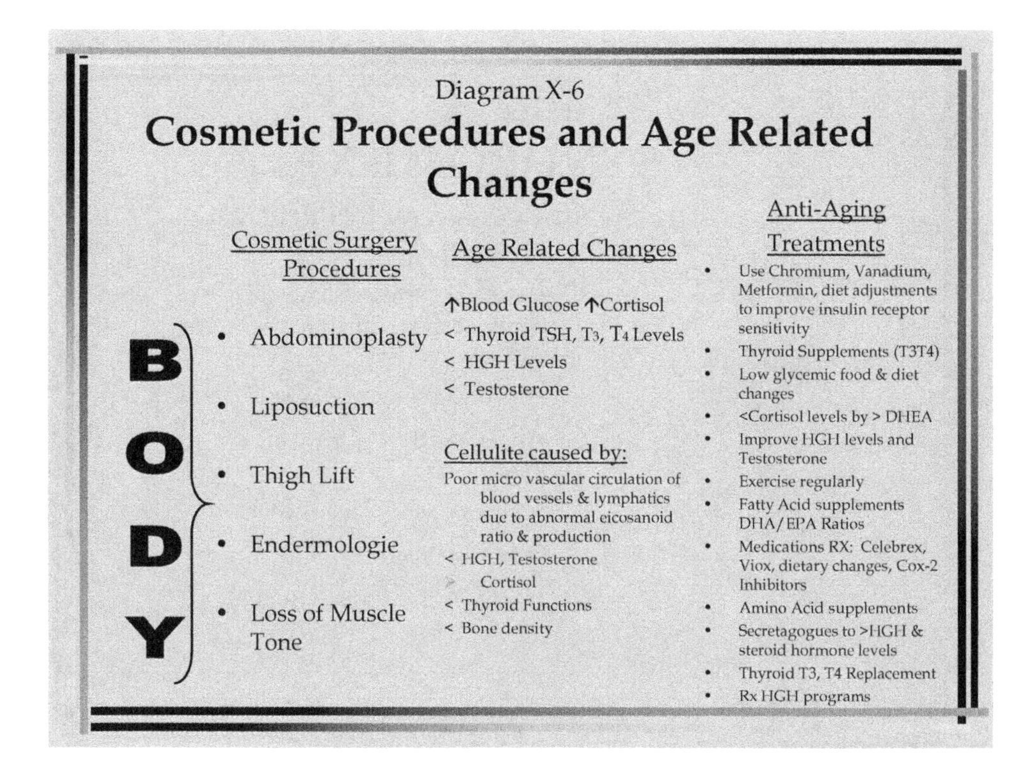

Maintaining appropriate thyroid function with adequate levels of triiodothyronine (T_3), thyroxine (T_4) and thyroid-stimulating hormone, or thyrotropin (TSH) levels is the initial step in helping to keep metabolic rates at optimal levels to burn body fat and maintain optimal body composition.

Finally, maintaining balanced ratios of estrogen and progesterone will help slow fat deposition.

The truncal obesity seen in most people over 40 years of age has been directly related to what has been called **syndrome X**[32,33] (Diagram X-7). Physical manifestations include a pear-shaped or oval-shaped abdomen in the mid-40s, with progressively worsening body composition as the years pass by. Also associated with this process, biochemically, are elevated low-density lipoprotein (LDL) levels, decreased high-density lipoprotein (HDL) levels, elevated cholesterol levels and elevated uric acid levels with decreased insulin receptor sensitivity.[34,35] The appearance of syndrome X not only is a harbinger to age-related physical changes in the body but also always heralds the onset of cardiovascular disease and the other signs of health risk that occur with this process (Diagram X-8). Most of this classic body configuration is due to omental fat accumulation.

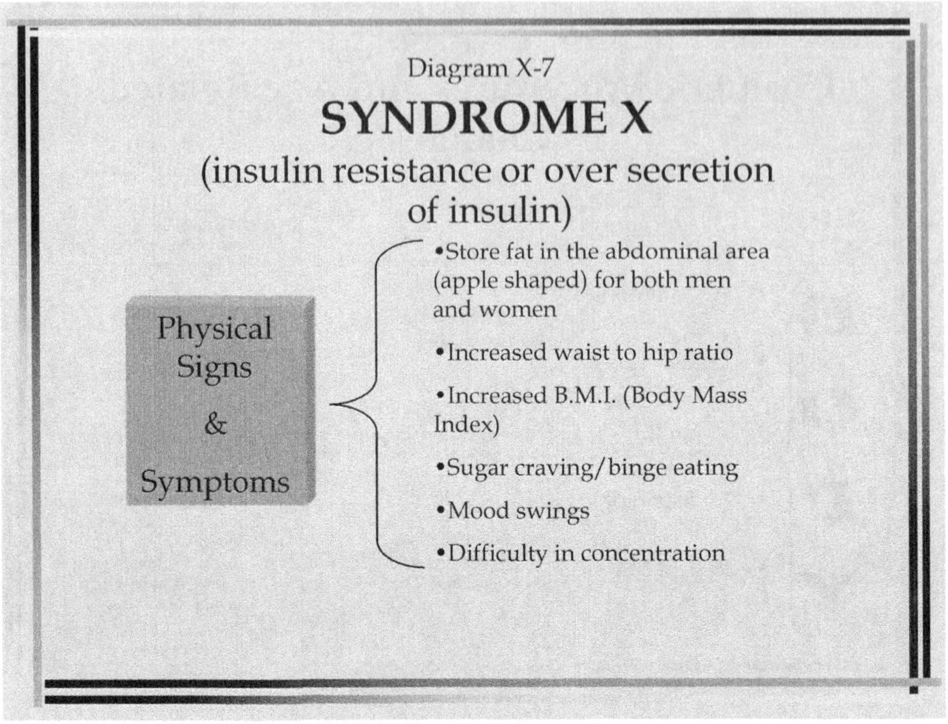

Diagram X-7

SYNDROME X

(insulin resistance or over secretion of insulin)

Physical Signs & Symptoms

- Store fat in the abdominal area (apple shaped) for both men and women
- Increased waist to hip ratio
- Increased B.M.I. (Body Mass Index)
- Sugar craving/binge eating
- Mood swings
- Difficulty in concentration

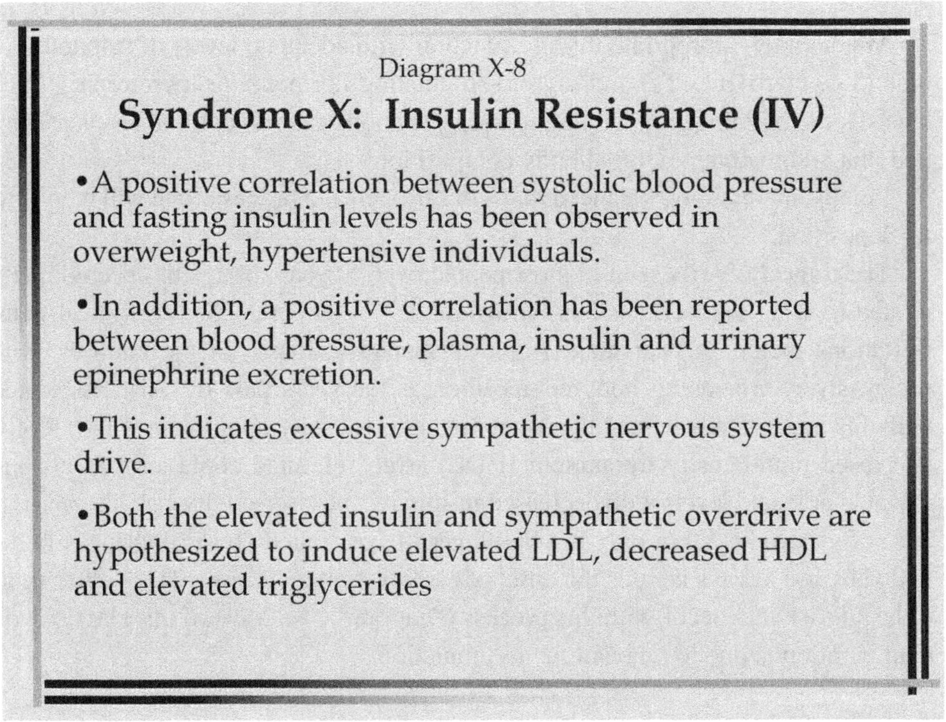

Diagram X-8

Syndrome X: Insulin Resistance (IV)

- A positive correlation between systolic blood pressure and fasting insulin levels has been observed in overweight, hypertensive individuals.
- In addition, a positive correlation has been reported between blood pressure, plasma, insulin and urinary epinephrine excretion.
- This indicates excessive sympathetic nervous system drive.
- Both the elevated insulin and sympathetic overdrive are hypothesized to induce elevated LDL, decreased HDL and elevated triglycerides

The Chest

Clinical Anti-Aging Treatment Concepts

Male Patients Within the breasts of the male patient, the presence of **gynecomastia** (excessive growth of the mammary glands) slowly appears after age 40. It is caused by an excess body fat deposition in the upper chest and breast area. This is also secondary to a lack of appropriate exercise and a decrease in muscle maintenance hormones, including testosterone and growth hormone. The cause of this physical age-related deformity is also directly linked to declines in testosterone levels. This occurrence is due to *conversion* of testosterone to estrogen as a result of the increase in aromatase enzyme[36,37] activity in the fat cells in the body or to an increase in total amount of body fat (Diagram X-9). This causes an increase in fatty deposition in the chest area of men.

The plastic surgical treatment of this age-related change is subcutaneous mastectomy or suction-assisted lipectomy. The anti-aging treatment consists of the use of supplemental zinc and chrysin to inhibit the aromatase enzyme complexes, decreasing body fat quantity and supplying key diet, exercise and insulin regulation regimens (Diagram X-10).

Female Patients In female patients, two types of breast-related changes can occur. The first category, breast enlargement, is frequently secondary to abnormal levels

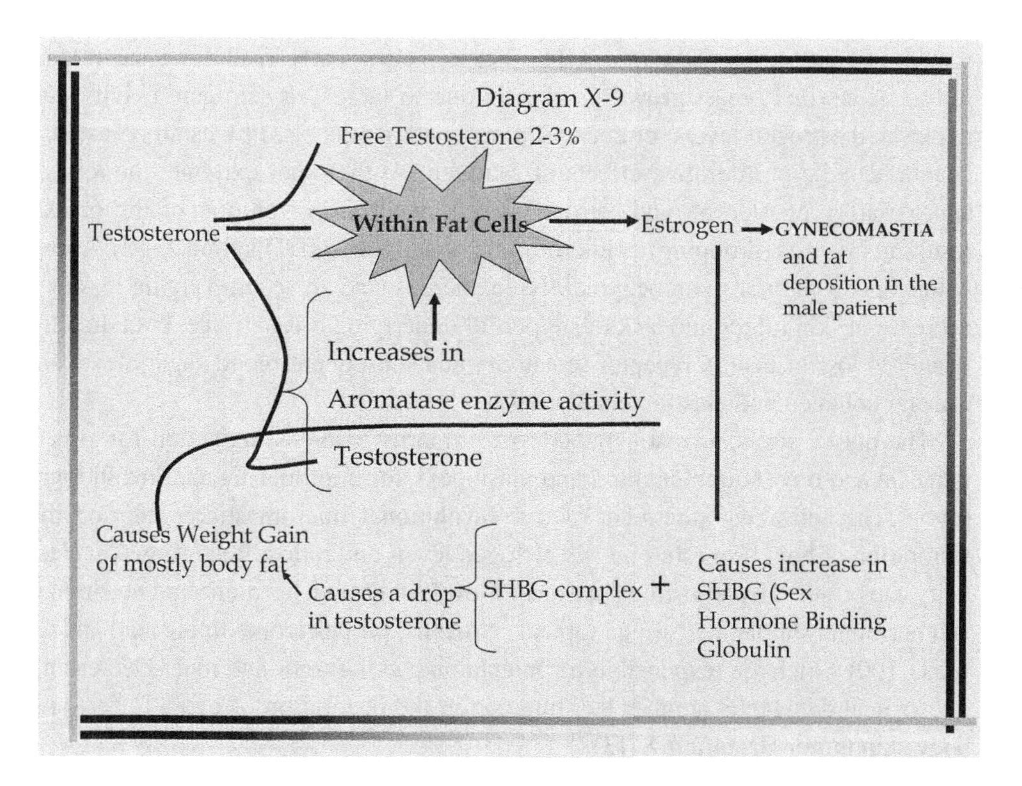

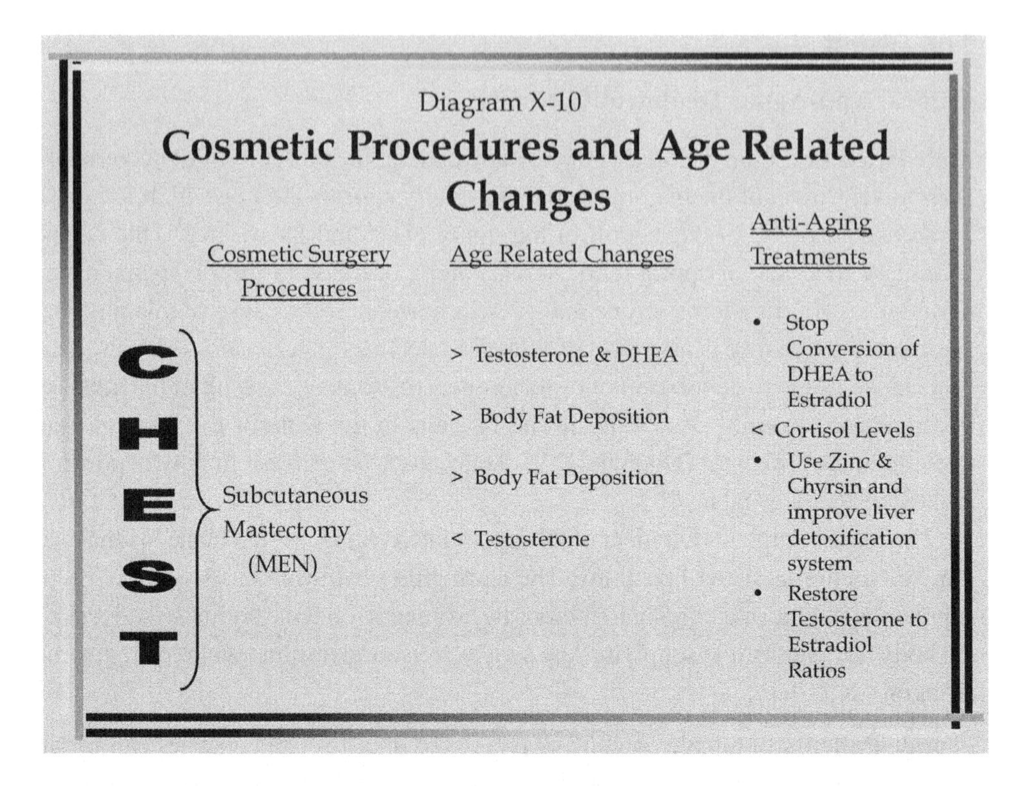

of either progesterone or estrogen, with inappropriate ratios producing continued growth of breast tissue (Diagram X-11). This can occur early in adolescence or later in life. Abnormal breast growth is usually due to increased estrogen activity due to elevated estrogen levels, or decreased progesterone levels. Progesterone usually counteracts the proliferative effects of estrogen. At the other extreme, the loss of appropriate progesterone and estrogen levels results in involution of the breast, resulting in ptosis (drooping) or micromastia (small breasts) (Diagram X-12). Along with changes in breast parenchymal tissue, there is also an accompanying loss or a decrease in skin turgor and lack of support of underlying breast tissue, both directly related to loss of insulin receptor sensitivity and subsequent blood sugar elevation, causing collagen and elastin degradation.

The plastic surgical treatments consist primarily of breast reduction for gigantomastia and breast augmentation and mastopexy for micromastia and involutional ptosis. The anti-aging treatment of true involutional micromastia centers on the restoration of both progesterone and estrogen levels and ratios. Restoring skin elasticity can be aided by adjusting appropriate levels of growth hormone and nutritional and hormonal supplements (e.g., topical estrogens, progesterone, lipoic acid and C-MED-100) which are responsible for maintaining skin turgor and tone. Decreasing free radical damage is another key function in the restoration of overall chest and body skin turgor (Diagram X-13).[16,18]

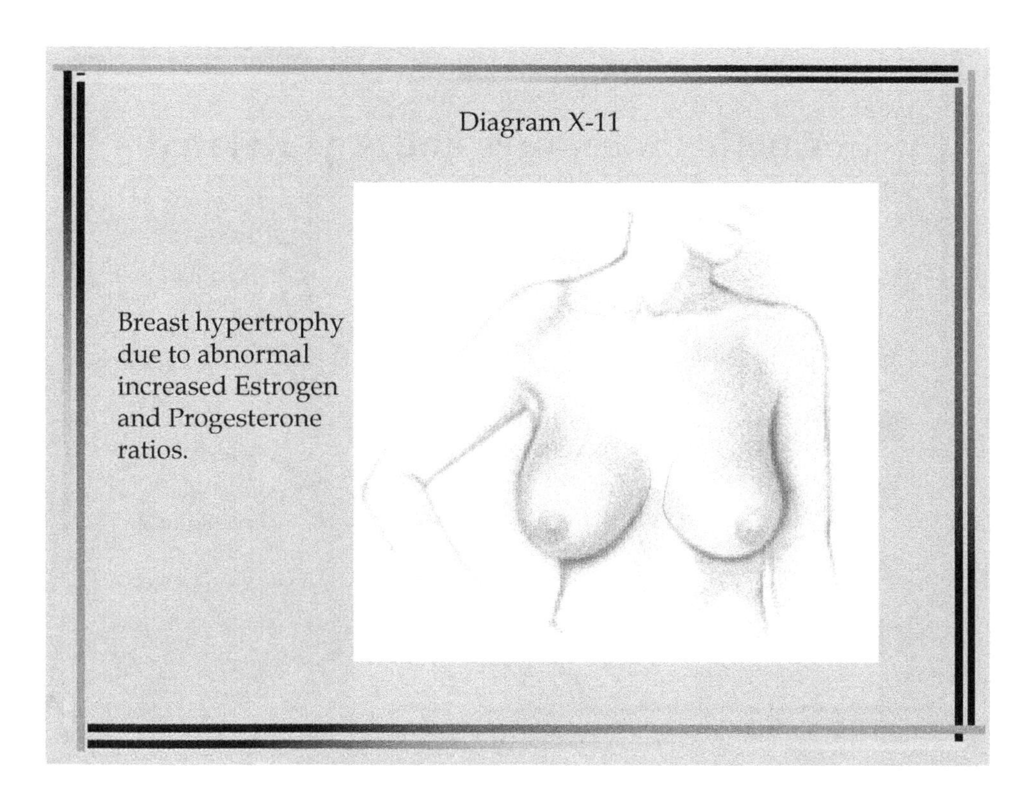
Diagram X-11
Breast hypertrophy due to abnormal increased Estrogen and Progesterone ratios.

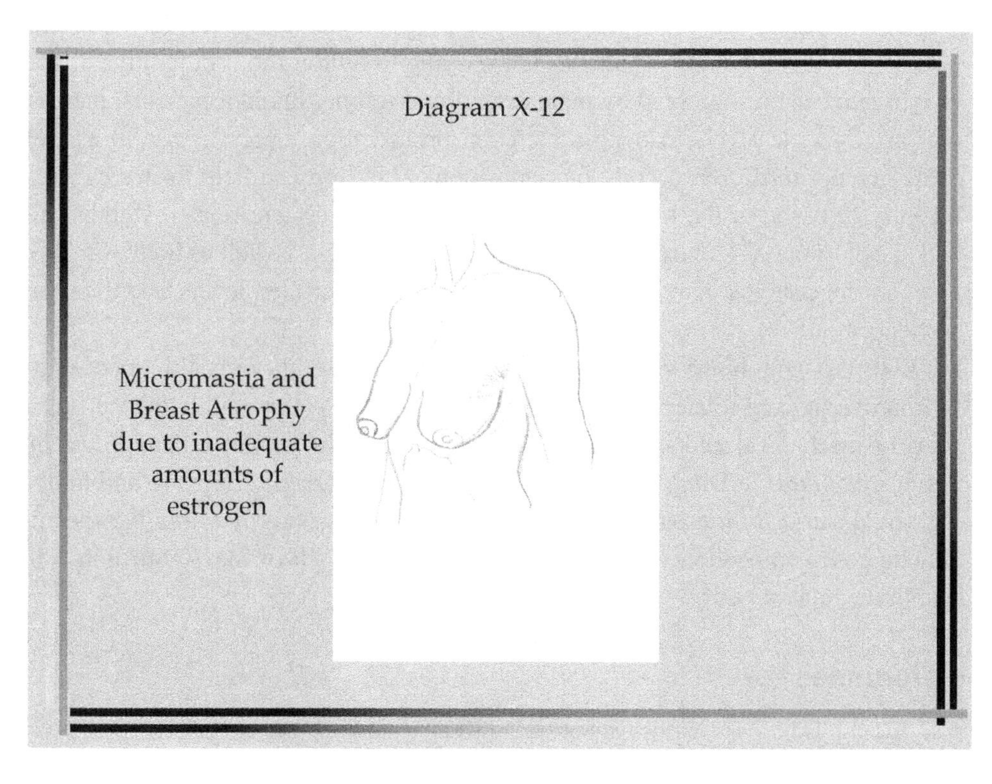
Diagram X-12
Micromastia and Breast Atrophy due to inadequate amounts of estrogen

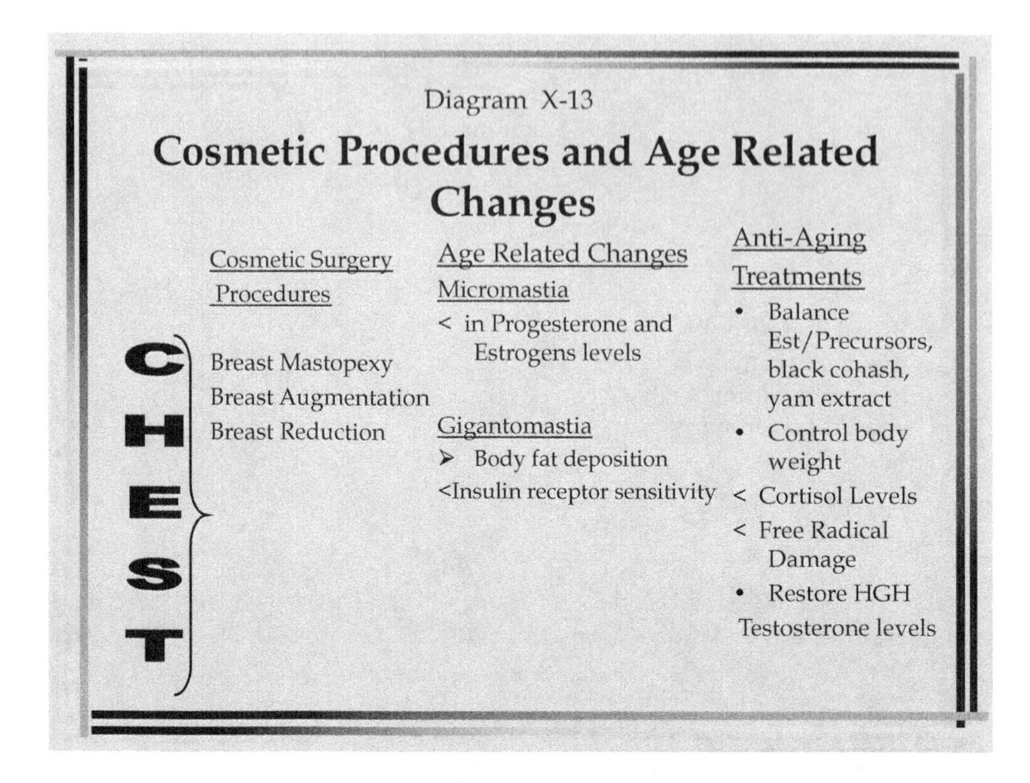

Gigantomastia treatment centers on restoring the balance of estrogen and progesterone levels in younger patients. Here trying to inhibit the proliferative effects of estrogen can be attempted by increasing progesterone. In older patients, increasing progesterone levels to oppose estrogen proliferation effects is useful as well as an attempt to decrease body fat deposition. This step can help to decrease the quantity of tissue in the breast area commonly seen with age-related weight gain and in syndrome X. Controlling cortisol and insulin levels as well as body weight is *essential* for patients with gigantomastia because uncontrolled levels lead to excess fat deposition.

With this new inner vision, or perspective, we have not only the opportunity to improve the age-related physical appearance of our patients utilizing our latest array of mechanical and surgical technology but also the power to *manipulate the aging equation* (see Diagram X-1). We can begin to markedly improve and inhibit the actual causes of these changes, causes that have their origin within the specific genetic codes embedded within the chromosomes of each of the 100 trillion cells within the human body.

References

[1] Bland J. Improving Genetic Expression in Prevention of the Diseases of Aging. Gig Harbor, WA: Health Communications Int., Inc.; Seminar Series, 1998.

[2] Rucker R, Tinker D. The role of nutrition in gene expression: A fertile field for the application of molecular biology. *J Nutr.* 1986; 116: 177–189.
[3] Berdanier CD. Nutrition and Genetic Diseases. In: Berdanier CD, Hargrove JL, eds. Nutrients and Gene Expression. Boca Raton, FL: CRC Press; 1996: 1–20.
[4] Goodman J. Histone tails wag the DNA dog, *Helix,* University of Virginia Health System, Vol 17, Spring, 2000.
[5] Wolff SP, Jiang ZY, Hunt JV. Protein glycation and oxidative stress in diabetes mellitus and ageing. *Free Radic Biol Med.* 1991; 10: 339–352.
[6] Lavrovsky Y, Chatterjee B, Clark RA, Roy AK. Role of redox-regulated transcription factors in inflammation, aging and age-related diseases [review]. *Exp Gerontol.* 2000; 35(5): 521–532.
[7] Christman JW, Blackwell TS, Juurlink BH. Redox regulation of nuclear factor kappa B: therapeutic potential for attenuating inflammatory responses [review]. *Brain Pathol.* 2000; 10(1): 153–162.
[8] Camhi SL, Lee P, Choi AM. The oxidative stress response [review]. *New Horiz.* 1995; 3(2): 170–182.
[9] Adler V, Yin Z, Tew KD, Ronai Z. Role of redox potential and reactive oxygen species in stress signaling. *Oncogene.* 1999; 18(45): 6104–6111.
[10] Gius D, Botero A, Shah S, Curry HA. Intracellular oxidation/reduction status in the regulation of transcription factors NF-kappa B and AP-1 [review]. *Toxicol Lett.* 1999; 106(2–3): 93–106.
[11] Maher P, Schubert D. Signaling by reactive oxygen species in the nervous system [review]. *Cell Mol Life Sci.* 2000; 57(8–9): 1287–1305.
[12] Sen CK, Packer L. Antioxidant and redox regulation of gene transcription [review]. *FASEB J.* 1996; 10(7): 709–720.
[13] Allen RG, Tresini M. Oxidative stress and gene regulation [review]. *Free Radic Biol Med.* 2000; 28(3): 463–499.
[14] Thannickal VJ, Fanburg BL. Reactive oxygen species in cell signaling [review]. *Am J Physiol Lung Cell Mol Physiol.* 2000; 279(6): L1005–L1028.
[15] Roy AK. Transcription factors and aging [review]. *Mol Med.* 1997; 3(8): 496–504.
[16] Nose K. Role of reactive oxygen species in the regulation of physiological functions [review]. *Biol Pharm Bull.* 2000; 23(8): 897–903.
[17] Finkel T. Redox-dependent signal transduction [review]. *FEBS Lett.* 2000; 476(1–2): 52–54.
[18] Gamaley IA, Klyubin IV. Roles of reactive oxygen species: signaling and regulation of cellular functions [review]. *Int Rev Cytol.* 1999; 188: 203–255.
[19] Slater AF, Nobel CS, Orrenius S. The role of intracellular oxidants in apoptosis. *Biochim Biophys Acta.* 1995; 1271(1): 59–62.
[20] Giampapa VC. *Anti-Aging Surgery: A Step Beyond Cosmetic Surgery. Advances in Anti-Aging Medicine.* Vol. 1. New York: Mary Ann Liebert, Publishers, Inc.; 1996: 57–60.
[21] Fu MX, Wells-Knecht KJ, Blackledge JA, et al. Glycation, glycoxidation, and cross-linking of collagen by glucose: kinetics, mechanisms, and inhibition of late stages of the Maillard reaction. *Diabetes.* 1994; 43(5): 676–683.
[22] Ennerit I. Free radicals and aging of the skin [review]. *EKS.* 1992; 62: 328–341.
[23] Fuchs J, Milbradt R. Antioxidant inhibition of skin inflammtion induced by reactive oxidants; evolution of the redox couple dihydrolipoate/lipoate. *Skin Pharmacol.* 1994; 7(5): 278–284.
[24] Miyata T, Hori O, Zhang JH, et al. The receptor for advanced glycation end products (RAGE) is a central mediator of the interaction of AGE-β_2 microglobulin with human mononuclear lymphocytes via an oxidant-sensitive pathway. *J Clin Invest.* 1996; 98(5): 1088–1094.
[25] Bierhaus A, Chevion S, Chevion M. Advanced glycation end products induced activation of NF-κB is suppressed by alpha lipoic acid in cultured endothelial cells. *Diabetes.* 1997; 46(9): 1481–1490.
[26] Horrobin DF. Loss of delta-5-desaturase activity as a key factor in aging. *Med Hypothesis.* 1981; 7: 1211–1220.
[27] Shiverick KT, Rosenbloom AT, eds. *Human Growth Hormone Pharmacology: Basic and Clinical Aspects.* Boca Raton, FL: CRC Press; 1995.

[28] Zimmet P, Baba S. Central obesity, glucose intolerance and other cardiovascular risk factors. *Diabetes Res Clin Proc.* 1990; 16: S167–S171.
[29] Stern MP, Haffner SM. Body fat distribution and hyperinsulinemia as risk factors for diabetes and cardiovascular disease. *Arteriosclerosis.* 1986; 6: 123–130.
[30] Tong P. Thomas T, Berish T, et al. Cell membrane dynamics and insulin resistance in non–insulin-dependent diabetes mellitus. *Lancet.* 1995; 345: 357–358.
[31] Reaven G. Role of insulin resistance in human disease. *Diabetes.* 1998; 37: 1495–1507.
[32] Yim MB, Yim HS, Lee C, et al. Protein glycation: creation of catalytic sites for free radical generation. *Ann NY Acad Sci.* Apr 2001; 928: 48–53.
[33] Lin Y, Rajala MW, Berger JP, et al. Hyperglycemia-induced production of acute phase reactants in adipose tissue. *J Biol Chem.* Nov. 9, 2001; 276(45): 42077–42083.
[34] Tchernof A, Lamarche B, Prud'Homme D, et al. The dense LDL phenotype: association with plasma lipoprotein levels, visceral obesity and hyperinsulinemia in men. *Diabetes Care.* 1996; 19: 629–637.
[35] Lukaczer D. *Functional Medicine: Adjunctive Nutritional Support for Syndrome X.* Gig Harbor, WA: Health Communications Int., Inc.; 1998.
[36] Giampapa VC. Anti-Aging: Testosterone/Estrogen Ratios. *Mind & Muscle Power Magazine,* April 2000.
[37] Shippen E, Fryer W. *The Testosterone Syndrome.* New York: M. Evan & Co., Inc.; 1998.

11

Anti-Aging Treatment: Overview

Vincent C. Giampapa, M.D., F.A.C.S.

Everything should be made as simple as possible, but not simpler.
Albert Einstein

Controlling the Aging Equation and Optimizing Gene Expression

As mentioned at the beginning of each chapter in this book, the fundamental process of aging can be viewed as an **aging equation** encoded within our 46 chromosomes. This **genetic code** controls the aging equation. These key processes of aging can be summarized with the understanding that the primary focus of treatment should center around control of (1) **glycation**, (2) **inflammatory processes** and (3) **oxidative stress** in the form of decreasing free radicals in order to avoid deoxyribonucleic acid (DNA) damage, to improve DNA repair and to improve the process of methylation of DNA.

Control of these three processes is the cornerstone for the successful clinical treatment of the aging process and results in a documentable improvement and reversal of key biomarkers of aging and improvement in age-related changes in the body and face. In essence, they are the building blocks over which other anti-aging treatments are laid (see chapter opening diagrams).

The chapter opening illustrations show that through a nutritional and pharmaceutical approach we can markedly affect gene expression and DNA at the *intracellular* level.[1–3] As we have a positive impact on these processes, they directly affect hormonal levels, which helps to restore balance and function to the endocrine, immune, digestive, and central nervous systems. It is at this level that these systems begin

The Principles and Practice of Antiaging Medicine for the Clinical Physician, 189–202.

to self-integrate among each other. A homeostasis effect, similar to that present in a youthful, healthy individual, is accomplished. Built upon these improvements are the observable total body changes that occur with exercise, diet and mind-body techniques.

It is for this reason that control of these key fundamental processes at the DNA and cellular levels is the essential focus in anti-aging medicine and age-management programs. If one can successfully improve glycation, inflammation, oxidation and methylation to a significant level, the other overlying components are positively affected as well.

For the cosmetic surgeon, this approach forms the basis of an age-management program. As noted within the chapter opening diagrams, each of these changes is directly related to changes in both the face and body and is based on the scientific facts that each of these four key components affects DNA function. This ultimately results in the physiological changes that we note in our patients as they age. It also affects our *stem cell pool reserves*, which affects virtually every organ in the body. These ideas form the basis of the new **aging paradigm** (Diagram XI-1).

In evaluating an overall treatment program, it is essential to focus on the following clinical goals (see earlier chapters). These goals are summarized as follows:

1. Decrease DNA damage.
2. Increase DNA repair.

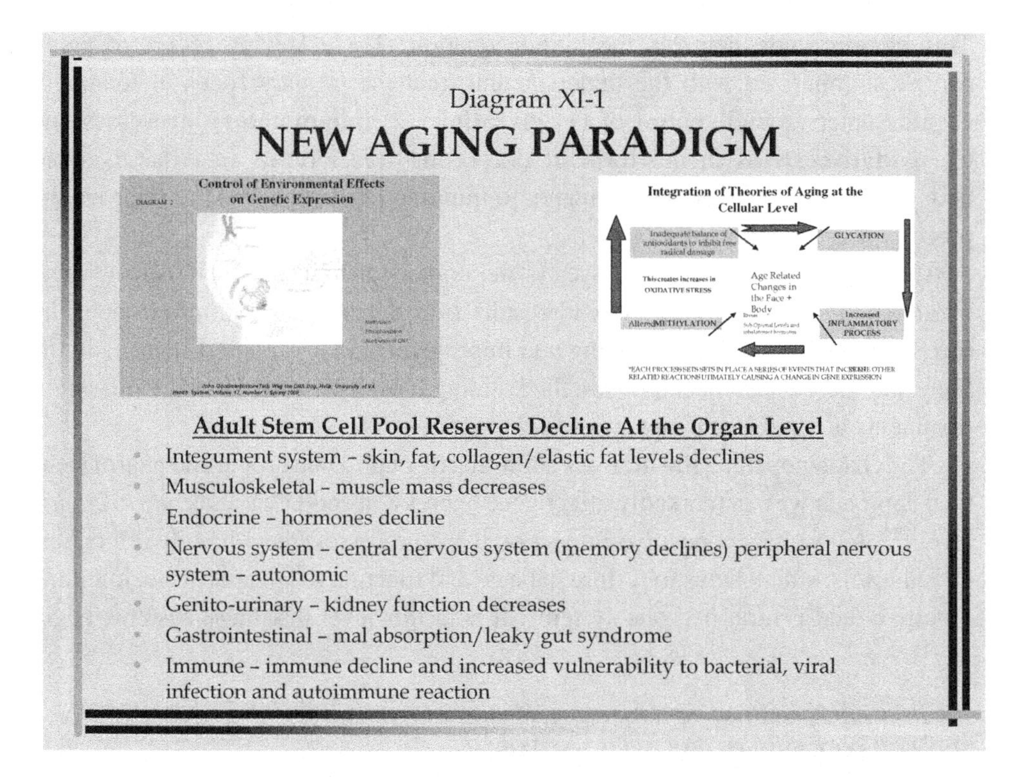

3. Augment immune function.
4. Optimize genetic expression.

Over the last few years, it has been documented that **nutraceuticals** (vitamins, minerals, phytochemicals, enzyme complexes) as well as prescription drugs *directly* influence gene expression.[2] That is, they may downregulate certain genes and *upregulate* other genes. If this is done in the correct fashion, they can *alter* cell signaling and thus directly affect hormonal levels. Hormones are ultimately responsible for changes in proteins and other growth factors, which have a direct impact on tissue regeneration and bodily changes on the physical level (Diagram XI-1).

Influencing Gene Expression and Altering Cell Signaling

Glycation For improving control of glycation, supplements that improve insulin sensitivity are recommended. These include alpha-lipoic acid (ALA), especially in the time-release form called Glucotize, along with vanadium, chromium, zinc, taurine, fenugreek and bitter gourd.

The prescription use of metformin or Glucophage, in microdoses at 125 mg two times a day with lunch and dinner, is an ideal way to control blood sugar elevation and loss of insulin receptor sensitivity.[3–5] It acts essentially as a caloric-mimic.

Aminoguanidine, 200 mg twice daily, can help inhibit protein crosslinking secondary to elevated glucose levels as well as avoid immune suppression.[6] Low-glycemic index diets and exercise are all essential parts of this process as well (Diagram XI-2).

An ideal macronutrient ratio (40, 30, 30) can directly affect the key hormones of aging (see Chapter V). Micronutrients, such as the essential amino acids, lysine, ornithine, citrulline, curcumin, vitamin E, and phytonutrients and coenzyme Q10, are also key compounds that improve glycation and dysglycemia.

Antioxidants An antioxidant program is essential to help combat the constant free radical production we experience as part of life. Antioxidant supplements containing vitamin A, vitamin C, and vitamin E, selenium and zinc in moderate doses and a host of natural compounds from vegetable-derived foods are essential to control free radical levels. In the chapter on DNA, it was shown that free radical levels are directly related to nuclear factor-κB (NF-κB) stimulation, or inhibition, which is one of the more important compounds in determining the rate of DNA repair. The rate and quantity of DNA damage are directly related to elevated free radical levels at both the nuclear and the cell membrane levels (Diagram XI-3). The wrong combination of mega doses of antioxidants can actually cause more free radicals and oxidative stress.

Deprenyl, also known as selegiline hydrochloride, has been shown in experimental studies to improve the levels of the intrinsic antioxidant compounds that we

Diagram XI-2

ANTI-AGING TREATMENT OVERVIEW

2. **IMPROVING GLYCATION AND DYSGLYCEMIA:**
   - Regular exercise
   - Low glycemic diet
   - Ideal macronutrient ratio
   - Improving insulin receptor sensitivity
     - Vanadium
     - Metformin
     - Micronutrients supplements, amino acids – lysine, arnithine, citrulline, curcumin, Vitamin E, phytonutrients, CQ^{10}
     - Chromium
     - Aminoguanadine
     - Crysolic acid
     - Carnosine

Diagram XI-3

ANTI-AGING TREATMENT OVERVIEW

5. IMPROVE ANTI-OXIDANT DEFENSE AT THE INTRA & EXTRACELLULAR LEVELS: to decrease free radicals
   - Complete Vitamins/Mineral supplement phytochemicals, pH regulating compounds
   - Deprenyl increases SOD/Catalase Levels at intracellular level (intrinsic anti-oxidants)
   - Inhibit NFκ-B with C-MED 100™

normally produce on our own.[7–9] These compounds, superoxide dismutase (SOD) catalase and glutathione peroxidase, are produced at the intracellular level and also at the level of the mitochondria, which is the site of maximum free radical production and free radical damage. **Glidosene**(TM)* and an oral form of SOD derived from melons has recently been shown to be effective as an oral agent.

Inflammation Control of the inflammatory process centers around the implementation of a hypoallergenic diet supplemented with digestive enzymes. These enzymes break down the protein peptides to amino acids and aid in absorption. Gastrointestinal support with such compounds (e.g., *Lactobacillus*) and amino acids (e.g., *L-glutamine*) along with these digestive enzymes markedly *decreases* inflammatory processes occurring at the cellular level.

Appropriate ratios of omega-6 and omega-3 essential fatty acids are one of the key ways of regulating the inflammatory pathways within the cells. Niacinamide, glucosamine sulfate and folic acid are all essential natural approaches.

Lipoic acid, boswellic acid, turmeric, gymnema, ginger and bitter gourd also directly affect the inflammatory process as well and act as natural Cox-2 inhibitors.[10]

Decreasing the toxic element load found in water and other factors such as foods and cooking style eliminates pro-inflammatory compounds (e.g., aluminum, copper and iron).

Prescription medications, also delivered in microdoses like celecoxib (Celebrex), 50 to 100 mg once or twice a day, act as pharmaceutical Cox-2 inhibitors and directly inhibit the pathways that are responsible for inflammatory eicosanoid production. Inhibiting NF-κB is also important in slowing aging at the cellular level and especially in the stem cell pool reserves (Diagram XI-4). C-MED-100 has been shown to inhibit both NF-κB and tumor necrosis factor alpha (TNF-α), key intracellular inflammatory compounds.[11,12]

Methylation Methylation is the process whereby certain genes are "turned on" and other genes are "turned off."[13] This process can be improved by additional supplements of vitamin B_6, vitamin B_{12}, folic acid and other methyl donors (e.g., betaine, trimethylglycine), and SAM-e (*S*-adenosyl-L-methionine) (Diagram XI-5).[13,14]

The use of antineoplastins synthesized by Dr. Stanley Burzynski has been shown to help regulate methylation patterns of DNA and to markedly decrease oncogenes activity. The use of antineoplastin (ATO)* is a valuable adjunct in reducing a large number of potential age-related cancers and should be considered an essential component of a comprehensive anti-aging program.

Decreasing cortisol levels also improves methylation. The easiest way to do this is by augmenting dehydroepiandrosterone (DHEA) levels, which decreases body

---

*Available through Optigenics, Inc.

*Available through the Burzynski Research Institute.

Diagram XI-4

ANTI-AGING TREATMENT OVERVIEW

3. IMPROVE INFLAMMATORY CONTROL BY DECREASING REACTIVE PROTEINS

(CRP, Amyloid -A + decreasing nitric oxide levels and TNF-A and NFκ-B levels)

- Lipoic acid, boswellic acid, tumeric, gymnema, bittergourd
- Decrease toxic element load (e.g. aluminum, iron, copper)
- Low allergy diet, lactobacillus supplementation
- L-glutamine, soy products
- Glucosamine sultate, folic acid
- Celebrex, VIOX
- Monoatomic elements (Platinium group compounds)
- DHA/EPA supplementation

Diagram XI-5

ANTI-AGING TREATMENT OVERVIEW

1. IMPROVING METHYLATION & ACETYLATION DEFECTS:

- B6
- B12
- Folate
- Betaine
- SAM-e (Sadenosylmethionine)
- Niacinamide
- MSM

fat levels as well as age-related brain cell death associated with related memory impairment.

Energy Another essential approach to anti-aging treatment is to improve mitochondrial oxidative function, or mitochondrial production of adenosine triphosphate (ATP). As mentioned earlier, ATP is the *essential* energy intermediate from which all cellular processes, as well as DNA repair, are accomplished. Keeping ATP production optimal is paramount in decelerating aging.

Improving mitochondrial oxidative function can be accomplished with additional levels of lipoic acid, *N*-acetylcysteine, niacinamide, coenzyme Q10, lipoic acid, L-carnitine and Microhydrin®, which is a microclustered silicon compound that acts as a strong hydrogen donor. Also important at this level are compounds such as taurine, vitamin E, and glutathione (Diagram XI-6).

Immunity Improving immune function can be accomplished with natural components from yeast cell extracts known as beta-1,3-glucans. The herbal compounds *Echinacea*, *golden seal* and *Astragalus* have all been documented to improve T-cell and immune functions.[15–17] Rishi mushroom extract and shitake mushroom concentrate are well-known immune stimulants as well. C-MED-100 is also a potent white blood cell stimulant and improves DNA repair within these cells (Diagram XI-7).

Diagram XI-6

ANTI-AGING TREATMENT OVERVIEW

6. IMPROVING MITOCHONDRIAL OXIDATIVE FUNCTION and ENERGY PRODUCTION:

- **Lipoic Acid, N-acetyl cysteine, niacinamide, CQ^{10}, lipoic acid, L-Carnitine and Microhydrin®**
- **Taurine, Vitamin E, Glutathione**

Diagram XI-7

ANTI-AGING TREATMENT OVERVIEW

9. IMPROVE IMMUNE FUNCTION:

- **Beta-1, 3 - glucans**
- **Echinacea**
- **Carboxy Alkyl Esters * (C-MED 100™)**
- **Golden seal**
- **Astragalus**

The most recent research with carboxy alkyl esters (C-MED-100) has documented a remarkable ability to use natural compounds to enhance immune function[16] as well as peripheral white blood cell counts. Clinical research at the University of Lund, Sweden, under the direction of Dr. Ronald Pero,[15–18] has scientifically documented these effects in both animal and human studies.

Improving water quality is also an essential component in age management. Since our cells are 98% water and because water is the medium in which all of these biochemical processes occur, it is essential to have the appropriate quantity and quality of water. Healthy water should be more alkaline, which helps to balance the intracellular and extracellular pH levels. An improvement in water surface tension and wetting ability, which are supplied by compounds like silica and microcluster compounds mentioned previously, enhances the cell's hydration capability and membrane permeability, therefore allowing cellular toxins to flow out of the cell and key essential micronutrients to flow into the cell along with key electrolytes (Diagram XI-8).

Improving gut function is another step in a program to improve aging. Simple maneuvers like chewing food well can help avoid "leaky gut syndrome," a common sequela in old age. Chewing well aids digestion because as we age, the quantity and quality of gastric acidity and digestive enzymes decrease. By having smaller pieces

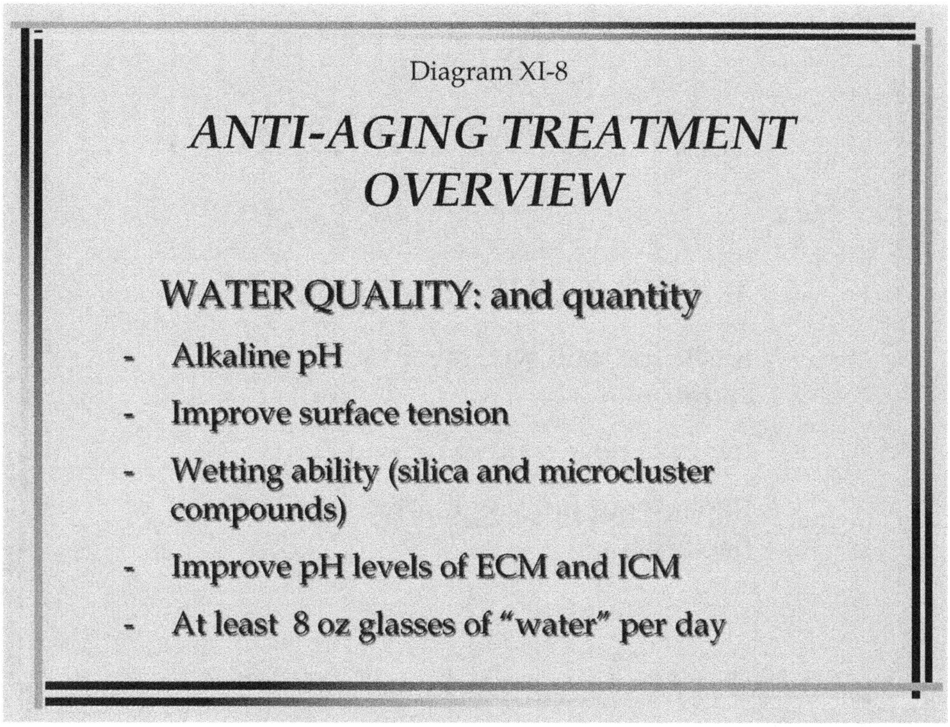

Diagram XI-8

ANTI-AGING TREATMENT OVERVIEW

WATER QUALITY: and quantity

- **Alkaline pH**
- **Improve surface tension**
- **Wetting ability (silica and microcluster compounds)**
- **Improve pH levels of ECM and ICM**
- **At least 8 oz glasses of "water" per day**

of food within the gastrointestinal (GI) tract, more of the food is made available in its essential components (e.g., carbohydrates, proteins and fats).

Increasing the fiber content in food is also important to help speed up transit time and to avoid the toxic effects of digestive enzymes and gastric juices on the gastrointestinal wall should they remain in contact with gut tissue for too long.

Replacing gut flora regularly with *Lactobacillus* and the media required for optimal gut flora to grow (Jerusalem artichoke or fructo-oligosaccharides) also improves GI function and the GI tract's ability to extract and metabolize key secondary vitamin and mineral complexes (Diagram XI-9).

Hormonal Replacement Protocols

Restoration of key hormonal levels should follow a strict regimen. Hormonal replacement in general and growth hormone restoration in particular should be accomplished in a fundamental stepwise process.

There is no place in anti-aging therapy for single-hormone replacement, In fact, single-hormone replacement with growth hormone or testosterone alone can be deleterious to the aging process and longevity. This is because the body has been designed to act in symphony and balance with all the key hormones previously discussed. Changes in one particular hormone alone can alter and create a cascade

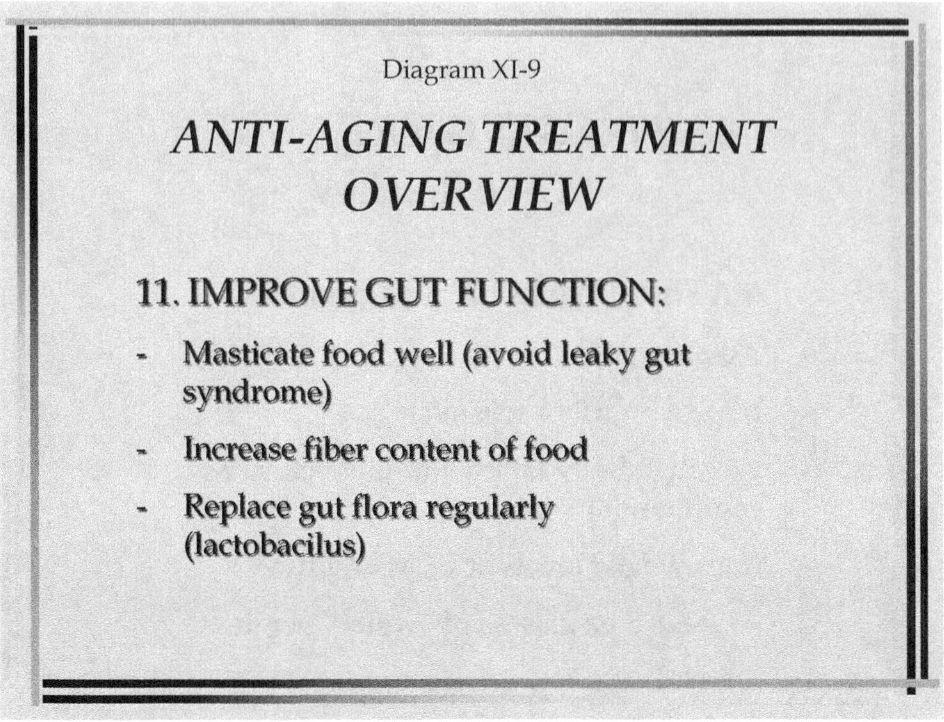

Diagram XI-9

ANTI-AGING TREATMENT OVERVIEW

11. IMPROVE GUT FUNCTION:

- **Masticate food well (avoid leaky gut syndrome)**
- **Increase fiber content of food**
- **Replace gut flora regularly (lactobacilus)**

of effects that upsets the delicate balance of the neuroendocrine system. Therefore, it is essential to treat all hormone levels in a stepwise and interrelated fashion (Diagram XI-10).[19]

Initially, a secretagogue (a compound that "signals" the body to produce more of its own natural hormones) is recommended. This is accomplished for the first 12 weeks to test the body's **functional organ reserve**, or the ability of the body's existing parenchymal tissues within the specific organ system or gland to produce more of its own hormones (Diagram XI-11). In this fashion, the body's normal feedback mechanisms are left intact, and we do not need to be concerned with overprescribing a specific hormone. A baseline hormonal screening precedes this approach.

Therefore, we can judge the effectiveness of this initial secretory trial as we monitor ongoing hormone levels. If the functional organ reserve levels are not age-appropriate, documented by a low initial baseline level and, secondarily, by the inability to maintain normal age-related levels of a specific hormone after a secretory trial, hormonal replacement treatments are begun.

Growth Hormone Replacement Therapy is initiated with the use of recombinant growth hormone, 0.5 to 1 unit/day. It is recommended that the hormone be administered Monday through Friday by subcutaneous injection. Weekends are left as

Diagram XI-10

ANTI-AGING TREATMENT OVERVIEW

7. IMPROVE HORMONAL LEVELS:

a) Regulate parasympathetic & sympathetic balance of autonomic nervous system

b) Secretagogues precursors to DHEA, Testosterone (steroid hormones) facilitate coordination of cell signaling and coordinate processes between the endocrine system and metabolic system to optimize repair and balance of physiological status. T3, T4 regulate metabolic rate.

c) Decrease Psychological Stress by decreasing time urgency, decrease Cortisol & elevate DHEA & other key hormonal biomarkers (i.e. C-AMP, C-GMP)

d) Balancing brain synchrony and improve brain wave activity

e) Improve 2nd messenger function (cordyceps sinensis)

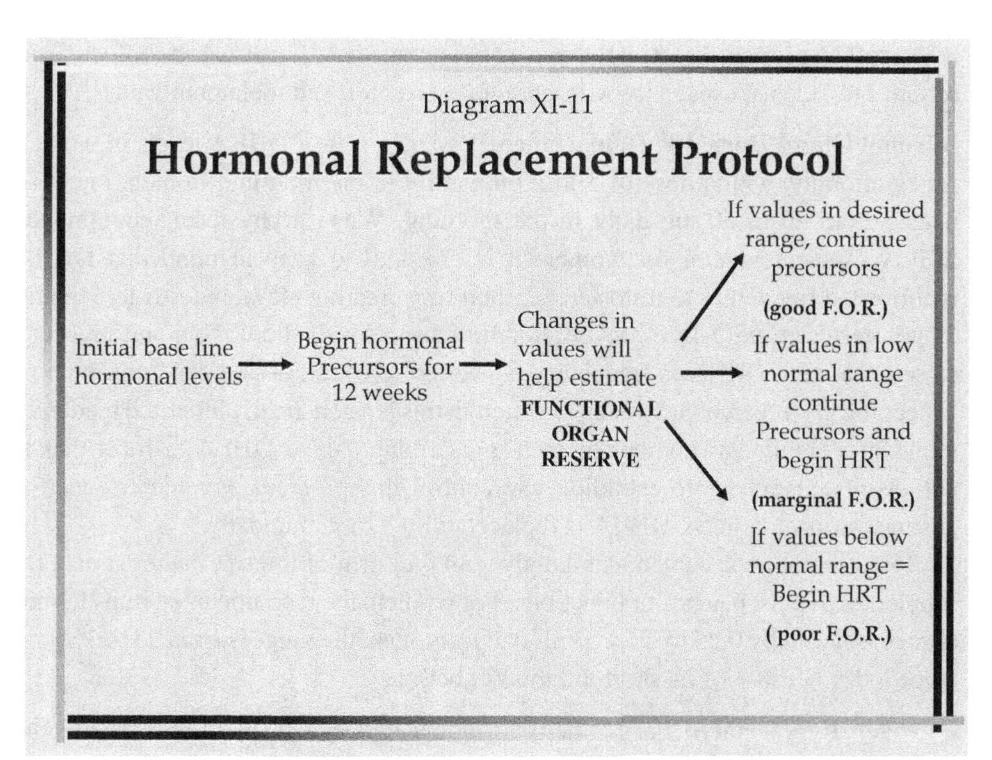

periods of nontreatment to allow the pituitary gland's normal feedback mechanisms to maintain equilibrium.

Thyroid Gland Hormone Replacement Thyroid supplementation is initiated with a dosage from 0.5 to 3 grains of triiodothyronine (T_3) and thyroxine (T_4), most commonly found in the form of Armour Thyroid,[19] an easily available prescription medication. This form of thyroid replacement is usually the most effective because it supplies both essential components of the thyroid hormone (T_3 and T_4). Supplying T_4 alone frequently results in inefficient T_3 levels because of depleted selenium levels in many individuals. Selenium is required to convert T_4 to T_3.[19]

Pineal Gland Hormone Replacement Restoration of pineal gland hormones is initiated by administration of melatonin, at 0.5 to 3 mg orally at night before bed. Appropriate melatonin levels have several advantages, which include:

1. DNA protection.
2. Improvement of immune function.
3. Body composition enhancement.
4. Regulation of general hormonal release patterns.
5. Improved sleep quality.

Melatonin levels can also be restored by prescribing tryptophan (1000 mg at bedtime). Tryptophan is converted to serotonin and then to melatonin. This regimen is less efficient than direct melatonin supplementation, but it can be effective in certain individuals, especially with both low serotonin and melatonin values.

Adrenal Gland Hormone Replacement Replacement of DHEA levels in women is begun orally, with a dose of 5 to 25 mg daily in the morning; in men, the dose ranges from 20 to 50 mg daily in the morning. When a physician recommends DHEA supplementation for women, it is essential to keep in mind that DHEA is converted frequently to testosterone, therefore creating elevated testosterone and dihydrotestosterone (5-DHT) levels. For this reason, levels should be monitored very closely. In men, DHEA is frequently converted to estradiol and can contribute to gynecomastia (breast enlargement). Although this is a genetically inherited tendency found in a significant number of men, prescribing 7-keto DHEA, a form that is not easily converted into estradiol, can control this problem. For menopausal or postmenopausal women, DHEA is replaced with 5 to 25 mg daily.

Testosterone replacement can be given in the form of transdermal creams, oral supplements, gels, patches or injections. For women, the recommended transdermal dosage is normally 0.25 to 2.5 mg/mL daily; for men, the range is from 25 to 160 mg twice a day because of its short duration of action.

Estrogen Replacement For women, hormonal replacement with the triestrogens is recommended to mimic the body's natural levels with a dose of 1.25 to 5 grams daily, twice per day in transdermal form until symptoms of deficiency are eliminated. The use of the lowest dose possible to eliminate symptoms is the ideal approach. Progesterone replacement should start with 20 to 200 mg B.I.D. in the form of a transdermal cream. Progesterone can also be given orally, 50 to 250 mg daily, (usual dose 100 mg) before bedtime. This is the preferred route.[20] If used, Transdermal creams are best prescribed in a twice-daily regimen (i.e., morning and evening) for optimal and stable blood levels throughout the day. Symptoms of progesterone deficiency are also monitored until subjective improvement is documented.

Dosage protocols[21] have been summarized in Chapter IX (see Charts 1 to 6). Hormone replacement therapy is a valuable component of an age management program for regulating and maintaining vital functions, such as bone mass maintenance, lipid profiles and memory and heart protection, even if the patient is initially asymptomatic.

References

[1] Rucker R, Tinker D. The role of nutrition in gene expression: a fertile field for the application of molecular biology. *J. Nutr.* 1986; 116: 177–189.

[2] Berdanier CD. Nutrition and genetic diseases. In: Berdanier CD, Hargrove JL, eds. *Nutrients and Gene Expresion.* Boca Raton, FL: CRC Press; 1996: 1–20.

[3] Goodman J. Histone tails wag the DNA dog. *Helix* (University of Virginia Health System). Spring 2000; vol. 17.

[4] Miyata T, Hori O, Zhang JH, et al. The receptor for advanced glycation end products (RAGE) is a central mediator of the interaction of AGE-β_2-microglobulin with human mononuclear phagocytes via an oxidant-sensitive pathway. *J. Clin Invest.* 1996; 98(5): 1088–1094.

[5] Iehara N, Takeoka H, Yamada Y, et al. Advanced glycation end products modulate transcriptional regulation in mesangial cells. *Kidney Int.* 1996; 50(4): 1166–1172.

[6] Tajiri Y, Myoller C, Grill V. Long-term effects of aminoguanidine on insulin release and biosynthesis: evidence that formation of advanced glycosylation end products inhibits B cell function. *Endocrinology.* 1997; 138(1): 273–280.

[7] Kitani K, Kanai S, Sato Y, et al. Chronic treatment of L-deprenyl prolongs the life span of male Fischer 344 rats. *Life Sci.* 1992; 52: 281–288.

[8] Milgram NW, Racine RJ, Nellis P, et al. Maintenance on L-deprenyl prolongs life in aged male rats. *Life Sci.* 1990; 47: 415–420.

[9] Carello MC, et al. L-Deprenyl increases activities of superoxide dismutase and catalase in certain brain regions in old male mice. *Life Sci.* 1994; 54: 975–981.

[10] Aggarwol et al. Spices as potent antioxidants with therapeutic potential. In: Cadenas and Packer, eds. *Handbook of Antioxidants*, ed 2. New York: Marcel Dekker; 2002: 445.

[11] Pero R. et al. Aliment. Water extracts inhibit NF-κB and prevent inflammation. *Pharmacol Ther.* 1998; 12: 1279–1289.

[12] Pero R, et al. Water extracts inhibit TNF alpha. *Free Radic Biol Med.* 2000; 29(1): 71–78.

[13] Chiang PK, Gordon RK, Tal J, et al. *S*-adenosyl methionine and methylation. *FASEB J.* 1996; 10(4): 471–480.

[14] Fenech MF, Dreosti IE, Rinaldi Jr. Folate, B_{12}, homocysteine status and chromosome damage rate in lymphocytes of older men. *Carcinogenesis.* 1997; 18(5): 1329–1336.

[15] Sheng Y, Bryngelsson C, Pero RW. Enhanced DNA repair, immune function and reduced toxicity of C-MED 100, a novel aqueous extract from *Uncaria tomentosa. J Ethnopharmacol.* 2000; 69: 15–16.

[16] Sheng Y, Pero RW, Wagner H. Treatment of chemotherapy-induced leucopenia in a rat model with aqueous extract from *Uncaria tomentosa. Phytomedicine.* 2000; 78: 137–143.
[17] Pero RW. Method of Preparation and Composition of a Water-Soluble Extract of the Plant Species *Uncaria* for Enhancing Immune, Anti-inflammatory, Antitumor and DNA Repair Processes of Warm-Blooded Animals. U.S. Patent No. 6,361,805B2.
[18] Akesson C, Pero RW, Ivars F. C-MED 100, a water extract of *Uncaria tomentosa*, prolongs survival in vivo. (Submitted 2001.)
[19] Yanick P, Giampapa VC. *Quantum Longevity.* Santa Barbara, CA: Christian Publishing Co.; 1998: 61–68.
[20] Colgan M. *Hormonal Health, Nutritional and Hormonal Strategies for Emotional Well-Being and Intellectual Longevity.* Vancouver, BC: Apple Publishing; 1996.
[21] Reiss U. *Natural Hormone Balance.* New York: Pocket Books, Simon & Schuster, Inc., 2001.

12

Introducing Anti-Aging Medicine into the Cosmetic Surgery Practice

Vincent C. Giampapa, M.D., F.A.C.S.
Oscar M. Ramirez, M.D., F.A.C.S.

Sit down before fact like a small child and be prepared to give up every preconceived notion, follow humbly wherever and to whatever abyss, nature may lead, or you will learn nothing.

T. H. Huxley

The Cosmetic Surgery Practice Today

The cosmetic surgery practice is *unique*, in many aspects, from other medical specialties and patient populations. First, it is for people seeking strictly *elective* services. The members of this group, in general, can afford the procedures that we as cosmetic surgeons and physicians offer. They are also extremely well informed. The demographics, which define this group, refer to patients ranging from 18 to 80 years of age. The largest group of patients falls within the "baby boomer" generation (Diagrams XII-1 and XII-2).[1] It is this group that is responsible for the skyrocketing number of cosmetic procedures over the last few years. In addition, the same group is expected to continue to be requesting our services in the future.

Of equal interest to this group are the concepts of "aging" and "aging optimally." This group of successful entrepreneurs and professionals is looking for a new "point of view" of aging and to prolong as much as possible the period of *middle life*, or the age 45- to 60-year categories (Diagram XII-3).[2] It is because of this recent shift in demographics, as well as the expendable income within this group, that the demands for anti-aging services have increased as well.

The Principles and Practice of Antiaging Medicine for the Clinical Physician, 203–215.

Diagram XII-1

Anti-Aging Market

There are 76 million Baby Boomers in the U.S., and they:

- have accepted the concept of preventive health maintenance to maximize "health span"
- spend over $12 billion on non-prescription vitamins and supplements annually
- want to know which nutraceuticals they really need and how to measure the effect on their health and longevity

Diagram XII-2

Key Trends

- Anti-aging is "hot" - hormone therapy, gene mapping and nutritional supplements are in the news
- Anti-Aging Medicine is rapidly gaining acceptance as a new medical specialty
- Over 6,000 physicians and others are members of the American Academy of Anti-Aging Medicine
- Consumer purchases of anti-aging nutraceuticals grew to over $2 billion in1998 in the U.S.
- Growth in consumer and physician interest in anti-aging medicine is sky-rocketing internationally

Diagram XII-3

A NEW "POINT OF VIEW" AND A NEW PARADIGM OF AGING

"We are not programmed to age and then die, we are programmed for self repair and longevity. "We are not reaching our full potential for health and longevity."

The present profile of a successful cosmetic surgical practice is one in which the cosmetic surgeon sees a regular number of patients entering his or her active patient database along with an ongoing continued number of earlier cosmetic surgical patients returning over a similar period of time (Diagram XII-4). This group of patients had left the practice and completed their initial cosmetic surgical services.

When cosmetic surgical patients have finished their initial relationship with the cosmetic surgeon, the present practice model allows for virtually 100% transition out of the practice. A small percentage of these patients will be peripherally tied to the cosmetic surgical practice through ancillary services that may be offered.

The most *successful* ancillary service that has kept patients in touch with the practice has been that of *skin care*. The ongoing success of skin care products and services can be viewed as the successful model for encouraging the patients to return to the physician who performed the initial surgery.

When considering the introduction of anti-aging medicine or age-management programs into the cosmetic practice, it is important to define the major differences between these two historic practice models.

The *present* model of cosmetic surgery practice is one in which, through the cosmetic surgical procedures and services, the "effects" of the aging process are treated successfully with both cosmetic surgical procedures and ancillary services (Diagram XII-5). These ancillary services, again, may include laser treatments,

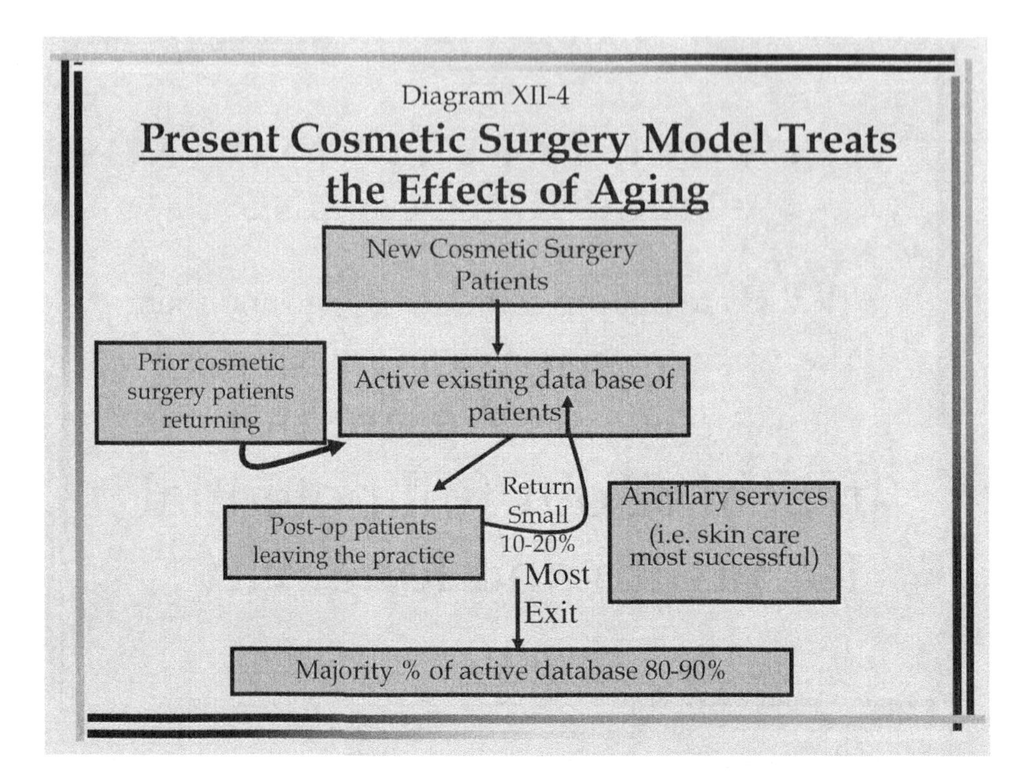
Diagram XII-4
Present Cosmetic Surgery Model Treats the Effects of Aging
New Cosmetic Surgery Patients
Prior cosmetic surgery patients returning
Active existing data base of patients
Return Small 10-20%
Ancillary services (i.e. skin care most successful)
Post-op patients leaving the practice
Most Exit
Majority % of active database 80-90%

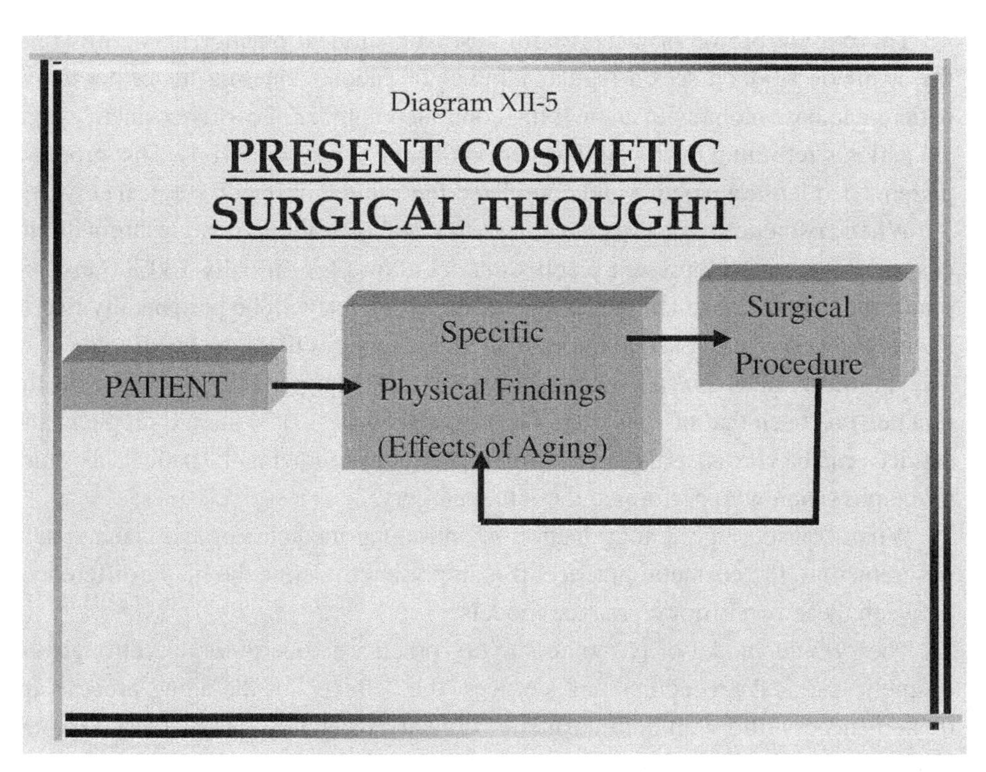
Diagram XII-5
PRESENT COSMETIC SURGICAL THOUGHT
PATIENT
Specific Physical Findings (Effects of Aging)
Surgical Procedure

skin care and office-based maintenance procedures, such as Botox and collagen injections.

The *new* model of cosmetic surgery should include an anti-aging medicine service. In this model, the cosmetic surgeon takes a leap forward to treat not only the *effects* of aging but also the *causes* (Diagrams XII-6 and XII-7). The new cosmetic patient enters the cosmetic surgeon's active database. Upon completing the surgical procedure, the patient exits the cosmetic surgical portion of the practice. The patient is then directed to the anti-aging medicine service in order to help extend the longevity of the surgical procedure originally performed and, more important, to receive the information, the products and the testing necessary to improve one's health span and quality of life as well (Diagrams XII-8 and XII-9).

Most present practices of cosmetic surgery should consider incorporating **age-management programs**. These programs can simply include "core" testing and "health" information centering on dietary, exercise and nutritional changes that improve gene expression and, ultimately, quality of health and life as well as the potential to increase longevity.

These programs require small amounts of the staff and physician's time, yet are highly effective. In a select number of practices with more physician time and interest, one can build a comprehensive anti-aging program and center. This system would involve extensive detailed biomarker testing and hormonal replacement

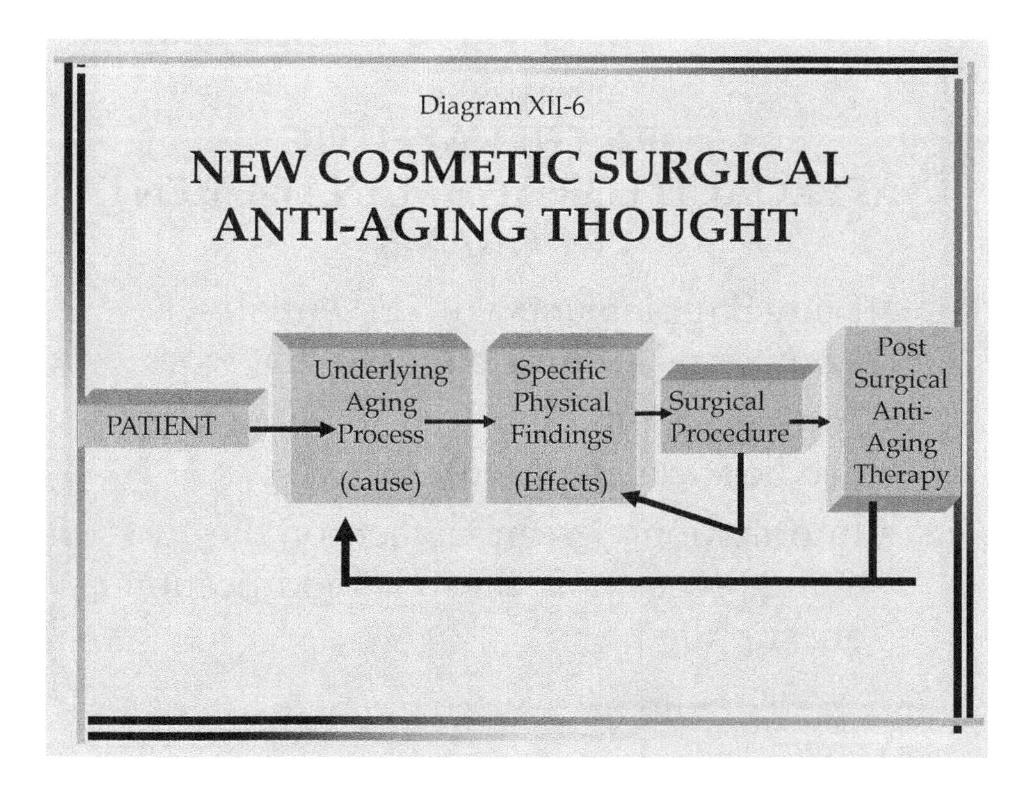

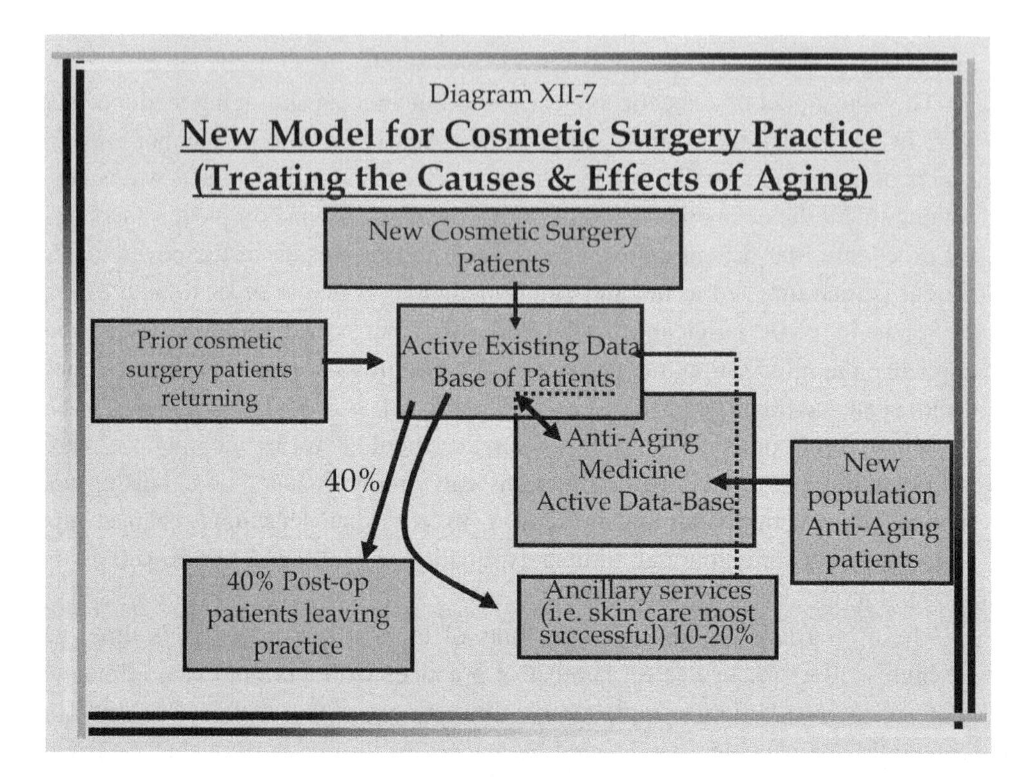

Diagram XII-8

CORE THERAPEUTIC APPROACH FOR AGE MANAGEMENT PROGRAMS

- Daily Supplements with Essential Compounds (nutraceuticals) that work
- Testing – Home Test Kit (no blood or needle sticks required)
- Information – Essential facts on the aging process and lifestyle modification are supplied.

Diagram XII-9

CORE THERAPEUTIC APPROACH FOR AGE MANAGEMENT PROGRAMS

HOME TEST KIT:

No blood work needed - (Urine Sample)

8 Hydroxy Guanosine = DNA Damage

8 EpiPgf-2a = Oxidative Stress

along with regular laboratory testing of previously treated biomarkers, including everything from deoxyribonucleic acid (DNA) studies to body composition levels. Extensive equipment is required at this level, as are well-trained staff and support personnel. This type of practice offers the most comprehensive and effective approach to optimal aging.

In a poll performed by Longevity Institute International in 1999 (Diagram XII-10), approximately 75% of cosmetic surgery patients who had recently undergone a surgical procedure indicated an interest in some form of anti-aging medical therapy. Patients were interested in maintaining their results and improving quality of life as well as longevity. The addition of the anti-aging medicine service or age-management program to the cosmetic surgery practice is also another source of attracting *new* patients for this service. In turn, a portion of them may become new cosmetic surgical patients as well.

Some of the key points to note when incorporating an anti-aging medical practice are the *perceptions* of the overall practice profile. With the addition of anti-aging medicine as part of a cosmetic surgery practice, the physician, for the first time, is presenting an opportunity to patients to be treated for not only the *effects of aging* but the *causes* as well. It is of primary importance to incorporate this new practice profile with all patients as they enter the practice. This practice profile contains a new view of the doctor-patient relationship as well. With an anti-aging medical service

Diagram XII-10

> 75% of plastic surgery patients want information on ANTI-AGING MEDICINE.

or age-management program within the practice, a lifelong relationship between the physician and the patient is emphasized and established, so that both the physician and the staff and office become a source of ongoing information about aging and how to age in the best manner.

This *change* in the doctor-patient relationship is one of the major positive effects within a practice that contains anti-aging medicine and helps to increase the sense of loyalty to the practice as well as a deeper level of medical respect for the physician.

One of the easiest ways of introducing anti-aging medicine to an existing patient population is to introduce the *concepts* upon the patient's first arrival to the office and to include presurgical treatments with key nutrients that will help with the healing process as well. This has been a successful model in many practices and improves not only the perceived value of the physician's practice but also the perceived value of the physician's interest in the patient.

A number of existing plastic surgery practices offer varying levels of anti-aging medicine participation for their patients to help them ease into this new concept and become more knowledgeable. A patient's initial introduction to anti-aging medicine may include the use of anti-aging supplements preoperatively to facilitate healing. These select nutrient combinations are continued postoperatively as part of the patient's introduction to age-management services. The use of supplements can be compared directly with the skin care model, which requires the patient's continued use of the products for a given overall effect on an ongoing basis. In this fashion, the expense is minimal, but the habit of taking anti-aging supplements and the interest and desire to investigate one's lifestyle and daily routines in general are introduced.

Introducing Biomarkers

Simple urine testing at this level, which consists of looking at DNA damage and oxidative stress (free radical levels),[3] is an easy way to obtain the patient's positive feedback on the effectiveness of an anti-aging supplement program. (Diagram XII-11). A thiol test* that measures DNA repair is expected to be available within the next year.[4]

Other practices offer more extensive testing programs for their postsurgical patients in order to help alleviate the recurrence of the original problem for which they may have presented to the cosmetic surgeon. For example, patients who have had gynecomastia (male breast enlargement) or lipodystrophy (fat deposits) or who are overweight in general can be directed into programs that focus on these specific problems caused by age-related changes in testosterone production, insulin receptor sensitivity and basal metabolic rate changes that occur with aging. Some practices have the diagnostic capability to measure full complements of biomarkers and can

Diagram XII-11

THE BASIC BIOMARKERS

To Follow for Cosmetic Surgery and Anti Aging Programs

1. 8OHD6 = Follows DNA Damage
2. 8 EpiPGT2x = Follow Lipid Peroxidation (i.e.free radical damage to the cell membrane)
3. THIOL TEST = To follow DNA repair capability

*Available through Campa Med, Inc.

create a custom-managed anti-aging program. Although this capability calls for more extensive training and a higher level of interest on the physician's part, this is the ultimate approach to age-evaluation and age-management programs that can be offered.

With the introduction of anti-aging medicine to a cosmetic practice, we must keep in mind that the practice offering these services is an essential information source for the patient as well. This information, which is available through the physician's cosmetic practice, will be an important factor in keeping patients interested in returning to the office in the future. Mailing short newsletters regarding key topics on aging is an excellent way to supply this ongoing information to patients.

Some other key ways to improve the success of an anti-aging medicine practice or a service within an existing cosmetic practice is to consider the use of seminars, either within the office setting or at a local hall for the public and invited patients, on topics of aging in general.

Usually, mixing the anti-aging topic with a focus on some cosmetic surgical procedure increases the number of audience members and participants. Regional seminars, tied in with local newspaper and radio advertising, can result in a major turnout and exposure for the surgeon who is looking to increase the cosmetic surgery practice as well as the anti-aging medicine patient database.

A suggested plan to increase exposure of anti-aging in the cosmetic surgeon's practice includes local magazine ads and radio announcements about the services and the uniqueness of these anti-aging programs. Cable and Web advertising are also effective sources of exposure.

Introducing the Concept of Anti-Aging Medicine and Age Management for Patients

With the appearance of a new cosmetic surgical referral to the physician's practice, it is important to *introduce the concept* of both cosmetic surgery and anti-aging medicine to the patient upon the first visit. This step, as mentioned previously, helps to improve the practice's overall profile and the patient's confidence in both the physician and the practice.

If the concept of a lifelong patient relationship is stressed and the surgery is as successful as it usually is, a greater sense of loyalty is established for the physician and his or her practice.

One method that has succeeded in introducing this concept to the cosmetic practice is to mention these concepts briefly at the time of the initial cosmetic surgical consultation (Diagram XII-12). This can be accomplished through written literature, through the use of a patient coordinator or by the physician. The concept of treating the effects of the aging process with cosmetic surgery and then offering the patient an opportunity to actually work on its underlying causes with the latest advances in

Diagram XII-12

The average cosmetic surgeon has a busy Anti-Aging Practice already existing within his own DATA BASE.

anti-aging medical science has been an extremely well received approach for those plastic surgeons already incorporating this system.

One of the more successful approaches has been to incorporate the use of an anti-aging supplement into the patient's preoperative preparatory program, usually approximately 2 weeks before the procedure. The patient takes a nutritional supplement that focuses on improving the aging process and helps to improve wound healing in general.

In my personal practice, anti-aging supplements are given to patients, and are included within the surgical fees. They are begun 2 weeks before the surgical procedure, and continued until 2 weeks after. Supplements are taken three times per day with breakfast, lunch and dinner. At approximately 2 weeks postoperatively, when the patients have finished a 1-month's supply, it is suggested that they continue the regimen for another 5 months in order to help maximize their healing capabilities during this important period of collagen deposition. I also emphasize that the supplements help improve overall results from the surgery. Most patients are happy to do this and, again, see this advice as increased personal attention and interest on the physician's part.

At about 2 months after the surgery, when the patient comes in for the follow-up visit, a courtesy anti-aging consultation with a patient coordinator or nurse, trained in the concepts of anti-aging medicine, takes place. It is during this time that the patient is introduced to the concepts of biomarker testing, hormonal replacement and maximizing the effects of the aging process. An anti-aging program designed for the patient's specific interest (e.g., weight loss, body composition management, or hormonal replacement) is also offered at this time. A significant percentage of cosmetic surgery patients who would be simply returning for surgical follow-up are now reentered into the active patient database as part of an anti-aging or age-management service. This includes product use and testing as well as other related testing options.

This approach has been highly successful in cosmetic surgical practices throughout the United States and focuses on utilizing the cosmetic surgical database and the physician's good will that has been established over a long period of time.

Within the busy cosmetic practice, there is virtually a built-in age-management program. This program can be accessed simply by sending an introductory letter to the patient and listing the available services. This in itself is an excellent way to bring back inactive patients who have not returned to the office for a while and to keep the patients from leaving the practice after completion of the initial surgery.

A significant amount of prepublished literature is available in the form of booklets and general information for the cosmetic surgeon considering an age-management program. This prepared information should focus on the essential therapy goals (Diagram XII-13).

Diagram XII-13

Age Management Therapy Goals

1) Control Environmental Effects -Environment defined as = diet, exercise, mind state, toxic elements, pollution, local radiation.
2) Improve the Function of the Aging Equation -Control Glycation, Inflammation, Oxidation/Free Radical Levels and Methylation at the cellular level.
3) Improve DNA Replication and Gene Expression -Improve the ratio and DNA Repair over DNA damage therefore resulting in less cell mutations and more accurate cell copies during cell replication. This preserves adult stem pools.

If any of the more involved efforts to establish a full anti-aging medicine practice are desired, a full-time patient coordinator/nurse/clinical specialist or an additional physician should be assigned to run the anti-aging program and to coordinate the patient's treatments, testing and programs so that patients are treated on an intimate, personal level and in a timely fashion.

References

[1] American Academy of Anti-Aging Medicine. Chicago: *Statistics on Aging Demographics*, 2000.

[2] Giampapa VC. Fifth International Symposium on Anti-Aging Medicine. Newark, NJ. August, 2000.

[3] Immunosciences Lab. Beverly Hills, CA.

[4] Pero RW, Hoppe C, Sheng Y. Serum thiols as a surrogate estimate of DNA repair correlates to mammalian lifespan. *J Anti-Aging Med.* 2000; 3: 242–249.

13

Anti-Aging Technologies

Present and Future Trends

Vincent C. Giampapa, M.D., F.A.C.S.

> Science is not only compatible with spirituality; it is a profound source of spirituality.
>
> Carl Sagan

"The first human being to live to be 200 years old is walking the planet today, and it may be one of our children, or perhaps even one of us!" It is important to discuss the possible timeline of present and future trends in anti-aging medicine and where it might lead (Diagram XIII-1).

As the 21st century dawns, it is apparent that we stand at the horizon of a revolution in anti-aging therapies and technologies. Within the next 5 to 10 years, a number of promising techniques should allow for a markedly expanded quality of health and extension of human longevity not previously thought to be possible. Breathtaking advances by major corporations in both the private sector (e.g., Advanced Cell Technology)[1] and the public sector (e.g., Affymetrix)[2] have documented the ability to take these new concepts and turn them into scientific reality.

Although it will be no more than a decade before many of these new technologies in anti-aging therapy arrive (over the next 10 years), it is absolutely essential to emphasize the fundamental element of these approaches based on our **human genetic code** — ***DNA*** (deoxyribonucleic acid). Even now we have the basic knowledge to improve the functions of this amazing molecule.

The technologies discussed in this chapter represent attempts to measure and improve what we have previously defined as **biomarkers**. To gauge the effectiveness of anti-aging medical therapies, the physician uses biomarkers instead of alleviating

The Principles and Practice of Antiaging Medicine for the Clinical Physician, 217–234.

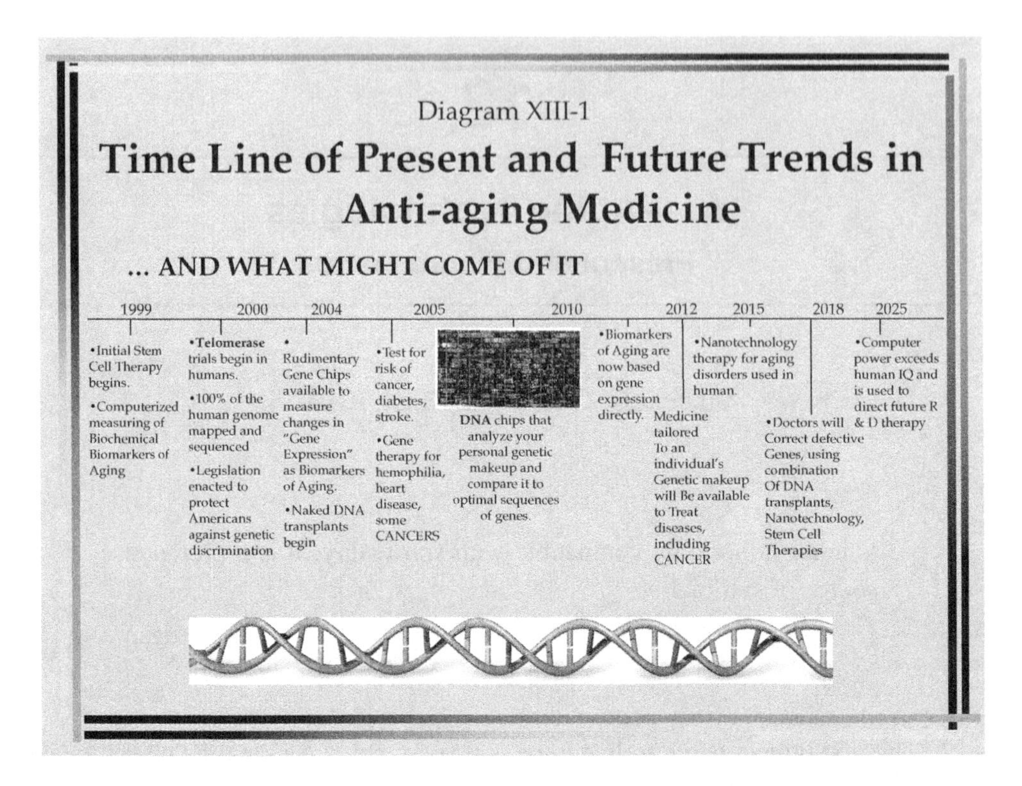

specific symptoms alone. In this fashion, we can *scientifically evaluate* the actual effects of an anti-aging program.

A new definition of the term biomarker is needed because the primary objective in age-management medicine (i.e., optimizing biological function while inhibiting the initiation of disease processes associated with age) has been redefined. This is in direct contradiction to the traditional approach of medicine at present. At the moment, our medical attempts are *interventional*, or designed to alter disease progression once it has started, but these attempts ignore the origin of the disease!

Most present biomarkers measure molecular and chemical changes in the body after specific treatment therapies have been utilized. These laboratory changes result from alterations in **gene expression** (e.g., proteins, hormones, and enzymes) (see previous chapters).

The *induction*, or *silencing*, of various genes is extremely important for human development both in utero and in early childhood.[3] The silencing process may play a causal role in the alterations of biomarkers associated with aging. New research by Dr. Stanley Burzynski of the Burzynski Research Institute in Houston, Texas has documented the importance of methylation as one of the key regulators of gene silencing and activation in aging, especially in regard to cancer. Therefore, one of the key goals in selecting a panel of biomarkers for aging should be to identify the age-related changes that we now know occur with specific gene expression alterations. These have recently been defined.[4,5]

At our Institute,* the present therapeutic approach centers on changing the cellular milieu (intercellular and intracellular fluids) and, therefore, *influencing gene expression* in a positive fashion to inhibit the causes of age-related disease. This is accomplished by utilizing a combination of various treatments (see earlier), including pharmaceuticals, hormone replacement therapy, micronutrient and nutraceutical supplements, regulating calories and macronutrient ratios, physical exercise regimens and mind-body techniques to induce relaxation.

Another way of stating the goal for our patients' therapy is to *optimize* physiological function while balancing biomarker levels to achieve or maintain values prevalent in the second and third decades of life. This is the period of optimal health and what can, therefore, be considered **optimal gene expression**.

The question arises: What is an optimal level of gene expression? One approach is to establish a **biomarker profile** that would be measured in early adulthood or in late adolescence, as is now being accomplished at our Institute. This allows the anti-aging practitioner to do more than make an educated guess based on what the "illusive" statistical human being would represent for a specifically chosen biomarker.

With the advent of **gene banking** today,[6] it is possible to selectively store and document our gene expression, or **profile**, at any specific age. The ideal time to do this is during late adolescence or in early adulthood. Although this may be somewhat late for many of us, it is important that we document and store our children's optimal gene expression, since it is the next generation who will probably be the ones to benefit most from future technologies.[6] Banking our own cells even at 40, 50, 60, and 70 years of age will still document our present gene expression and without a doubt will prove to be extremely useful with the advent of future technologies.

Key Technologies

In the coming years, a number of key technologies will give rise to a more comprehensive and detailed array of biomarkers that will be able to measure and more effectively treat the aging process. The essential technologies that will drive this revolution are:

1. Gene chips.
2. Genomics and Proteomics.
3. Gene transfection (plasmid DNA transfers).
4. Stem cell therapy (autologous regenerative medicine).
5. Nanotechnology and artificial intelligence.

*The Giampapa Institute for Anti-Aging Therapy and Age Management.

Gene Chips and Genomics With rapid development of **gene chips**, gene expression will easily be measured in a person. This gives the physician a more complete detailed picture of what is happening in the body with the aging process at a *specific* point in time. At present, *molecular* changes (objective biomarkers) are routinely documented; however, the physician cannot discern whether these changes are at the intimate level of gene expression or are due to factors that lie immediately after **gene transcription** (or **translation**) or during protein formation. Gene chips will make possible the true definition of optimal gene expression.

The ability to make this statement stems from the recent completion of the **Human Genome Project**, which has now provided the basic code for life as well as disease. Constant updates of human genome findings are available on several Web sites.[7,8]

It is a major benchmark, in the history of life on this planet, that we have discovered and can now read the blueprints to design and build a human organism (Diagram XIII-2). The simultaneous appearance of the computer and the gene chip has allowed this information to have acute clinical relevance. Thus, the newly deciphered information within the 23 pairs of human chromosomes will be a primary force in the evolution of biomarkers and will represent a new branch of science called **genomics**.

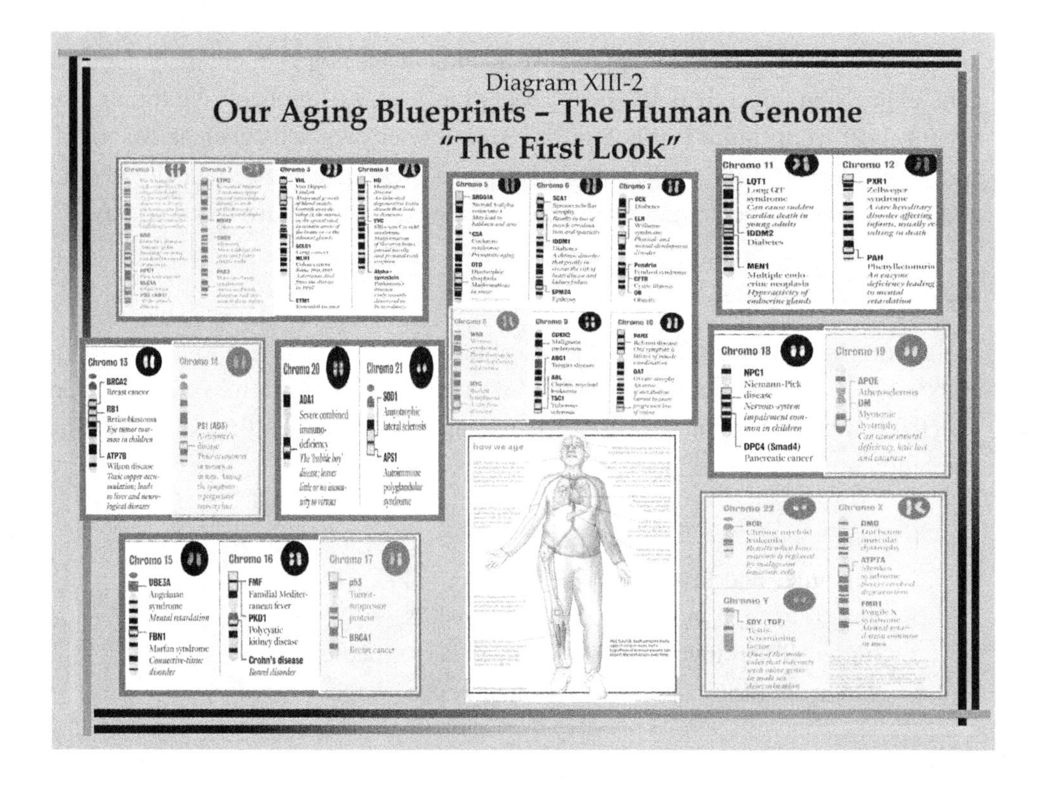

Diagram XIII-3

PERSONAL GENE CHIPS REVEAL YOUR RISK FOR DEADLY DISEASES

How to Speed-Read Genes

HIGH-TECH CHIPS read your genetic blueprints to forecast what diseases you could get, or help determine which treatments would work best for a disease you already have. To decode the answers, single strands of known segments of DNA--the chemical code that forms the blueprints for life--attach to unknown DNA samples.

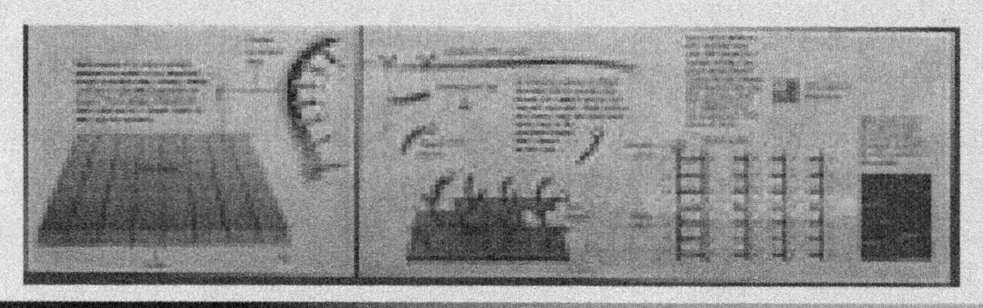

The pattern and recognition capacity of a computerized gene chip will be a significant advance toward the ability to define the basic gene expression pattern responsible for the healthy period of life that we designate as early adulthood.

How does a gene chip work (Diagram XIII-3)? The chips are composed of pre-selected *sequences* of human genes, which can be chosen to monitor the expression of key genetic sequences now known to be involved in the aging process. RNA is isolated from the subject's cells and is tagged with a material that fluoresces when the RNA binds to its complementary DNA segment. A spectrophotometer scans the chip, and a computer catalogs the fluorescent areas and thus identifies which genes are being expressed. This technology is easily adopted to determine the effects of anti-aging therapies and gene expression over the short term and the long term. Furthermore, basic research on aging has already profiled the key genes that are altered during the aging process.[5] Many of these detrimental genetic patterns do not occur during calorie restriction and life span extension experiments documented in animals.[9–15] These patterns have also been shown to be inhibited in primates that have been treated under the same conditions of caloric restriction and improved nutrient density (i.e., decreased calorie intake and increased essential vitamin, mineral and cofactor levels).

The major family of these genes[5] are involved in regulating:

1. Glucose.
2. Modification (methylation) of DNA.
3. Expression of inflammatory processes intracellularly.
4. Control of oxidation, among others.

These genes also control the same processes that occur on a cellular level (see discussion of the aging equation in Chapter I).

In addition to these improvements in diagnostic capacity, both **plasmid DNA transfers** (Diagram XIII-4) and **stem cell therapy** (Diagram XIII-5) will have a radical effect in the effort to slow the aging process.

DNA Transfection (Plasmid DNA Transfer) Currently, attempts are being made with transfer plasmid vectors containing DNA sequences. Plasmids are simply non-harmful bacteria-like structures. With this technique, select genes can be placed into a subject involving only the simplest genetic engineering technology. Although this work is being carried out only in animals and in the sickest patients, future techniques will be easily applied to helping individuals for the purpose of optimizing key biomarkers and managing the aging process as well.

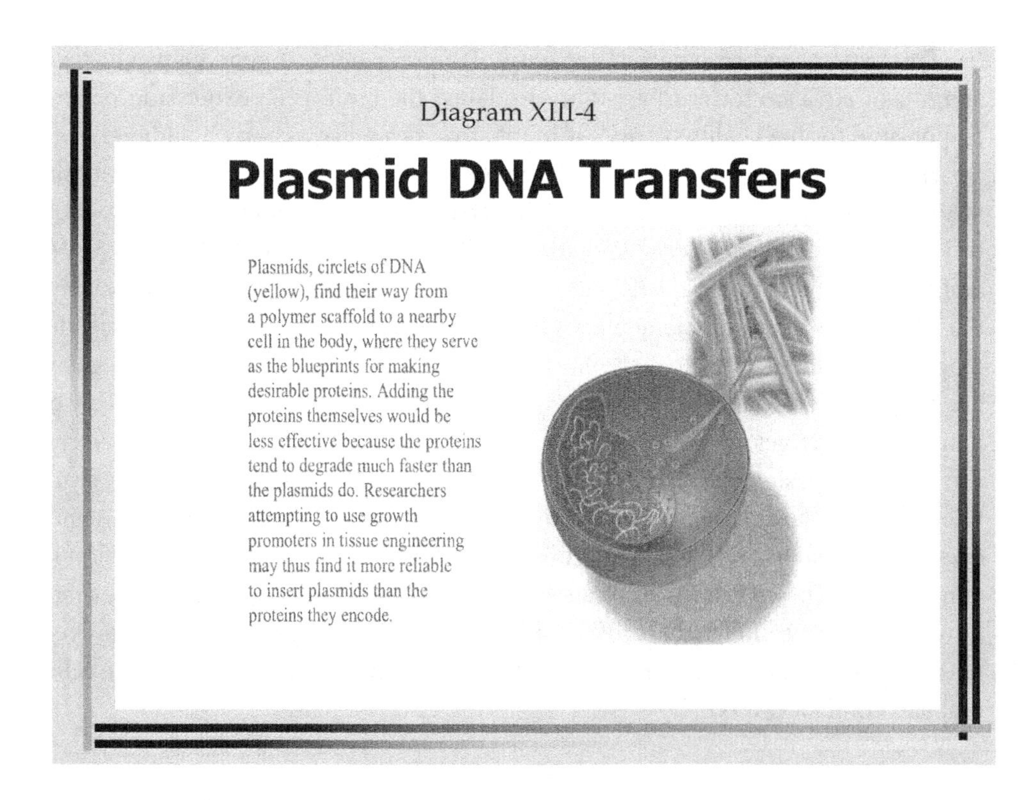

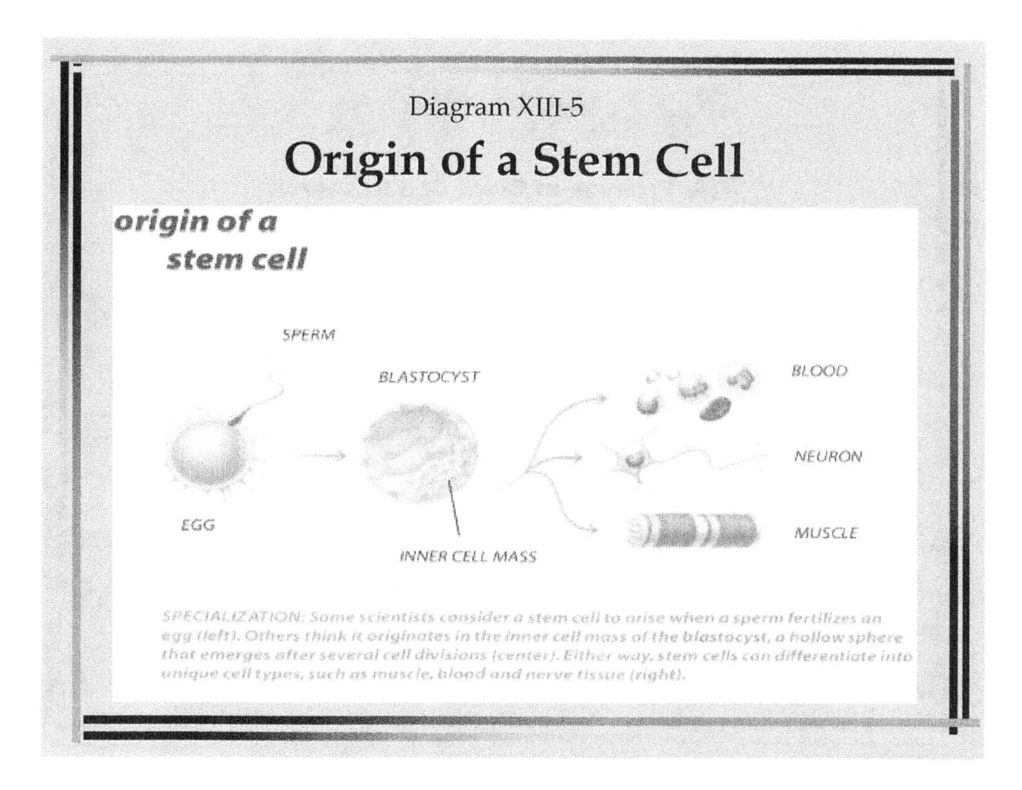

For example, in the immediate future, it will be possible to transfect DNA into muscle cells or to perhaps even directly inject RNA via phospholipid encapsulation to help control oxidation of free radical damage. Many studies support the observation that high levels of **intrinsic antioxidant enzymes**, such as superoxide dismutase, catalase and glutathione peroxidase are correlated with an increased life span.[16] These enzymes will most likely be the first compounds for delivery via this technology.

Additional studies in animals have also demonstrated that striated muscle cells take up the genes coding for these antioxidant enzymes and express functional enzymes. This current technology makes such therapies viable in principle and highly likely in practicality.

Focusing on improving the key intrinsic antioxidant systems with DNA transfer technology will be one of the most straightforward ways to slow aging, to limit free radical damage and, therefore, to optimize the aging process. This therapeutic approach will become available within our lifetime.

Stem Cells The use of bioengineering and *cloning* of stem cells is being intensely studied in many laboratories around the world. Experiments such as the cloning of Dolly the sheep clearly point to the possibility of differentiating adult DNA (e.g., a buccal cell DNA sample from present gene banks) into enucleated stem cells. Once these stem cell colonies are available, they may allow us to selectively regenerate

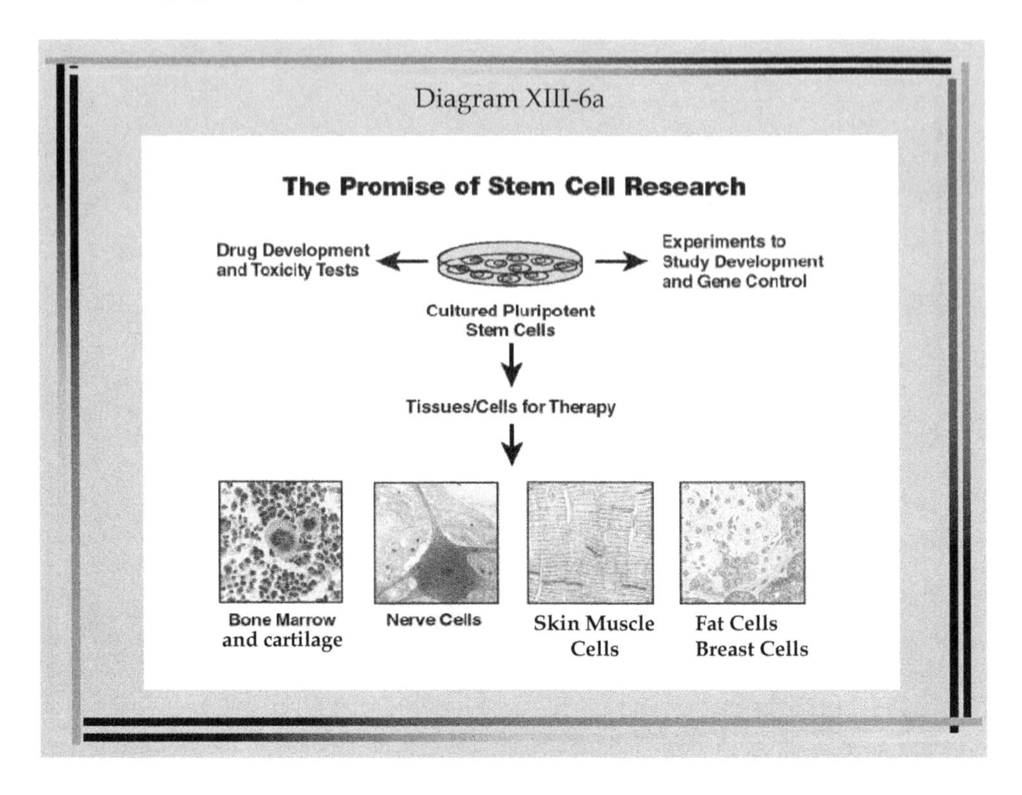

any component of our aging human body (Diagram XIII-6a). Presently utilized stem cell therapy, as practiced by Dr. William Rader,[17] involves introducing a new source of cells in embryological form (fetal stem cells) that do not trigger immune rejection even though they are composed of different DNA (the donor's DNA). According to Dr. Rader, this is because they are not differentiated enough, developmentally, to "generate an antigenic response."

Human Therapeutic Cloning and Regenerative Medicine

As the techniques of molecular biology become more sophisticated, sources of cells with specific embryological characteristics will be produced from cells from the patient's own body (Diagram XIII-6b). Recently, the world was stunned by the announcement of the successful creation of the first cloned human embryo by Advanced Cell Technology, in Worster, Massachusetts, under the leadership of Dr. Michael West. Dr. West refers to this new approach to tissue regeneration as **autologous regenerative medicine**. It is all based on **human therapeutic cloning**.[18–21]

With this technique, many of the effects of aging we presently treat with surgical and laser technologies will fall by the wayside to make room for actual cell replacements cloned from our *own* cells (Diagram XIII-7). Three basic approaches

Diagram XIII-6b

Stem Cells Types, Sources and Daughter Tissues

Stem Cell Type	Source	Daughter Tissue
Embryonic	Embryo or fetal Tissue	All types
Hematopoietic	Adult bone	Blood cells, brain marrow
Neuronal	Fetal brain	Neuro, glia blood
Mesenchymal	Adult bone marrow	Muscle, bone, cartilage, tendon

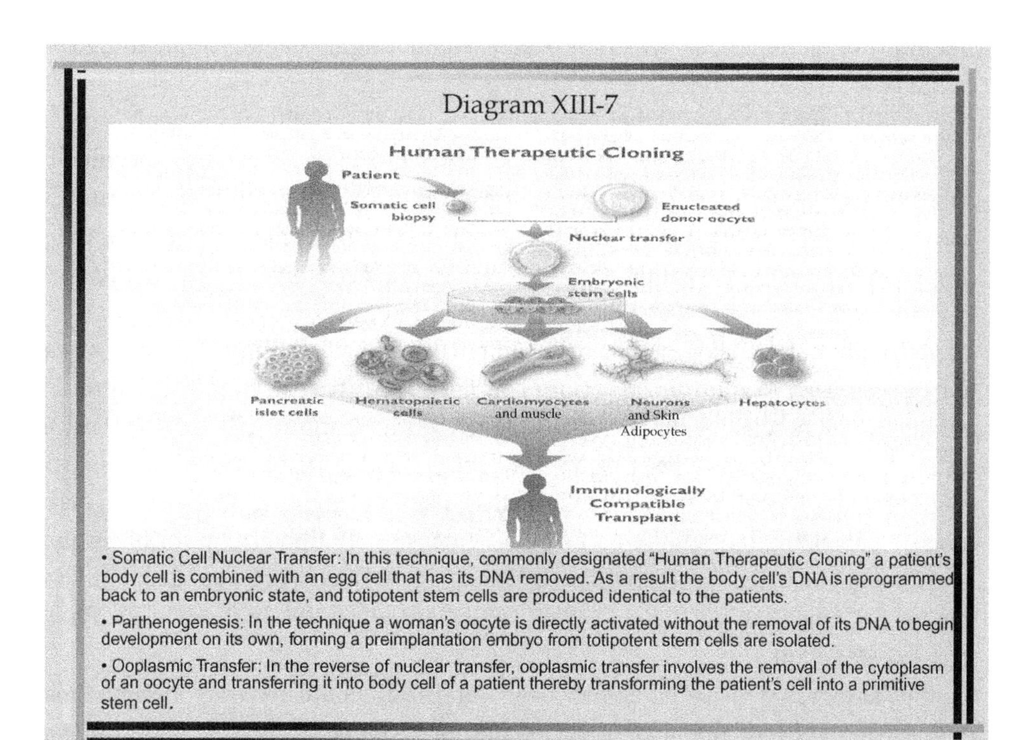

exist to manufacture embryonic cells identical to human adult cells:

- Somatic cell nuclear transfer
- Parthenogenesis
- Ooplasmic transfer

The use of stem cells should also help to answer the hypothesis that **mitochondria** are key players in the aging process — an area of intense speculation in recent research. New stem cells that are introduced to a donor will bring with them new **virgin mitochondria** into an aging body. This should help to upregulate or increase adenosine triphosphate (ATP) production. Many recent experiments have indicated that during aging, there is a loss of mitochondrial and nuclear DNA repair capacity along with their key enzymes. This may be avoided with new mitochondrial elements, transplanted from stem cells, and thus resulting in the improvement of these intracellular hallmarks of aging.

Mitochondria are crucial for maintaining the body in a state of **negative entropy**; that is, mitochondria are responsible for maintaining cellular organization and continuity of structure. It is through enhanced cellular energetics that processes such as cyclic adenosine monophosphate (cAMP)–mediated second messenger amplification of hormonal signals are accomplished. Although it has not yet been proven, these transplanted embryonic stem cells most likely will positively affect and improve the biological function of the neighboring and aging cells throughout the body.

More advanced stem cell therapy will also be able to selectively target key organ systems impaired or destroyed during the aging process and even to specifically repair damage within key organ sites themselves. Depending on which germ cell line is expanded (e.g., ectoderm, endoderm, or mesoderm), organ-specific anti-aging therapy will be possible within the next decade and, certainly, within the life span of those reading this text (see Diagrams XIII-6a and b).

An important focus of the same research in the immediate future will be restoration of central nervous system function. This has already been accomplished through intrathecal stem cell injections.[17] Cells of ectodermal origin will also be used to restore key brain centers responsible for glucose and cortisol regulation (i.e., the ventral medial lateral or dorsal medial lateral nuclei of the hypothalamus). Therefore, by enhancing the regulation of blood glucose with tighter control of insulin, cortisol and glucagon, important biomarkers of aging, such as glycation and inflammation, are more likely to be controlled and balanced, similar to those levels found in a younger adult.

Furthermore, stem cell treatment for improved central nervous system function will provide treatment options for the most feared aspect of aging: *mental*

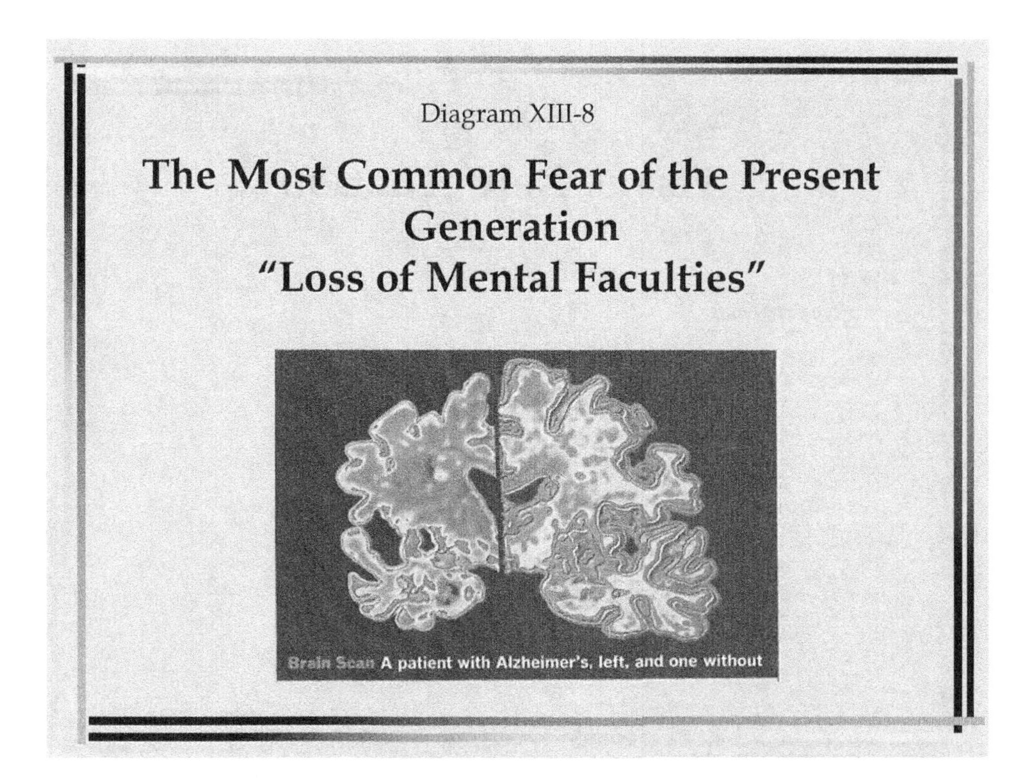

deterioration, age-related memory impairment and **Alzheimer's disease** (Diagram XIII-8). Yet another approach to stem cell therapy is being developed by Aastrom Biosciences, which utilizes *adult stem cells* from bone marrow (Diagram XIII-9).

Shortly after these technologies become mainstream as anti-aging therapies, the ultimate anti-aging process, **nanotechnology**, may be ready for clinical application (Diagram XIII-10).

In its early developmental form, nanotechnology will be applied on a macroscopic level to treat anatomical defects. Most likely, surface damage to the skin will be the first area of experimentation. As it rapidly matures, this technology will sharpen its resolution and will focus on the cellular and molecular levels for applications such as DNA repair. These **Biobots**,[22] well described in Drexler's book, *Engines of Creation*, will be designed to *repair* and *reconfigure* DNA; thus it will be possible to reverse DNA damage and inherited genetic lesions as well as to deliver chemotherapy agents to specific cell sites[23] (see Diagram XIII-11). The application of these technologies will allow for optimal gene expression and repair of either inherited or environmentally initiated defects. With this ability, the medical community will be able to push the limits of optimal life span and health span to ranges not imagined at present.

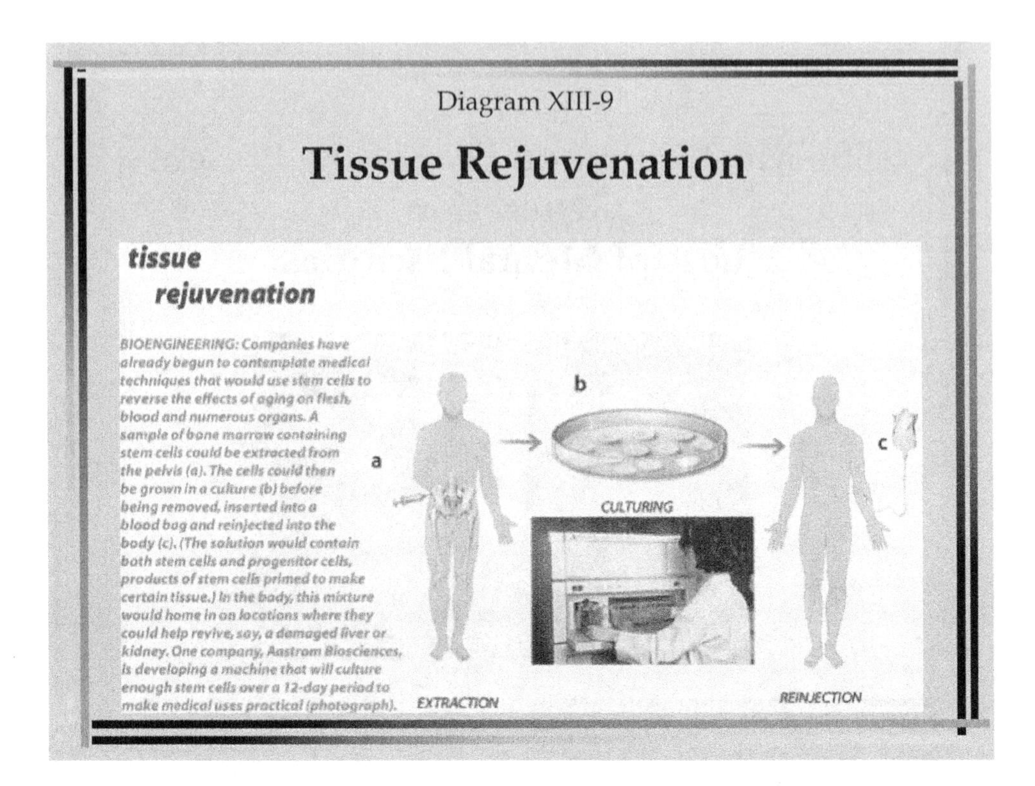
Diagram XIII-9
Tissue Rejuvenation
tissue
rejuvenation
BIOENGINEERING: Companies have already begun to contemplate medical techniques that would use stem cells to reverse the effects of aging on flesh, blood and numerous organs. A sample of bone marrow containing stem cells could be extracted from the pelvis (a). The cells could then be grown in a culture (b) before being removed, inserted into a blood bag and reinjected into the body (c). (The solution would contain both stem cells and progenitor cells, products of stem cells primed to make certain tissue.) In the body, this mixture would home in on locations where they could help revive, say, a damaged liver or kidney. One company, Aastrom Biosciences, is developing a machine that will culture enough stem cells over a 12-day period to make medical uses practical (photograph).
a
b
c
CULTURING
EXTRACTION
REINJECTION

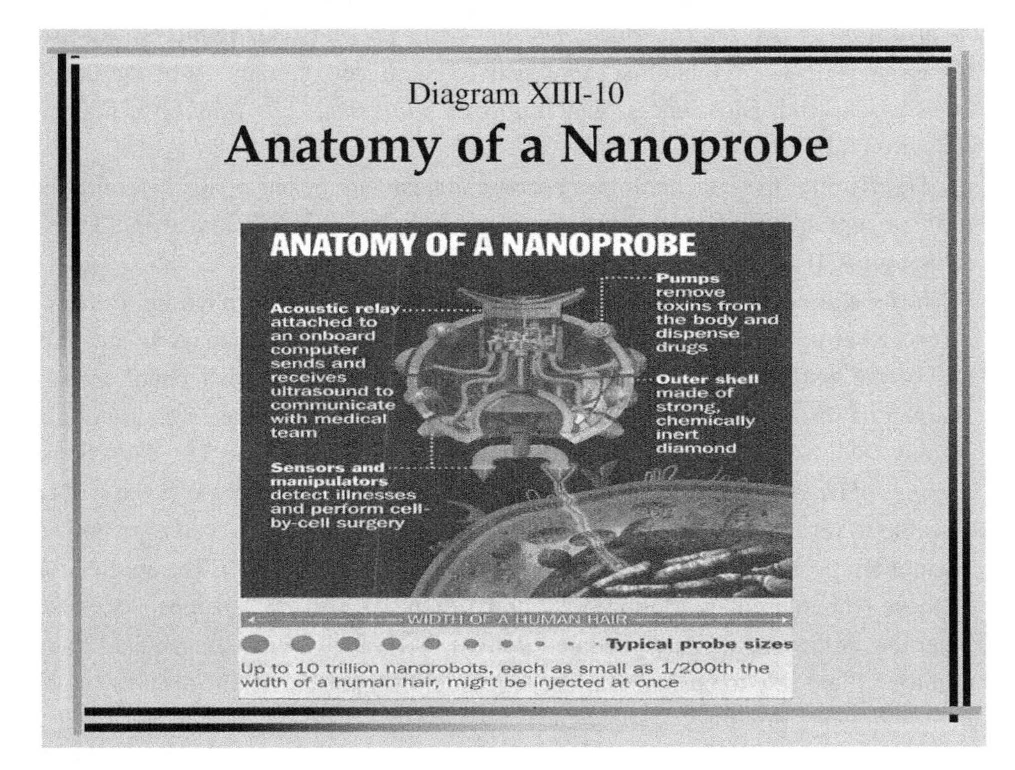
Diagram XIII-10
Anatomy of a Nanoprobe
ANATOMY OF A NANOPROBE
Pumps remove toxins from the body and dispense drugs
Acoustic relay attached to an onboard computer sends and receives ultrasound to communicate with medical team
Outer shell made of strong, chemically inert diamond
Sensors and manipulators detect illnesses and perform cell-by-cell surgery
WIDTH OF A HUMAN HAIR
Typical probe sizes
Up to 10 trillion nanorobots, each as small as 1/200th the width of a human hair, might be injected at once

Diagram XIII-11

Nanomotors

- **Create a hybridized motor to deliver cells and/or mechanical devices to perform a specific function**
  - pioneered at Cornell University
  - inspired by the protein rotors that convert cellular fuel into motion-generating energy in bacterial organisms
  - graft parts of bacterial motors onto a metal nanostructure
  - light-harvesting mechanisms, inspired by photosynthesis, create ATP for the nanomotor, eliminating the need to provide an external fuel source

Application: arrive at tumor cells to synthesize and deliver chemotherapy agents on-site, reducing toxicity to healthy tissue

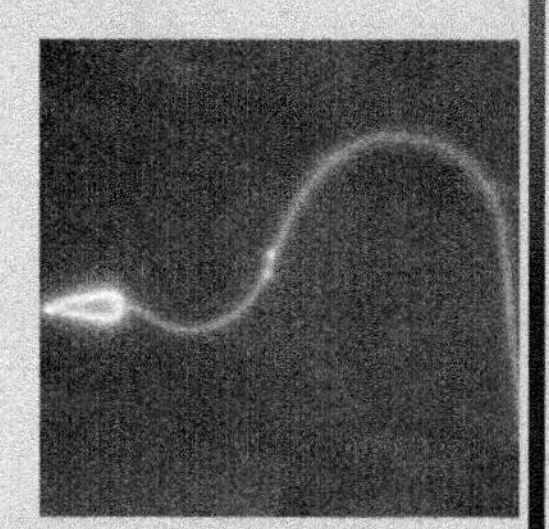

When evaluating anti-aging medicine techniques, note that the following points are documented as fact in this early part of the 21st century:

1. Presently, an individual's genetic sequences are unknown at the time of treatment.
2. Although a person's gene expression is unknown at the time of treatment, it can be inferred on the basis of laboratory tests and subjective questionnaires.
3. Anti-aging interventions presently utilized range from pharmaceutical to nutraceutical to lifestyle modification.
4. Measuring a collection of biomarkers, both subjective and objective, validates present anti-aging therapies and documents that *they do work* in improving quality of life and health span; augmentable life span effects have yet to be evaluated but appear highly probable.
5. Computer programs, such as the Biomarker Matrix Profile, allow for **pattern recognition** to analyze large amounts of data and to document different levels of aging improvement (Diagram XIII-12).

These five key points describe the anti-aging therapy, as currently practiced at our Institute, utilizing the patented Biomarker Matrix Profile program, as well as

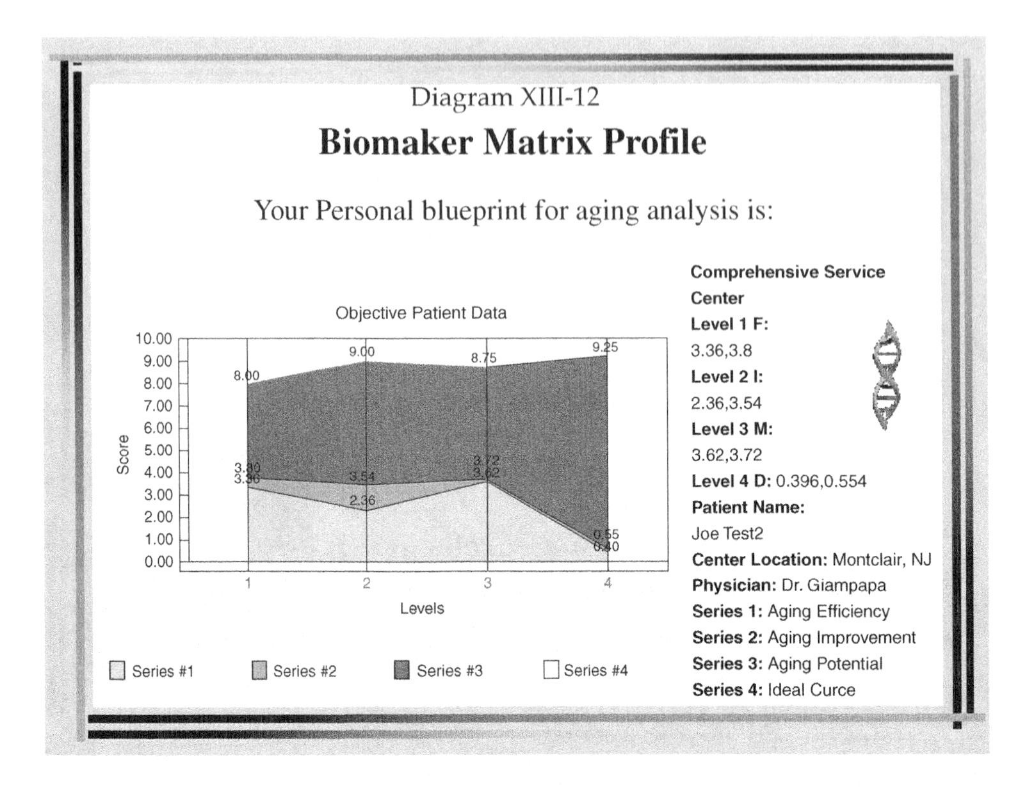

at other established anti-aging centers.[24] In essence, we have been documenting improvements in physiological, cellular and molecular and DNA function, which most likely reflect the result of improved gene expression.

Altered gene expression cannot be directly documented without incorporating a gene chip into the profile. More important, the current arsenal of anti-aging therapies attempts to *optimize* biological functioning without *altering* the structure or content of our genes. In essence, we have learned to make the genes we have work better — a major accomplishment! Nonetheless, optimizing biomarkers undoubtedly makes it possible to document changes resulting in improved quality of life and biological function.

The panels of biomarkers presently used from Immunosciences Labs Inc. (Beverly Hills, California) have been correlated with positive effects on health span and the quality of life as tested in humans over the last several years. Although extensions in human life span are impossible to prove at present, data from animal models, both nonprimate and primate, strongly implicate alterations in certain classes of genes in the suppression of age-related diseases, which should lead to longer life spans. According to the English biogerontologist Mr. Alex Comfort, the greater the number of biomarkers positively affected by a treatment regimen, the greater the probability of human life extension.

Anti-Aging Therapies in the Near Future

The Next Five to Ten Years Anti-aging therapies will soon undergo a dramatic transformation. The anti-aging clinic of the future will read a patient's genetic sequences and measure gene expression by extracting small samples of DNA and ribonucleic acid (RNA), most likely from the white blood cells, to be compared with the optimal age-related gene patterns, soon to be defined by the Human Genome Project. This will allow genetic defects to be detected and gene expression patterns to be monitored early in life.

Anti-aging therapeutics in the early part of the 21st century will combine today's therapeutic regimens and will include the new technologies, DNA-plasmid transfers, stem cell therapy, balancing DNA repair versus DNA damage and possible telomerase therapy. Furthermore, these therapies will be directed at *improving* the poor expression of specifically *inherited* genes as well as suboptimal biomarkers.

Ultimately, the therapeutic end-points will be an increased health span, an improved quality of life and a slower appearance of age-related diseases as well as of aging body and face contours. With the use of these new technologies, we will be able to document the effectiveness of anti-aging therapeutics on gene expression directly involved in the aging process.

The Next Ten to Twenty Years The anti-aging therapies in the more distant future — within the next 10 to 20 years — will involve direct manipulation of genetic sequences within living human beings. Starting with a sample of the patient's DNA and RNA, more advanced and highly selective gene chips will focus on the content and expression of the patient's nuclear and mitochondrial DNA. This will allow us to solve the puzzle of cellular energetics (ATP production) and DNA replication.

Genomics, by this time, will be a mature science and will provide a more detailed analysis of inherited and environmentally acquired genetic defects. A vast array of genes, by then, will have been analyzed, such as those involved in important processes like hormone production, signal transduction and signal amplification at the intracellular level, genomic stability, oxidative phosphorylation (ATP energy production), DNA repair, transcription, translation, protein turnover and methylation.

The expanding universe of tools for information concerning gene manipulation and nanotechnology will make it possible to identify defects, and solutions will be proposed to correct them using custom-designed, self-directed molecular machines. These novel nanomachines will be designed to carry out intracellular repairs on defective genes according to the information harvested from gene chips.

This therapy will be easily administered in the doctor's office of the future. A computerized biomarker profiler, incorporating gene chip information, will document the results. This approach should make it feasible to extend the life span and health span and to push the limitations and definitions beyond the present human genetic capacity.

The ability to document changes in gene expression and repair of DNA sequences will transform anti-aging therapeutics into a discipline that will be custom-tailored for each age-related disease process. This technology, coupled with a logarithmic growth of computer intelligence over the next few decades, will allow computers to surpass human intelligence scores and to begin to direct future research on aging as well as age-management treatment protocols. Advances in software programs recently designed to mimic human neuronetworks, as described by Ray Kurtzweil in *The Age of Spiritual Machines*,[25] will vastly increase our ability to understand complex relationships between different biological systems as we age. The merging of man and machine has already taken place at John Hopkins, where implanted chips are now being used to control a computer by way of a patient's thoughts (Diagram XIII-13).[26]

The 21st century will usher in a new era of medicine that finally treats the *causes* and not the *effects* of aging, right down to the genetic level. We will be able to control, repair and recreate our genetic inheritance and, most likely, to direct the quality of our lives and longevity as well as our physical appearance.

Shortly, we will look back on these last few decades and realize that we have ideologically progressed from the concept of **spontaneous germ theory of disease**

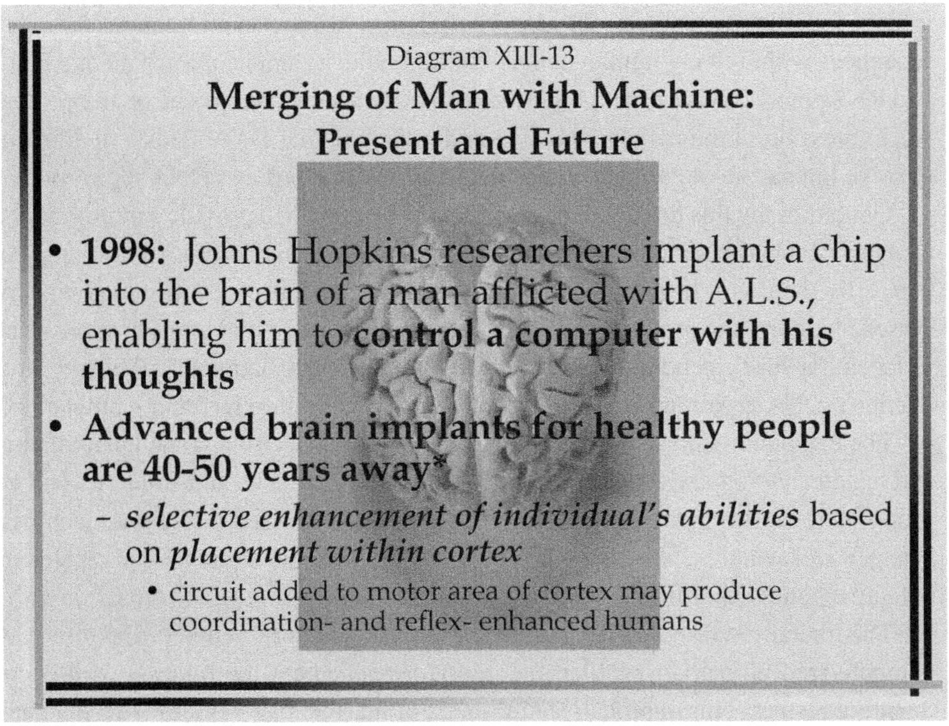

Diagram XIII-13

Merging of Man with Machine: Present and Future

- **1998:** Johns Hopkins researchers implant a chip into the brain of a man afflicted with A.L.S., enabling him to **control a computer with his thoughts**
- **Advanced brain implants for healthy people are 40-50 years away***
  - ***selective enhancement of individual's abilities*** based on ***placement within cortex***
    - circuit added to motor area of cortex may produce coordination- and reflex- enhanced humans

to deciphering and enhancing our own genetic makeup and our aging codes. We humans, a *carbon-based* life form, will soon come to realize that we have created a new *silicon-based* life form called *computers* having an intellectual capacity far beyond our own. In the immediate future, computers will aid us in our quest for enduring health and eternal youth at a speed much beyond our human capability.

For almost a century, as cosmetic surgeons and physicians we have been trained in the art and science of enhancing our physical appearance, but now we have the opportunity to be among the first in medicine to utilize this new information and perspective on aging. The coming decades will redefine what it means to be a cosmetic surgeon and physician of the 21st century.

References

[1] Advanced Cell Technology, Worcester, MA.

[2] Affymetrix, Inc., San Francisco, California.

[3] Jones PA, Takai D. The role of DNA methylation in mammalian epigenetics. *Science.* 2001; 293: 1068–1070.

[4] Cerami A. Hypothesis: glucose as mediator of aging. *J Am Gerontol Soc.* 1985; 33: 626–634.

[5] Lee, C, Klopp RG, Weindruch R, Prolla TA. Gene expression profile of aging and its retardation by caloric restriction. *Science.* 1999; 285: 1390–1393.

[6] Youth Cell Technologies, Inc., Greensboro, NC. Available at *www.youthcell.com.*

[7] Bioinformatics. Available at *weizman.ac.il/cards.*

[8] Available at *http://www.ncbi.nlm.nih.gov/omim/.*

[9] Bodkin NL, Ortmeyer HK, Hansen BC. Long-term dietary restriction in older-aged rhesus monkeys: effects on insulin resistance. *J Gerontol Biol Sci Med Sci.* 50: B142–B147.

[10] Cefalu WT, Wagner JD, Wang ZQ, et al. A study of caloric restriction and cardiovascular aging in Cynomolgus monkeys: a potential model for aging research. *J Gerontol Biol Sci Med Sci.* 1997; 52: B98–B102.

[11] Duffy PH, Reuers RJ, Leakey JA, et al. Effect of chronic caloric restriction on physiological variables related to energy metabolism in male Fischer 344 rat. *Mech Aging Dev.* 1989; 48: 117–133.

[12] Fernades G, Friend P, Yunis EJ, Good RA. Influence of dietary restriction on immunologic function and renal disease in (NZB x NZW) F mice. *Proc Natl Acad Sci U S A.* 1978; 75: 1500–1504.

[12a.] Green RM. *The Human Embryo Research Debates: Bioethics in the Vortex of Controversy.* New York: Oxford University Press; 200l.

[13] Holehan AM, Merry BJ. The experimental manipulation of ageing by diet. *Biol Rev.* 1986; 61: 329–368.

[14] Kim MJ, Roecher EB, Weindruch R. Influences of aging and dietary restriction on red blood cell density profiles and antioxidant enzyme activities in rhesus monkey. *Exp Gerontol.* 1993; 28: 515–527.

[15] Lee DW, Yu BP. Modulation of free radicals and superoxide dismutase by age and dietary restriction. *Aging.* 1991; 2: 357–362.

[16] Fossel M. *Reversing Human Aging.* New York: William Morrow & Co, Inc.; 1996: 37–40.

[17] Rader W, Schiller A (Personal communication). Leonardis Clinic, Medra, Inc.; 2001.

[18] Cibelli J, Lanza R, West M. First cloned human embryo. *J Regen Med.* November 25, 2001.

[19] Lanza RP, Cibelli JB, West MD. Human therapeutic cloning. *Nat Med.* 1999; 5(9): 975–977.

[20] Lanza RP, Cibelli JB, West MD. Prospects for the use of nuclear transfer in human transplantation. *Nat. Biotechnol.* 1999; 17(12): 1171–1174.

[21] Lanza RP, et al. The ethical validity of using nuclear transfer in human transplantation. *JAMA*. 2000; 284(24).
[22] Drexler E. *Engines of Creation: The Coming Era of Nanotechnology*. New York: Bantam, Doubleday Dell Publishing Group, Inc; 1998.
[23] Nanomedicine nears the clinic. *Technology Review*. 2000; January/February: 60–65.
[24] Giampapa VC, Klatz MR, DiBernardo EB, Kovarik AF. *Biomarker Matrix Protocol, Advances in Antiaging Medicine*, Vol. 1. Larchmont, NY: Mary Ann Liebert Publishers, Inc.; 1996: 221–223.
[25] Kurtzweil R. *When Computers Exceed Human Intelligence: The Age of Spiritual Machines*. New York: Penguin Putnam, Inc.; 1999.
[26] Sergio Naveda, Director of Intelliwise Research and Training, Sao Paulo, Brazil. In *Interactive Week*, January 10, 2000, p 34.

14

Stem Cell Utility in Anti-Aging Medicine:

Focus on the Tissue Microenvironment

Steven J. Greco PhD[1] and Pranela Rameshwar[2]

[1] *1Department of Medicine, Division of Hematology/Oncology, University of Medicine and Dentistry of New Jersey-New Jersey Medical School, Newark, NJ, USA*
[2] *2UMDNJ-New Jersey Medical School, MSB, Rm. E-579, 185 South Orange Ave, Newark, NJ, USA; rameshwa@umdnj.edu*

The utility of stem cells in regenerative medicine has been ongoing for several decades, primarily in the form of bone marrow (BM) transplants to treat various hematological disorders and other immune-related diseases. This type of treatment has provided valuable information on the influence of the recipient tissue's microenvironment, or niche, on stem cell behavior following implantantion. More recently, stem cells have been examined as a potential therapy for numerous diseases, tissue repair and aging disorders. Of note, stem cells have more recently been explored as anti-aging therapies and have application in skin regeneration. One consideration that poses a formidable task for the successful clinical application of stem cells in these modalities is the impact of the host tissue microenvironment on the desired therapeutic outcome. *In vitro*, stem cells exist in surroundings directly controllable by the researcher to produce the desired cellular behavior. *In vivo*, the transplanted cells are exposed to a dynamic host microenvironment laden with soluble mediators and immunoreactive cells. In this chapter, we focus on the possible contribution by microenvironmental factors, and how these influences can be overcome in anti-aging therapies utilizing stem cells. Although the information could be extrapolated to any stem cell, we focus on mesenchymal stem cells (MSCs) since these can be selected

The Principles and Practice of Antiaging Medicine for the Clinical Physician, 235–244.

from autologous sources with reduced ethical concern. These stem cells can also cross an allogeneic barrier. Specifically, we examine three ubiquitous microenvironmental factors, IL-1α/β, TNFα and SDF-1α, and consider how inhibitors and receptor antagonists to these molecules could be applied to increase the efficacy of MSC anti-aging therapies while minimizing unforeseen harm to the patient.

Stem Cells, Anti-Aging Medicine and the Tissue Microenvironment

The emergence of stem cells therapies has sparked excitement among scientists, clinicians and patients alike regarding the potential treatment of previously untreatable conditions. However, while stem cells are close to approval for a few indications and have begun to be utilized for cosmetic purposes, the implementation of many stem cell therapies in patients may still be years away. When considering translating these therapies into patients, there are two principal concerns that must be resolved: I. Can the stem cells efficiently produce the desired therapeutic outcome, albeit tissue replacement or repair, *in vitro*?; and II. Can the *in vitro* studies be replicated *in vivo*, both short- and long-term, with increased confidence?

A vast number of tissue types have been generated from both embryonic (ESCs) and adult stem cells. ESCs are pluripotent cells derived from the inner cell mass of the blastocyst, which hold tremendous potential in generating specified tissue types (Lerou and Daley, 2005). However, the potential for immune rejection, together with the possibility of tumor formation has caused their application in humans to proceed with caution (Lerou and Daley, 2005). Adult stem cells tend to be tissue-specific cells with limited differentiation potential as compared to ES cells. Adult stem cells are clinically attractive therapies due to their reduced risk of tumorigenesis and ability to expand with relative ease (Cheng et al., 2004).

Among the many types of adult stem cells, those resident to the bone marrow (BM), particularly mesenchymal stem cells (MSCs), have gained extensive interest among scientists and clinicians (Deans and Moseley, 2000). MSCs are mesodermal cells primarily resident to the adult BM, which undergo lineage-specific differentiation to generate bone, fat and cartilage among other tissue types (Bianco et al., 2001). MSCs have also been reported to transdifferentiate into defined ectodermal and endodermal tissues *in vitro*, thus alluding to their inherent plasticity (Jeon et al., 2007; Eberhardt et al., 2006; Choi and Panayi, 2001; Ong et al., 2006; Greco and Liu et al., 2007; Greco and Zhou et al., 2007; Cho et al., 2005; Trzaska et al., 2007). MSCs are available for autologous therapies, have a unique ability to bypass immune rejection and are inherently migratory (Potian et al., 2003). Whereas tissues derived from ESCs or other types of stem cells may be rejected when transplanted, MSCs offer the potential for allogeneic transplantation and a readily available source of "off-the-shelf" stem cells for personalized therapies.

Recently, MSCs as well as their mesodermally-related brethren, adipose-derived stem cells (ADSCs), have been applied in cosmetic/anti-aging medicine for regeneration of skin and underlying connective tissues (Barret et al., 2009; Yang et al., 2010). Given the relative ease with which MSCs and ADSCs can be harvested and expanded from the patient, the application of a patient's own autologous stem cells is an ideal scenario. Luckily, even if the patient is unable to donate the BM or fat tissue for harvesting the cells, the unique immune properties of these adult stem cells allows their potential for an allograft as well. Unfortunately, these unique immune properties alone do not guarantee that the cells will produce the desired therapeutic outcome, regardless of the donor cell source.

In vitro, a MSC's growth conditions can be closely monitored to favor stem cell growth and/or differentiation. *In vivo*, the transplanted MSCs are exposed to local immune cells and soluble mediators that could influence the cells' behavior, either positively or negatively regarding the desired outcome. This concept of the tissue microenvironment has become a growing concern, and may be an important factor in deciding whether a stem cell therapy succeeds or fails (Phinney and Prockop, 2007; Greco et al., 2004; Yan et al., 2006; Grassel and Ahmed, 2007).

A prototypical example of a tissue microenvironment affecting stem cell behavior is observed among hematopoietic stem cells (HSCs) and their niche within the BM. HSCs are relatively quiescent cells located close to the BM endosteum at relatively low oxygen concentration (Greco et al., 2004). As HSCs differentiate, the maturing immune cells migrate towards the central sinus of the BM under progressively higher oxygen concentrations (Greco et al., 2004). The change in oxygen is a key determinant in the maturation of the immune cells before they leave the BM and migrate into the peripheral circulation (Greco et al., 2004). In contrast, MSCs are located close to trabecular bone near the central sinus of the BM (Bianco et al., 2001). As MSCs migrate towards the endosteum under progressively lower concentrations of oxygen, the stem cells differentiate into stromal fibroblasts, which form the principal support structure for immune cell maturation (Bianco et al., 2001).

This example demonstrates that local microenvironmental changes in variables such as oxygen concentration can drastically affect the behavior of MSCs. Since MSCs have been shown to generate a vast number of tissues, they have clinical implications in a wide array of diseases and disorders. Among possible transplantation sites are tissues such as dermal, adipose, cardiac, neural, pancreatic and bone. Each tissue provides a unique local microenvironment that can affect the success of the therapy. The problem facing researchers and clinicians is accurately developing *in vitro* models to recapitulate the tissue microenvironment so that cellular behavior can be observed prior to transplantation. This is no easy task considering the dynamic nature of the microenvironment.

For example, in regenerative skin therapy, transplantation of MSCs or ADSCs alone will generate a local immune response, even if the cells are tolerated. This

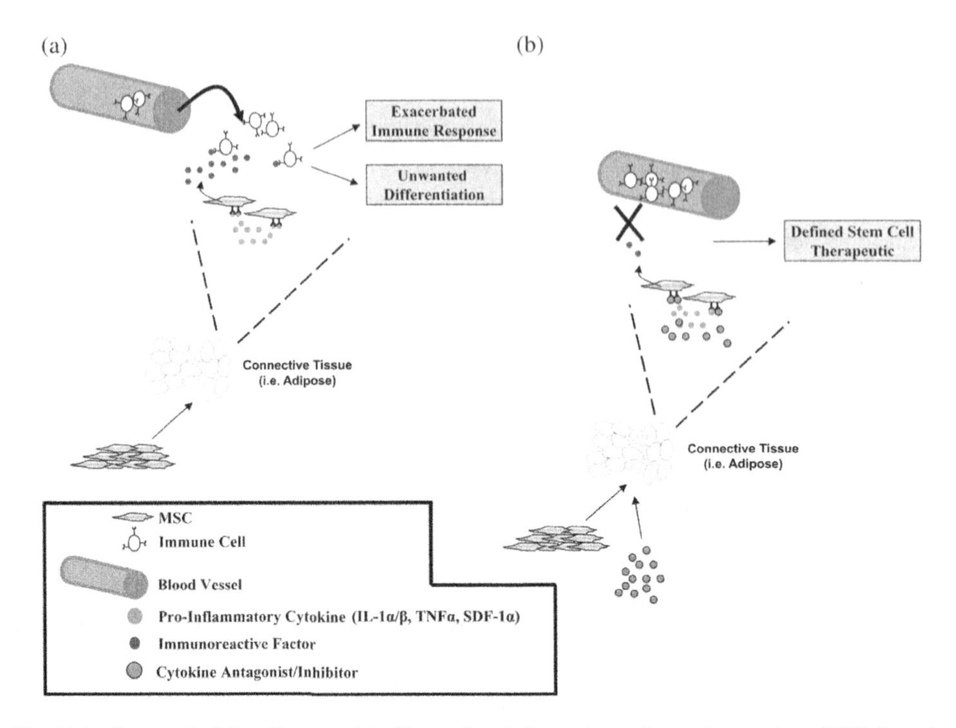

Fig. 14.1 Cartoon depicting the potential effects of an inflammatory microenvironment on MSC therapies. (a) A recipient tissue, such as adipose or dermis, if injured is laden with inflammatory cytokines, such as IL-1α, IL-1β, TNFα or IL-6, within the microenvironment. MSCs introduced into the microenvironment can respond to the inflammatory stimuli by synthesizing and releasing immunoreactive factors, such as SP or other neuropeptides. Excessive production of SP would lead to an exacerbated immune response and infiltration of additional immune cells into the injured tissue. Additionally, the inflammatory stimuli could have an untoward effect on MSC differentiation, for example differentiation into fibroblasts rather than adipocytes. (b) Alternatively, delivery of MSCs together with an inflammatory cytokine antagonist/inhibitor may abrogate immunoreactivity and allow the desired stem cell therapeutic to proceed unhindered.

modest release of soluble factors could disrupt homeostasis within the dermal milieu by causing release of inflammatory mediators, such as cytokines. MSCs express receptors for many cytokines, thus demonstrating their potential to respond to local microenvironmental changes (Bianco et al., 2001; Greco and Rameshwar, 2007). Excessive cytokine release within the dermal transplantation site could lead to the production of other soluble factors by the MSCs themselves. If these factors are immunoreactive, then other immune cells could infiltrate the tissue and cause an exacerbated immune response, rejection of the transplant or differentiation of the MSCs into an undesirable phenotype (Figure 14.1a). On the other hand, MSCs have been shown to be a potent source of trophic factors (Phinney and Prockop, 2007). These findings indicate that MSCs could also be used to aid endogenous tissue repair, perhaps even more so than in cell replacement.

If MSCs or ADSCs are found to negatively impact the dermal microenvironment through exposure to soluble mediators, there are still approaches to augment their use as an effective therapeutic. When considering the example presented in Figure 14.1a, which provides an example of MSC implantation into fat tissue, inclusion of specific cytokine receptor antagonists or inhibitors could suppress the untoward effects of the host microenvironment on the transplanted MSCs, thereby leading to a well-defined therapeutic outcome (Figure 14.1b). It is stressed that clinicians evaluate these types of approaches before implanting stem cells for cosmetic/ anti-aging purposes.

Throughout the remainder of this chapter we will address feasibility of using similar pharmacologic approaches in MSC/ADSC transplants to address the above concerns. More specifically we will focus on three ubiquitous microenvironmental factors: IL-1α/β, TNFα and SDF-1α, and examine how receptor antagonists or inhibitors to these factors, whether FDA approved or in development, may limit the potential negative influences of the tissue microenvironment.

Interleukin-1α/β

IL-1α and IL-1β are members of the IL-1 superfamily of cytokines. These pro-inflammatory mediators are primarily synthesized by macrophages, monocytes and dendritic cells, and are responsible for immune defense against infection (Table 14.1) (Dinarello, 1994). IL-1α and IL-1β are also key regulators of hematopoesis and the inflammatory process (Table 14.1) (Dinarello, 1994). Both cytokines are found throughout the body, thus they are expected to be present within the local microenvironment of the dermis.

MSCs express IL-1RI, which is the principal receptor for both IL-1α and IL-1β (Greco and Rameshwar, 2007). IL-1α was found to alter the behavior of undifferentiated MSCs and neurons differentiated from MSCs (Greco and Rameshwar, 2007). Specifically, stimulation of MSCs with IL-1α caused production of the neurotransmitter, substance P (SP), by undifferentiated and differentiated cells (Greco and Rameshwar, 2007). Similar effects were not observed in cells stimulated with IL-1β (Greco and Rameshwar, 2007). SP has involvement in various physiological functions, such as the perception of pain, however the peptide also has a stimulatory effect on immune cell development and function (Greco et al., 2004). The excessive levels of SP could lead to immune cell infiltration and an exacerbated immune response, which may cause rejection of the transplant.

To counter the negative effects of IL-1α present within a tissue microenvironment, co-therapies utilizing MSCs and specific IL-1R antagonists or inhibitors may be successful. The IL-1R antagonist (IL-1ra) is naturally occurring and binds to the IL-1RI. IL-1ra competes for binding to the IL-1RI with IL-1α and IL-1β, however binding of this ligand does not result in an intracellular signal (Dinarello, 1994). A commercially available IL-1RI antagonist is Kineret®, also known as Anakinra®, which is a recombinant, non-glycosolated version of human IL-1ra (Table 14.1) (Bresnihan and Cunnane, 1998; Gabay, 2000; Arend, 1993; Hannum et al., 1990).

Table 14.1. Microenvironmental factors implicated in the outcome of MSC therapies.

Cytokine/ Chemokine	Source	Physiological Function	Receptor Expression on MSCs	Inhibitors/ Antagonists	Reference
IL-1α/β	macrophages, monocytes, dendritic cells	immune response, inflammation, hematopoiesis	yes	IL-1ra, Kineret®	Bresnihan and Cunnane, 1998; Gabay, 2000; Arend, 1993; Hannum et al., 1990
TNFα	macrophages	immune response, inflammation, proliferation, differentiation, tumorigenesis, viral replication	yes	Remicade®, Humira®, Enbrel®	Knight et al., 1993; Choy and Panayi, 2001
CXCR4	tissue-specific stromal cells	immune response, inflammation, hematopoiesis, chemotaxis, tumor metastasis	yes	Mozobil®, T134, tannic acid	Cashen et al., 2007; Arakaki et al., 1999; Chen et al., 2003

Listed microenvironmental factors are ubiquitously expressed throughout the body and have known effects on MSCs. Co-therapy with pharmacologics, such as receptor antagonists or specific inhibitors, may improve the desired therapeutic outcome.

The drug has been used in the treatment of inflammatory conditions such as rheumatoid arthritis.

Tumor Necrosis Factor α

TNFα is a pro-inflammatory cytokine principally synthesized by macrophages, which is involved in the acute phase of systemic inflammation (Locksley et al., 2001). More specifically, TNFα mediates immune cell homing, proliferation and differentiation, as well as tumorigenesis and viral replication (Table 14.1) (Locksley et al., 2001). TNFα is ubiquitously found throughout the body, and is another important factor present within local tissue microenvironments.

MSCs express the principal receptor for TNFα, TNF-R1 (Table 14.1) (Greco and Rameshwar, 2007). Recent studies have shown that undifferentiated MSCs incubated with TNFα have a greater ability to migrate in the presence of chemokines compared to cells incubated without TNFα (Ponte et al., 2007; Schmal et al., 2007). Increased chemotaxis of MSCs would be clinically important if proper homing to the site of tissue injury became more efficient. However, since MSCs are inherently chemotactic, increased sensitivity to chemokine gradients could cause continuous mobilization within a tissue, and impede proper homing and delivery of the desired therapeutic.

To offset any deleterious effects of microenvironmental TNFα co-thereapies with existing TNFα inhibitors may be beneficial. Currently there are three approved TNFα inhibitors, infliximab (Remicade®), adalimumab (Humira®) and etanercept (Enbrel®), which are primarily used to treat inflammatory and autoimmune disorders (Table 14.1) (Knight et al., 1993; Choy and Panayi, 2001). Infliximab and adalimumab are monoclonal antibodies that bind TNFα and block signaling through the TNF-RI. Etanercept is a large molecular weight, soluble recombinant TNFα receptor fusion protein that binds TNFα and prevents signaling through membrane-bound TNF-RI. Administration of these pharmacologics in combination with MSC therapies may negate any untoward effects of TNFα on the desired therapeutic outcome.

Stromal Cell-Derived Factor-1α

SDF-1α, also known as CXCL12, is a chemokine produced by stromal fibroblasts, which mediates inflammation and the immune response through modulating lymphocyte chemotaxis (Table 14.1) (Bleul et al., 1996). Additionally, SDF-1α regulates hematopoeisis and has a role in tumor metastasis (Table 14.1) (Bleul et al., 1996). Like IL-1α/β and TNFα, SDF-1α is ubiquitously found throughout the body, and is an important soluble mediator of the tissue microenvironment.

Within the BM, the primary lineage-specific progeny of MSCs are stromal fibroblasts (Bianco et al., 2001; Deans and Moseley, 2000). It is not surprising then that SDF-1α is important in the biology of MSCs. Expression of the principal SDF-1α receptor, CXCR4, has been demonstrated on MSCs, where it has been shown to mediate site-directed homing of MSCs in models of tissue engineering (Schantz et al., 2007). In the BM, SDF-1α is vital to the hematopoietic supportive function that MSCs exert to maintain proper hematopoeisis (Van Overstraeten et al., 2006).

As mentioned earlier, preconditioning MSCs with cytokines such as IL-1β and TNFα increased the migratory capacity of the cells (Ponte et al., 2007; Schmal et al., 2007). However, these enhancing effects were shown to be independent of SDF-1α, and were instead mediated by other chemokines (Ponte et al., 2007).

In general, the clinical relevance of SDF-1α in the success of MSC therapies is positive, since SDF-1α gradients help MSCs home to sites of tissue injury (Bleul et al., 1996). In theory, MSCs could be administered systemically and allowed to respond to SDF-1α gradients for proper delivery to the target tissue, although this is less relevant for cosmetic approaches where target tissues are topical and superficial. SDF-1α is also a potent lymphocyte chemoattractant (Bleul et al., 1996). Excess production of SDF-1α within the microenvironment could potentially lead to increased immune cell infiltration and transplant rejection.

To counteract any negative influences of local microenvironmental SDF-1α on MSC therapies, there are several available pharmacologics that inhibit the SDF-1α/CXCR4 interaction. Plerixafor, also known as Mozobil® or AMD3100, is a partial antagonist of CXCR4 (Table 14.1) (Cashen et al., 2007). Current *in vivo* studies have shown that pre-treatment of MSCs with AMD3100 significantly prevented stem cell migration to the injured rat brain (Wang et al., 2007). Similar approaches may be beneficial in order to prevent non-specific MSC migration. T134, a small molecule CXCR4 inhibitor, and tannic acid, a water-soluble polyphenol widely distributed within the plant kingdom that acts as a selective CXCR4 antagonist, both inhibit the SDF-1α/CXCR4 interaction and may have future application as approved pharmacologics (Table 14.1) (Arakaki et al., 1999; Chen et al., 2003). Administration of these compounds would most likely have to be locally delivered to the target tissue, since systemic inhibition of SDF-1α/CXCR4 could disrupt BM homeostasis or lead to excessive HSC mobilization into the peripheral circulation. This local approach is most relevant for skin regeneration.

Summary

This chapter addresses the concept of the tissue microenvironment in cosmetic medicine, particularly in regenerative skin therapies, and examines its clinical importance in MSC and ADSC transplantation. By accurately developing *in vitro* models that mimic the local tissue milieu, the behavior of MSCs or their differentiated progeny can be observed prior to transplantation into patients. Through like approaches, ill effects from soluble mediators or other cell types, which can impede the desired therapeutic outcome, can be assessed. In many cases, co-therapy with a pharmacologic such as a cytokine receptor antagonist may negate the deleterious effects of the microenvironment and optimize the therapeutic potential of the stem cells. Here, we have focused on three ubiquitous microenvironmental factors with known effects on MSC function, and addressed how local delivery of inhibitors to these factors could improve the stem cell therapy.

References

[1] Arakaki R, Tamamura H, Premanathan M, et al. 1999. T134, a small-molecule CXCR4 inhibitor, has no cross-drug resistance with AMD3100, a CXCR4 antagonist with a different structure. *J Virol*, 73:1719–1723.

[2] Arend WP. 1993. Interleukin-1 receptor antagonist. *Adv Immunol*, 54:167–227.

[3] Barret JP, Sarobe N, Grande N, Vila D, Palacin JM. 2009. Maximizing results for lipofilling in facial reconstruction. *Clin Plast Sur*, 35:487–92.

[4] Bianco P, Riminucci M, Gronthos S, et al. 2001. Bone marrow stromal stem cells: nature, biology, and potential applications. *Stem Cells*, 19:180–192.

[5] Bleul CC, Fuhlbrigge RC, Casasnovas JM, et al. 1996. A highly efficacious lymphocyte chemoattractant, stromal cell-derived factor 1 (SDF-1). *J Exp Med*, 184:1101–1109.

[6] Bresnihan B, Cunnane G. 1998. Interleukin-1 receptor antagonist. *Rheum Dis Clin North Amer*, 24:615–628.

[7] Cashen A, Nervi B, DiPersio J. 2007. AMD3100: CXCR4 antagonist and rapid stem cell-mobilizing agent. *Future Oncol*, 3:19–27.

[8] Chen X, Beutler JA, McCloud TG, et al. 2003. Tannic acid is an inhibitor of CXCL12 (SDF-1alpha)/CXCR4 with antiangiogenic activity. *Clin Cancer Res*, 9:3115–3123.

[9] Cheng N, Janumyan YM, Didion L, et al. 2004. Bcl-2 inhibition of T-cell proliferation is related to prolonged T-cell survival. *Oncogene*, 23:3770–3780.

[10] Cho KJ, Trzaska KA, Greco SJ, et al. 2005. Neurons derived from human mesenchymal stem cells show synaptic transmission and can be induced to produce the neurotransmitter substance P by interleukin-1 alpha. *Stem Cells*, 23:383–391.

[11] Choi KS, Shin JS, Lee JJ, et al. 2005. In vitro trans-differentiation of rat mesenchymal cells into insulin-producing cells by rat pancreatic extract. *Biochem Biophys Res Commun*, 330:1299–1305.

[12] Choy EH, Panayi GS. 2001. Cytokine pathways and joint inflammation in rheumatoid arthritis. *N Engl J Med*, 344:907–916.

[13] Deans RJ, Moseley AB. 2000. Mesenchymal stem cells: biology and potential clinical uses. *Exp Hematol*, 28:875–884.

[14] Dinarello CA. 1994. The interleukin-1 family: 10 years of discovery. *FASEB J*, 8:1314–1325.

[15] Eberhardt M, Salmon P, von Mach MA, et al. 2006. Multipotential nestin and Isl-1 positive mesenchymal stem cells isolated from human pancreatic islets. *Biochem Biophys Res Commun*, 345:11670–1176.

[16] Gabay C. 2000. IL-1 inhibitors: novel agents in the treatment of rheumatoid arthritis. *Expert Opin Invest Drug*, 9:113–127.

[17] Grassel S, Ahmed N. 2007. Influence of cellular microenvironment and paracrine signals on chondrogenic differentiation. *Front Biosci*, 12:4946–4956.

[18] Greco SJ, Corcoran KE, Cho KJ, et al. 2004. Tachykinins in the emerging immune system: relevance to bone marrow homeostasis and maintenance of hematopoietic stem cells. *Front Biosci*, 9:1782–1793.

[19] Greco SJ, Liu K, Rameshwar P. 2007. Functional similarities among genes regulated by OCT4 in human mesenchymal and embryonic stem cells. *Stem Cells*, 25:3143–3154.

[20] Greco SJ, Rameshwar P. 2007. Enhancing effect of IL-1alpha on neurogenesis from adult human mesenchymal stem cells: implication for inflammatory mediators in regenerative medicine. *J Immunol*, 179:3342–3350.

[21] Greco SJ, Zhou C, Ye JH, et al. 2007. An interdisciplinary approach and characterization of neuronal cells transdifferentiated from human mesenchymal stem cells. *Stem Cells Dev*, 16:811–826.

[22] Hannum CH, Wilcox CJ, Arend WP, et al. 1990. Interleukin-1 receptor antagonist activity of a human interleukin-1 inhibitor. *Nature*, 343:336–340.

[23] Jeon SJ, Oshima K, Heller S, et al. 2007. Bone marrow mesenchymal stem cells are progenitors in vitro for inner ear hair cells. *Mol Cell Neurosci*, 34:59–68.

[24] Knight DM, Trinh H, Le J, et al. 1993. Construction and initial characterization of a mouse-human chimeric anti-TNF antibody. *Mol Immunol*, 30:1443–1453.

[25] Lerou PH, Daley GQ. 2005. Therapeutic potential of embryonic stem cells. *Blood Rev*, 19:321–331.

[26] Locksley RM, Killeen N, Lenardo MJ. 2001. The TNF and TNF receptor superfamilies: integrating mammalian biology. *Cell*, 104:487–501.

[27] Ong SY, Dai H, Leong KW. 2006. Inducing hepatic differentiation of human mesenchymal stem cells in pellet culture. *Biomaterials*, 27:4087–4097.

[28] Phinney DG, Prockop DJ. 2007. Concise review: mesenchymal stem/multipotent stromal cells: the state of transdifferentiation and modes of tissue repair–current views. *Stem Cells*, 25:2896–2902.

[29] Ponte AL, Marais E, Gallay N, et al. 2007. The in vitro migration capacity of human bone marrow mesenchymal stem cells: comparison of chemokine and growth factor chemotactic activities. *Stem Cells*, 25:1737–1745.

[30] Potian JA, Aviv H, Ponzio NM, et al. 2003. Veto-like activity of mesenchymal stem cells: functional discrimination between cellular responses to alloantigens and recall antigens. *J Immunol*, 171:3426–3434.

[31] Schantz JT, Chim H, Whiteman M. 2007. Cell guidance in tissue engineering: SDF-1 mediates site-directed homing of mesenchymal stem cells within three-dimensional polycaprolactone scaffolds. *Tissue Eng*, 13:2615–2624.

[32] Schmal H, Niemeyer P, Roesslein M, et al. 2007. Comparison of cellular functionality of human mesenchymal stromal cells and PBMC. *Cytotherapy*, 9:69–79.

[33] Trzaska KA, Kuzhikandathil EV, Rameshwar P. 2007. Specification of a dopaminergic phenotype from adult human mesenchymal stem cells. *Stem Cells*, 25:2797–2808.

[34] Van Overstraeten-Schlögel N, Beguin Y, Gothot A. 2006. Role of stromal-derived factor-1 in the hematopoietic-supporting activity of human mesenchymal stem cells. *Eur J Haematol*, 76:488–493.

[35] Wang Y, Deng Y, Zhou GQ. 2007. SDF-1alpha/CXCR4-mediated migration of systemically transplanted bone marrow stromal cells towards ischemic brain lesion in a rat model. *Brain Res*, in press.

[36] Yan J, Xu L, Welsh AM, et al. 2006. Combined Immunosuppressive Agents or CD4 Antibodies Prolong Survival of Human Neural Stem Cell Grafts and Improve Disease Outcomes in ALS Transgenic Mice. *Stem Cells*, 24:1976–1985.

[37] Yang JA, Chung HM, Won CH, Sung JH. 2010. Potential application of adipose-derived stem cells and their secretory factors to skin: discussion from both clinical and industrial viewpoints. *Expert Opin Biol Ther*, 10:495–503.

Resources

Glossary

Adenosine (A) One of the four nucleotides that make up the chemistry of DNA and RNA. Adenosine is a purine; the other purine base is guanosine (G). Adenosine always pairs with thymidine (T).

Adenosine triphosphate (ATP) The energy-rich currency used by cells to drive metabolic reactions.

Aging The process of cellular breakdown, resulting primarily from DNA damage.

Anabolism/anabolic A series of enzymatic reactions that build larger molecules from smaller ones.

Anti-aging Processes that delay, stop or retard normal aging at the cell, organ or general body level.

Antioxidants Any of a large class of substances that neutralize free radicals before they cause damage.

Apoptosis Programmed cell death, or a protective process of cellular self-destruction. Cells that have outlived their usefulness accept a signal to self-destruct, leaving younger, healthier cells to take up their functions.

Autoantibody Antibody that is formed in response to normal cellular constituents within the body. Autoantibodies are active in autoimmune disorders.

Autoimmune disorder Attack of the immune system against itself.

Base A nitrogen-containing molecule that, when combined with deoxyribose sugar and phosphate, is a building block of DNA called a nucleotide. Pyrimidine and purine are the nucleotides that make up the structure of DNA and RNA.

Base pair Two nucleotides, always a pyrimidine with a purine, that are paired and held together by a hydrogen bond.

Biological age A person's age according to one's metabolic function, as opposed to one's chronological age.

Carboxy alkyl esters (CAEs) The water-soluble active elements in cat's claw, *Uncaria tomentosa*. The term CAE is a general chemical name indicating that all of its members contain an acid functional group (carboxy) chemically bonded to an alcohol group.

Carcinogen An agent that causes cancer.

Catabolism/catabolic A series of enzyme-catalyzed reactions that break down complex molecules (fats, carbohydrates, proteins) into energy-rich smaller molecules such as adenosine triphosphate (ATP).

Cell cycle The orderly events in cellular reproduction. DNA replication is central to this activity. Any defects in DNA can result in production of daughter cells that cannot function properly.

Centromere A compact region contained in a chromosome where sister chromatids (the two exact copies of each chromosome) are formed after replication. This is the only place where DNA strands are joined together.

Chromosomes Tightly coiled strands of DNA and associated proteins that carry the genome.

Chronological age A person's age in calendar years.

Clone Cells produced by repeated division of a single common cell or organism that contains a single parent copy of DNA. This kind of reproduction is considered "asexual" because it does not involve two parents.

Cloning The process of generating exact copies of cells containing only one set of parent chromosomes; a form of asexual reproduction.

Codon A sequence of three nucleotides in a DNA or a messenger RNA molecule that represents instructions for incorporation of a specific amino acid in a protein strand that is or should be built.

Coenzyme A small nonprotein molecule that attaches to an enzyme made of protein and that is required for its activation. Niacin (niacinamide) in its coenzyme forms, NAD and NADP, are coenzymes for more than 200 enzymatic reactions.

C-reactive protein (CRP) A protein found in early stages of cardiovascular disease and some cancers.

Cytokine An extracellular signaling protein or peptide that acts as a local mediator in cell-to-cell communication. Cytokines are important signaling proteins in immune responses and nerve and brain cell proliferation.

Cytosine (C) A pyrimidine base that always pairs with guanosine (G) in DNA strands.

Deoxyribonucleic acid (DNA) A double-stranded helix that carries genetic code.

Disease genes Genes that encode for abnormal proteins that cause disease.

DNA adducts Chemical changes in the DNA molecule that have not been specified by the four bases. DNA adducts are usually removed by excision repair.

DNA repair capacity The ability of natural cellular processes to repair mistakes that occur in DNA.

DNA sequencing Determining the order of nucleotides in a DNA molecule. The primary goal of the Human Genome Project, DNA sequencing allows DNA to be studied in new ways. See **Proteomics.**

DNA transcription Copying of one strand of DNA onto a complementary copy of DNA or RNA. DNA transcription allows expression of the genome.

Electron donor An atom or molecule that gives an electron to another atom or molecule and is thereby oxidized. Fatty acids are one kind of molecule that donates electrons to oxygen. As a result, the fatty acid becomes oxidized and cannot function properly. "Reactions such as these lead to oxidative stress.

Electron transport The movement of electrons along a chain of carrier molecules, going from higher-energy molecules to lower-energy molecules and producing ATP in the process. During electron transport, some oxygen free radicals are created.

Entropy The degree of randomness in a system. Within the body, metabolism is normally well ordered. Oxidative stress induces entropy or disorder in metabolic processes.

Enzyme A protein that catalyzes a reaction between two or more atoms or molecules. Enzymes control all metabolic processes within the body.

Excision repair The most common kind of DNA repair. The damaged section of DNA is removed and replaced with the correct bases and sequence. The final step seals and strengthens the repaired area.

Free radical An atom or molecule with an unpaired electron; it lacks chemical stability and pulls electrons from stable molecules, setting off a chain reaction — rogue molecules that disrupt normal cell function by stealing electrons to gain molecular stability; produced mainly at the mitochondrial level or from by-products of the immune system reactions.

Free Radical Theory of Aging The idea that, over time, free radicals cause the progressive deterioration of biological systems that we call aging. This is the predominant theory of how we age.

Gene sequencing Mapping the specific order of nucleotide base pairs occurring in genes.

Genetic Control Theory of Aging The idea that how we age is encoded in our genetic material.

Genome The sequence of genes on DNA that spell out our individual identity.

Glycation The crosslinking of proteins at the cellular and genetic levels is caused by unregulated glucose levels, insulin surges and insulin receptor insensitivity. This directly affects gene expression and protein synthesis. Glycosylation of the immunoglobulin (Ig) molecule results in modified configuration of the immunoglobulins, which then have altered function and may contribute to autoimmune reactions.

Glycosylation The formation of linkages with glycosyl groups, as between D-glucose and the hemoglobin chain, to form the fraction hemoglobin A_{Ic}, whose level rises in association with the raised blood d-glucose concentration in poorly controlled or uncontrolled diabetes mellitus.

Guanine (G) A purine base that always pairs with cytosine (C) in DNA and uracil (U) in RNA.

Histones Specialized structural proteins that suppress certain segments of DNA so that they are not normally transcribed. Histones are the essential structures that regulate gene expression by acetylation and phosphorylation.

Inflammation Initiated by acute-phase proteins, this process creates an increase in key cytokines (TH_1, C-reactive protein, serum amyloid A, IL-1a, IL-1b, inside and outside the cell) and initiates aging changes in the vascular tissues, brain, joints and gastrointestinal tract and modifies gene expression and posttranslational protein formation.

Junk DNA Sections of DNA that do not contain genes.

Lipids Fatty oil-soluble substances. Biological membranes are made of lipids, a combination of phospholipids and long-chain polyunsaturated fatty acids.

Messenger ribonucleic acid (mRNA) An RNA molecule that specifies the amino acid sequence of a protein.

Metabolic disorders Disorders that involve a genetic error in sequencing an enzyme or enzymes involved in a specific metabolic pathway.

Methylation The addition of a CH_3 (a methyl group) controls the masking portion of specific regions of DNA and the unmasking of others, usually of the promoter regions. This alters how genetic messages are translated by turning the molecular genetic switch from "on to off" or from "off to on," thereby modifying genetic expression.

Molecular medicine A branch of medicine that focuses on abnormalities in specific metabolic pathways involved in a disease.

Necrosis A process in which cells die because they are injured or damaged in some way. Necrosis occurs from outside damage; apoptosis (programmed cell death) occurs because of internal cellular signals.

Neurohormone A hormone that stimulates the brain and nerves.

Neurotransmitter A small signaling molecule that relays chemical messages between the brain and nerve cells (neurons).

Nicotinamide Vitamin B_3 (niacin, nicotinic acid), a precursor of the coenzymes NAD and NADP.

Nicotine adenine dinucleotide (NAD) A coenzyme form of niacin (nicotinamide) that participates in oxidation-reduction (redox) reactions and produces energy.

Nicotine adenine dinucleotide phosphate (NADP) A niacin coenzyme that participates in biosynthetic pathways.

Nuclear factor kappa B (NF-κB) Key nuclear transcription factor involved in DNA repair at the intranuclear level (active form) and intracellular level (inactive form). A nuclear transcription factor that regulates numerous genes encoding proteins important in apoptosis, inflammation and cellular growth. Normally, NF-κB activates apoptosis to combat the free radical hazards of oxidative stress.

Nucleic acid The base pairs that make up RNA and DNA.

Nucleoside A base, either a purine or pyrimidine, that is chemically bonded to a sugar and phosphate.

Nucleotide Originally a combination of a (nucleic acid) purine or pyrimidine, one sugar (usually ribose or deoxyribose) and a phosphoric group; by extension, any compound containing a heterocyclic compound bound to a phosphorylated sugar by an N-glycosyl link (e.g., adenosine monophosphate, NAD^+).

Nutraceuticals Substances derived from food or herbs that have pronounced pharmacological and medicinal effects.

Oxidation *(specific definition)* Loss of an electron by an atom or molecule, typically through the process of oxygen reduction (redox) reactions.

Oxidation *(generalized definition)* The amount of free radical damage produced at the intracellular and extracellular levels. Oxidation directly affects genetic structure and function and cell membrane and organelle function.

Oxidative stress The total body effect of molecules that have been damaged by oxygen free radicals.

Phospholipids Lipid molecules used to construct biological membranes that act as biological soaps because they have water-soluble and lipid-soluble ends.

Polyunsaturated fatty acids (PUFAs) Essential fatty acids that are elongated and desaturated (i.e., double bonds between carbon atoms) to form the long-chain PUFAs (poly-unsaturated fatty acids) found in biomembranes.

Protein The major macromolecular constituent of cells, of which enzymes are one kind.

Proteomics The study of how proteins carry out the genetic orders of DNA. This is the fastest-growing area to be generated from the Human Genome Project. The protein complement generated from DNA is vastly more complicated than unraveling the genome because there are numerous proteins, and defining what each does and how it works is challenging to the new frontier of medical research.

Purine The larger of the two bases that make up DNA and RNA. Adenosine and guanine are purines. They bond only with pyrimidines.

Pyrimidine The smaller of the two bases that make up DNA and RNA. Cytosine and thymidine are pyrimidines. A pyrimidine and a purine bond to form a base pair.

Recombinant DNA (rDNA) Any DNA molecule that is formed by joining a DNA segment from different sources. rDNA is widely used in cloning of genes and genetic modification organisms (GMOs).

Reduction Adding an electron, the opposite of oxidation; occurs in oxidation reduction (redox) reactions. Can be the loss of an oxygen molecule or the gain of a hydrogen atom.

Replication Reproductive and exact duplication of DNA.

Ribonucleic acid (RNA) Nucleic acid material that encodes for proteins. RNA receives its instructions by pairing with a single strand of DNA. Then it carries the genetic code into the ribosome, where it can be translated into synthesis of new proteins.

Ribosome A cellular particle where RNA synthesizes proteins.

Saturated fatty acids Lipids with no double bonds in the carbon backbone. Saturated fats are solid at room temperature. Most animal fats are saturated.

Signal transduction The initiation and carrying of messages between cells.

Stem cell research The study of stem cell reproduction and its applications.

Stem cells Cells that carry the genetic code but have not differentiated into specialized cells that perform a particular function.

T cell A type of lymphocyte responsible for cell-mediated immunity. Subsets include helper cells, suppressor cells, cytotoxic cells, natural killer cells and memory T cells. Also known as CD3, CD4 (helper), and CD8 (cytotoxic cells) because of the receptor proteins on their surfaces.

Thymidine (T) One of the pyrimidine bases found in DNA and RNA. Thymidine always bonds with the purine base adenosine (A).

Transcription factors Proteins that initiate transfer of genetic information.

Tumor necrosis factor alpha (TNF-α) A protein produced by white blood cells that initiates killing of tumor cells. TNF-α also promotes inflammation, activates NF-κB and increases oxidative stress.

Uracil (U) A pyrimidine base that replaces cytosine in RNA.

Compounding Pharmacies/Anti-Aging Formularies

ApotheCure, Inc.
13720 Midway Rd., Suite 109
Dallas, TX 75244
(214) 960-6601
Fax: (800) 687-5252
www.apothecure.com

California Pharmacy and
Compounding Center, Inc.
307 Placentia Ave., #0102
Newport Beach, CA 92663
(800) 575-7776
Fax: (949) 642-0725

College Pharmacy
3505 Austin Bluffs Parkway, Suite 101
Colorado Springs, CO 80918
(800) 888-9358
Fax: (800) 556-5893
www.collegepharmacy.com

Homelink Pharmacy
2650 Elm Ave., Suite 104
Long Beach, CA 90806
(310) 988-0260
(800) 272-4767
www.kahealani.com

Hopewell Pharmacy
1 West Broad Street
Hopewell, NJ 08525
(800) 792-6670
(609) 466-1960
Fax: (800) 417-3864
Fax: (609) 466-8222
www.hopewellrx.com

International Anti-Aging Systems
P.O. Box 2995
London, England N10 2 NA
44-181-444-8272
www.antiagingsystems.com

Medical Center Pharmacy
10721 Main Street
Fairfax, VA 22030
(703) 273-7311

Wellness Health & Pharmaceuticals,
Inc.
2800 South 18th Street
Birmingham, AL 35209
(800) 227-2627
Fax: (800) 369-0302
www.wellnesshealth.com

Web Sites

General Anti-Aging Information Sources

www.antiaginginfosite.com
www.anti-aginginfosource.com
www.worldhealth.net

Information on Each Human Gene Mapped or Sequenced

www.bioinformatics.weizmann.ac.il/cards

Anti-Aging Product Sources

Neutraceuticals, Hormones (C-MED 100)

www.antiaginginfosite.com
www.lef.org
www.optigene-x.com
www.worldhealth.net

Skin Products

www.giampapainstitute.com
(Physician-
and pharmacy-compounded skin
care products for optimal effects)
www.nvperriconemd.com
(A comprehensive skin care line for
anti-aging for the face and body)

Anti-Aging Programs and Products

www.cenegenics.com
www.giampapainstitute.com (A full array of anti-aging programs from basic programs with home testing to the most comprehensive testing available. All programs focus on DNA therapy, hormonal replacement and brain health and hormone growth replacement therapy.)

Isologen Technologies
2500 Wilcrest
Houston, TX 77042
(713) 780-4754
www.cosmeticskin.com/dr_boss.shtml

Anti-Aging Surgery

www.youthfulneck.com (Web site of Plastic Surgery Center International for the most advanced anti-aging surgical "procedures available anywhere. Surgical procedures are used intimately with anti-aging therapies to maximize "long-term results and initial healing.)

Anti-Aging Laboratories

Aeron Lifecycles Laboratory
1933 Davis Street, Suite 310
San Leandro, CA 94577
(800) 631-7900
(510) 729-0375
Fax: (510) 729-0383

American Medical Testing Laboratories (AMTL)
One Oakwood Blvd., Suite 130
Hollywood, FL 33020
(954) 923-2990
(800) 881-AMTL
Fax: (954) 923-2707
www.alcat.com

Douglas Laboratories
600 Boyce Road
Pittsburgh, PA 15205
(800) 245-4440
(412) 494-0122
Fax: (412) 494-0155
www.douglaslabs.com

Genox Laboratories, Inc.
1414 Key Highway
Baltimore, MD 21230
(410) 347-7616
(800) 810-5450
Fax: (410) 347-7617

Great Smokies Diagnostic Laboratories
63 Zillicoa Street
Asheville, NC 28801
(800) 522-4762
Fax: (828) 252-9303
www.gsdl.com

Kronos Science Laboratory
2222 East Highland, Suite 220
Phoenix, AZ 85016-4872
(877) 576-6675

Fax: (602) 667-5623
www.kronoscentre.com

Immunosciences Labs, Inc.
8730 Wilshire Blvd., Suite 305
Beverly Hills, CA 90211
(800) 950-4686
www.shasta.com

Optigene-X Lab
Biomarkers of Aging Panels
151 New Monmouth Road
Middletown, NJ 07748
(732) 275-1175
(866) OPTIGNX
Fax: (732) 275-1727
www.optigene-x.com

Quest Diagnostic Labs
One Malcolm Avenue
Teterboro, NJ 07608-1070
(201) 393-5000
(800) 222-0446
www.questdiagnostics.com

Anti-Aging Diagnostic Equipment

Biological Technologies International, Inc.
P.O. Box 560
Payson, AZ 85547
(928) 474-4181
Fax: (928) 474-1501
www.bioterrain.com

Bioanalogics, Inc.
7909 S.W. Cirrus Drive, Bldg. 27
Beaverton, OR 97008
(800) 327-7953
www.bioanalogics.com

H-Scan
Richard Hochchild
Hoch Company
2915 Pebble Drive
Corona Del Mar, CA 92625
(949) 759-8066
http://informagen.com

Anti-Aging Journals

Anti-Aging Bulletin
IAS Ltd.
Les Autelets
Sark GY9 OSF
Channel Islands, Great Britain

Anti-Aging Medical News
2415 N. Greenview
Chicago, IL 60614

Journal of Anti-Aging Medicine
Mary Ann Liebert, Inc., Publishers
2 Madison Avenue
Larchmont, NY 10538
(914) 834-3100
Fax: (914) 834-3689
www.liebertpub.com

Life Extension Magazine
P.O. Box 229120
Hollywood, FL 33022-9120
(800) 841-5433
www.lef.org/newshop/cgi-shop/subscribe.cgi

Anti-Aging Centers

Cenegenics Medical Institute
851 Rampart Boulevard
Sir Williams Court Complex, Suite 100
Las Vegas, NV 89145
(888) YOUNGER

www.888younger.com
www.cenegenics.com

Giampapa Institute
89 Valley Road
Montclair, NJ 07042
(973) 783-6868
Fax: (973) 746-3777
www.giampapainstitute.com

Anti-Aging and Cancer Treatment Centers

Burzynski Research Institute, Inc.
Stanley R. Burzynski, M.D., Ph.D.
9432 Old Katy Road
Houston, TX 77055
(713) 335-5697
Fax: (713) 335-5699
www.volmed.com

Educational Seminars

The American College for Advancement in Medicine
23121 Verdugo Drive
Laguna Hills, CA 92653
Fax: (949) 455-9679
www.acam.org

American Academy of Anti-Aging Medicine A^4M
2415 N. Greenview
Chicago, IL 60614
(773) 528-4333
Fax: (773) 528-5390
www.worldhealthnet/resources

Giampapa Institute
89 Valley Road
Montclair, NJ 07042
(973) 783-6868
Fax: (973) 746-3777
www.giampapainstitute.com

Clinical Creations, LLC
Nicholas V. Perricone, M.D.
377 Research Parkway
Meriden, CT 06450
(888) 823-7837
Fax: (203) 379-0817
www.clinicalcreations.com
www.nvperriconemd.com

Mind Body Centers

Monroe Institute
62 Roberts Mountain Road
Faber, VA 22938
(434) 361-1252
Fax: (434) 361-1237
www.monroeinstitute.org/voyagers/voyages

Sample Core Age Management Program

For the plastic surgeon, a core age management program can be organized according to the format shown here. It has been shown to be highly effective, as well as concise and easy to understand from both the patients' point of view and the medical staffs' as well. This format has been utilized at the Giampapa Institute for the last four years with a high degree of success. Alterations or changes in the organizational content can be made on a practice-by-practice and doctor-by-doctor basis.

Age Management Program
Lab Results

Patient Name: ______________________ Sex: __M__ Age: __54__

Date Sample Received: __09/10/01__ Date Testing Completed: __09/21/01__

Patient DNA Damage Levels*

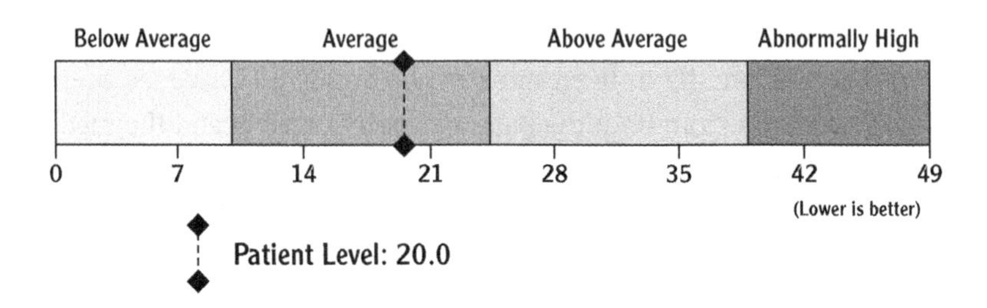

*DNA Damage Levels (reflected as 8-OHdG values). Scientific research has shown that preserving and repairing DNA contributes to optimal health and longevity.

Patient Free Radical Levels†

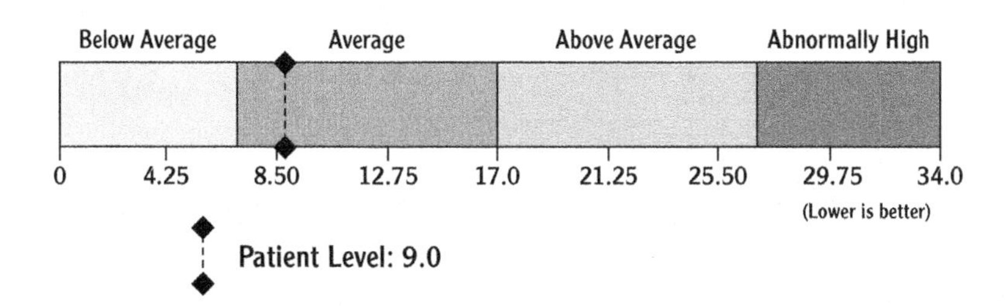

†Elevated Free Radical Levels (reflected as 8-epi-$PGF_{2\alpha}$ values) are directly related to increased damage to human cell membranes and DNA which are the key factors responsible for the aging process we experience as we grow older.

Age Management Program
Lab Results

Patient Name: ____________________ Sex: _M_ Age: _54_
Date Sample Received: _10/11/01_ Date Testing Completed: _10/18/01_

Patient DNA Damage Levels*

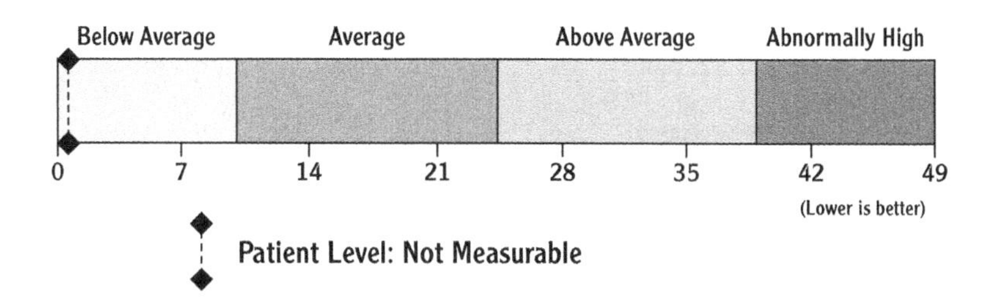

*DNA Damage Levels (reflected as 8-OHdG values). Scientific research has shown that preserving and repairing DNA contributes to optimal health and longevity.

Patient Free Radical Levels†

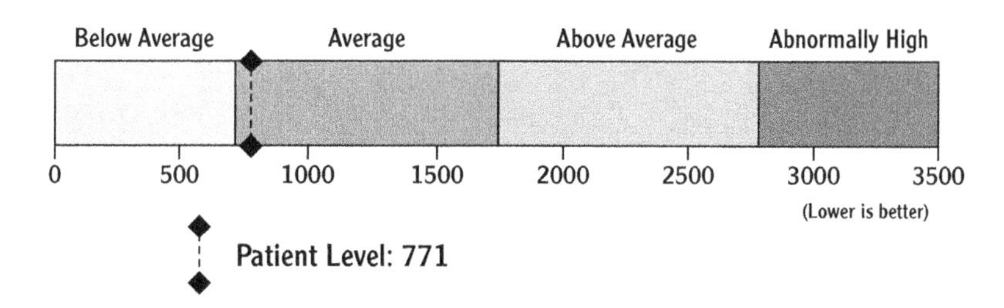

†Elevated Free Radical Levels (reflected as 8-epi-$PGF_{2\alpha}$ values) are directly related to increased damage to human cell membranes and DNA which are the key factors responsible for the aging process we experience as we grow older.

Subjective Questionnaire: Part I

Patient Name:________________________ Date: ____________

Month (Circle): Month 1 Month 4 Month 8 Month 12

Please rate each item by placing a circle around the appropriate number.

			Negative Change			Start of Program			Positive Change		
Mental Functions											
Energy Levels Daily (daily NRG Levels)	−5	−4	−3	−2	−1	0	1	2	3	4	5
(exercise tolerance)	−5	−4	−3	−2	−1	0	1	2	3	4	5
Increased Mental Alertness											
Sense of Well Being (increased/decreased)	−5	−4	−3	−2	−1	0	1	2	3	4	5
Anxious	−5	−4	−3	−2	−1	0	1	2	3	4	5
Depressed	−5	−4	−3	−2	−1	0	1	2	3	4	5
Memory (recent)	−5	−4	−3	−2	−1	0	1	2	3	4	5
(old)	−5	−4	−3	−2	−1	0	1	2	3	4	5
Mentally Focused (mental clarity)	−5	−4	−3	−2	−1	0	1	2	3	4	5
Improved Recall (remembering names …etc.)	−5	−4	−3	−2	−1	0	1	2	3	4	5
Overall Stress Level (more/less relaxed)	−5	−4	−3	−2	−1	0	1	2	3	4	5
Integument											
Age Spots on Hands and Face	−5	−4	−3	−2	−1	0	1	2	3	4	5
Skin Thicker	−5	−4	−3	−2	−1	0	1	2	3	4	5
Body Skin More Supple (less dry)	−5	−4	−3	−2	−1	0	1	2	3	4	5
Face Skin (less lined & improved tightness)	−5	−4	−3	−2	−1	0	1	2	3	4	5
Nails Stronger or Weaker	−5	−4	−3	−2	−1	0	1	2	3	4	5
Nails Growing Faster or Slower	−5	−4	−3	−2	−1	0	1	2	3	4	5
Hair Growing Faster or Slower	−5	−4	−3	−2	−1	0	1	2	3	4	5
New Hair in Bald Spots	−5	−4	−3	−2	−1	0	1	2	3	4	5
Ruptured Blood Vessels (black and blue spots)	−5	−4	−3	−2	−1	0	1	2	3	4	5
Body Hair Loss or Gain	−5	−4	−3	−2	−1	0	1	2	3	4	5
Immune System											
Allergic Reaction & Symptoms (itching, swollen eyes, etc.)	−5	−4	−3	−2	−1	0	1	2	3	4	5
Healing (gums, cuts, etc.—faster or slower)	−5	−4	−3	−2	−1	0	1	2	3	4	5
Colds & Flu	−5	−4	−3	−2	−1	0	1	2	3	4	5

Subjective Questionnaire: Part II

Measurements

Weight _________lbs.

Chest _________in.

Waist _________in.

	Negative Change					Start of Program	Positive Change				
Digestive System											
Digestive Problems (gas, accident)	−5	−4	−3	−2	−1	0	1	2	3	4	5
Regular Bowel Movements	−5	−4	−3	−2	−1	0	1	2	3	4	5
Bulkier Stools (sink in water/float in water)	−5	−4	−3	−2	−1	0	1	2	3	4	5
Skeletal System											
Bone and Joint Pain	−5	−4	−3	−2	−1	0	1	2	3	4	5
Flexibility	−5	−4	−3	−2	−1	0	1	2	3	4	5
Joint Swelling & Redness	−5	−4	−3	−2	−1	0	1	2	3	4	5
Urinary System											
Continentcy (loss of urine control)	−5	−4	−3	−2	−1	0	1	2	3	4	5
Frequent	−5	−4	−3	−2	−1	0	1	2	3	4	5
Getting Up at Night	−5	−4	−3	−2	−1	0	1	2	3	4	5
Urgency	−5	−4	−3	−2	−1	0	1	2	3	4	5
Body Composition											
Weight Gain	−5	−4	−3	−2	−1	0	1	2	3	4	5
Weight Loss	−5	−4	−3	−2	−1	0	1	2	3	4	5
Stronger Abdominal Muscles	−5	−4	−3	−2	−1	0	1	2	3	4	5
Firmer Thigh Muscles	−5	−4	−3	−2	−1	0	1	2	3	4	5
Youthful Body Contour	−5	−4	−3	−2	−1	0	1	2	3	4	5
Love Handles & Bells Disappearing	−5	−4	−3	−2	−1	0	1	2	3	4	5
More Endurance When Exercising	−5	−4	−3	−2	−1	0	1	2	3	4	5
More Strength When Exercising	−5	−4	−3	−2	−1	0	1	2	3	4	5
Under Upper Arms	−5	−4	−3	−2	−1	0	1	2	3	4	5
Abdomen	−5	−4	−3	−2	−1	0	1	2	3	4	5
Thighs	−5	−4	−3	−2	−1	0	1	2	3	4	5
Sag Under Chin	−5	−4	−3	−2	−1	0	1	2	3	4	5
Facial Skin	−5	−4	−3	−2	−1	0	1	2	3	4	5

Subjective Questionnaire: Part III

	Negative Change					Start of Program	Positive Change				
Sexual Status											
Libido Level	−5	−4	−3	−2	−1	0	1	2	3	4	5
Firmer Erection (men)	−5	−4	−3	−2	−1	0	1	2	3	4	5
Frequency of Sexual Relations	−5	−4	−3	−2	−1	0	1	2	3	4	5
Increased Sensitivity	−5	−4	−3	−2	−1	0	1	2	3	4	5
Menopausal Symptoms (hot/cold flashes, headaches, etc.)	−5	−4	−3	−2	−1	0	1	2	3	4	5
Period: Regular	−5	−4	−3	−2	−1	0	1	2	3	4	5
Irregular	−5	−4	−3	−2	−1	0	1	2	3	4	5
Sleep Function											
Fall Asleep Faster	−5	−4	−3	−2	−1	0	1	2	3	4	5
Sleep Interruption	−5	−4	−3	−2	−1	0	1	2	3	4	5
Sleep Throughout Night	−5	−4	−3	−2	−1	0	1	2	3	4	5
More Dreaming Sensation	−5	−4	−3	−2	−1	0	1	2	3	4	5
Dietary Habits											
Sugar or Sweet Cravings	−5	−4	−3	−2	−1	0	1	2	3	4	5
Appetite (more hungry)	−5	−4	−3	−2	−1	0	1	2	3	4	5
(less hungry)	−5	−4	−3	−2	−1	0	1	2	3	4	5
Frequency of Meal Habits (less or more)	−5	−4	−3	−2	−1	0	1	2	3	4	5

Specific Food Cravings:
Name Them: ____________________

Thirst											
Drink More Water or Less	−5	−4	−3	−2	−1	0	1	2	3	4	5
Crave Salt More or Less	−5	−4	−3	−2	−1	0	1	2	3	4	5
Body Temperature (body feels)											
More Cold	−5	−4	−3	−2	−1	0	1	2	3	4	5
More Hot	−5	−4	−3	−2	−1	0	1	2	3	4	5
Flashes Hot or Cold	−5	−4	−3	−2	−1	0	1	2	3	4	5

Note Other Comments or Concerns ____________________

____

____

____

____

Suggested Therapy Schedule for: ________________ Date: __________

Nutraceuticals	Hormonal Precursors	Prescriptions	Hormonal Replacement (HRT)
Upon Arising			
Breakfast			
Mid-Morning			
Lunch			
Mid-Afternoon			
Dinner			
Before Bedtime			

Giampapa Institute™
Personal Exercise Recommendations
for: ___________________

The enclosed stretching exercises are designed to keep all major muscle groups and joints flexible and mobile. They are also combined with deep breathing to help improve pulmonary function and at the same time to begin to quiet and center the mind.

This simple 12-minute procedure performed daily will markedly improve overall body function and structure and introduce you to the need and importance for the establishment of routines or regimens in your life. These exercises can be done first thing in the morning before your day begins or in the evening before retiring. Many patients find that doing this twice a day helps them to begin their day in a quiet, centered and focused manner and to end their day feeling stress free and thus getting a better quality of sleep.

The illustrations that follow will help you visualize this simple sequence of exercises. Please note that you should not be fooled or led to believe that these simple exercises will not have a major benefit in your overall sense of well-being and body function. They are based on a series of Tibetan yoga procedures that have endured for over 4,000 years and are actively utilized even today.

The normal wear and tear of aging can cause structural disorders and a lack of flexibility in joints and muscles. These biomarkers of the aging process respond to exercise, stretching and relaxation techniques — all things you can control.

There are five anti-aging breathing and stretching exercises that activate the powerful anti-aging centers of the body. These exercises improve Level 1 biomarkers (increased flexibility in joints, improved pulmonary function, etc.) and oxygenate body tissues while allowing them to get rid of excessive acidity (pH-balancing effects).

In addition, these exercises relax the nervous system, leaving us in a calm mental state. Every position sends a molecular message to the core centers of our cellular environment and excites different molecular functions of our cellular soup. These different positions activate acupuncture points and vital energy centers of the body and radically diminish the chance that you'll be grimacing with pain with advancing age.

Be sure to take a series of deep breaths before and after each positional exercise.

*To calculate Maximum Heart Rate (MHR) subtract your age from 220; that is: 220–50 years old = 170 heartbeats per minute (MHR)

Stretching	**Cardiovascular training 55–65% of MHR**	**Resistance exercise**	**Stretching & deep breathing**
See instructions below. *Duration 5 to 10 min.*	☐ Rapid walking ☐ Rapid walking w/hand weights ☐ Swimming ☐ Biking *Duration 20 min.*	☐ Calisthenics ☐ Light weight training ☐ Pilates ☐ Yoga *Duration 40 min.*	See instructions below. *Duration 5 to 10 min.*

Stretching Exercise Postures & Breathing

(2 deep, slow breaths between each exercise; 10 repetitions each)

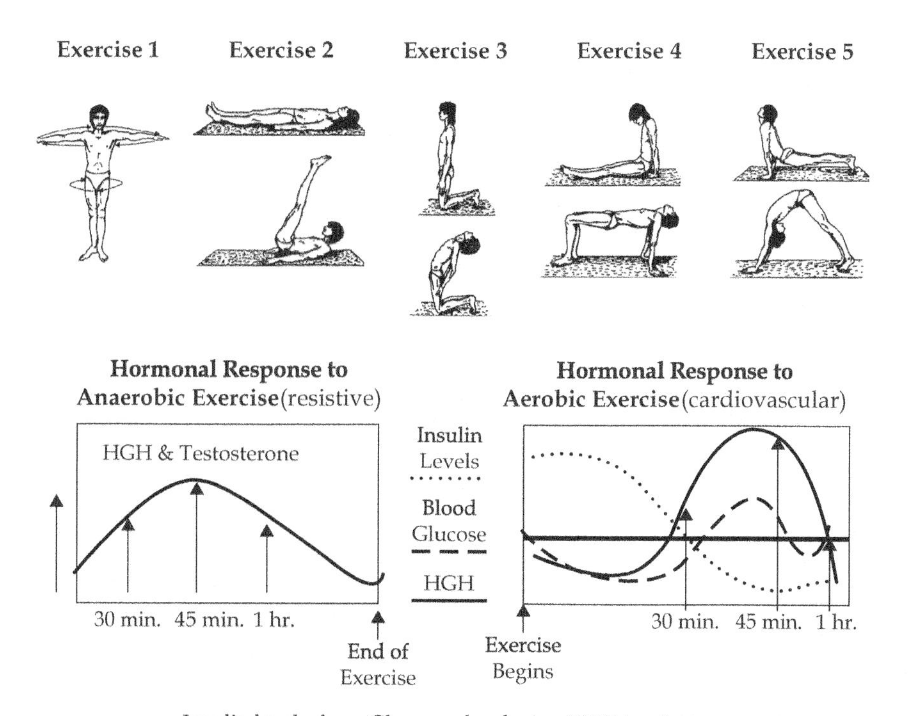

Insulin levels drop. Glucagon levels rise. HGH levels rise.

Skin-Fold Measurements

Patient Name:____________________

Women: X1 = hip bone skin-fold thickness (mm) and X2 = back of arm skin-fold thickness (mm).

Men: X1 = anterior thigh skin-fold thickness (mm) and X2 = back of arm skin-fold thickness (mm).

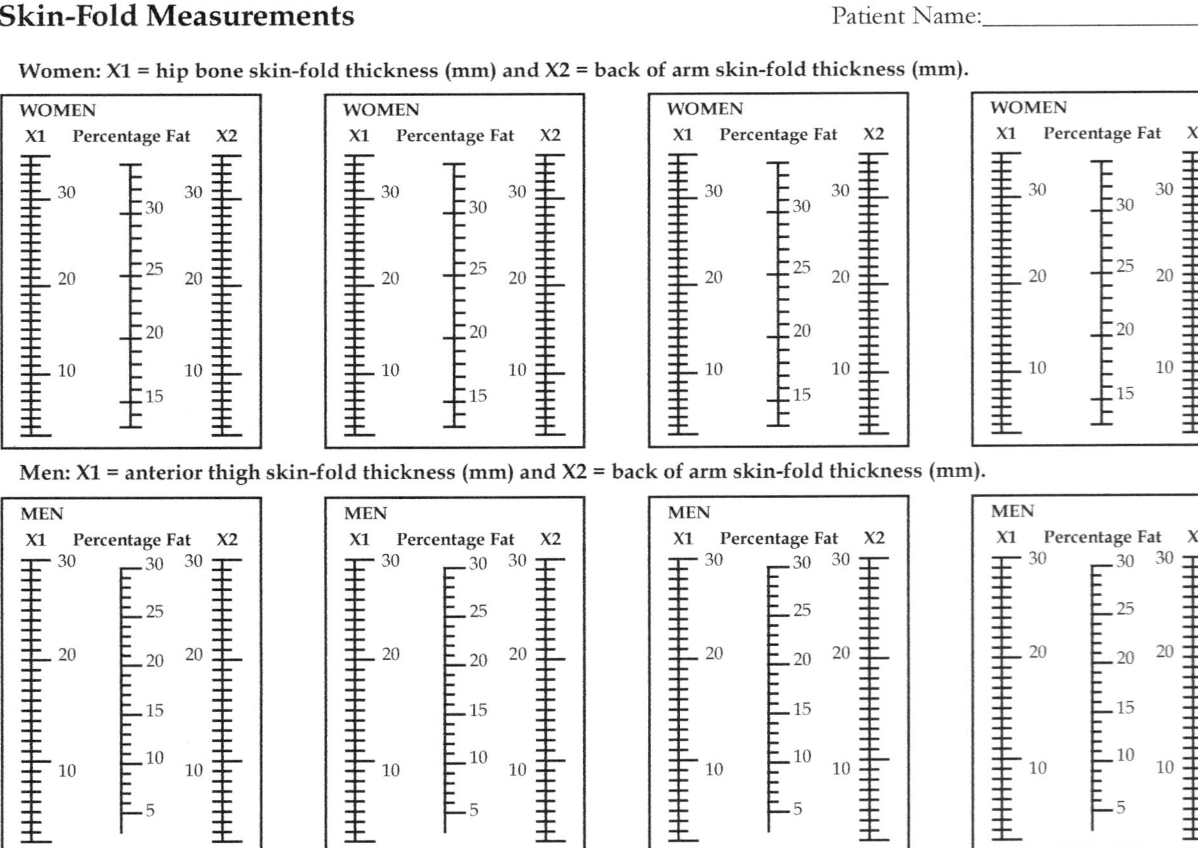

Body Mass Index (BMI) Patient Name:__________________

A BMI of 27 or over means your weight increases your chance of developing health problems like diabetes, heart disease and certain cancers. Find your height on the chart, then look across that row and find the weight nearest your own. The number at the top of the column is your BMI.

BMI	19	20	21	22	23	24	25	26	27	28	29	30	35	40
											RISK ZONE			
Height							**Weight in Pounds**							
5′	97	102	107	112	118	123	128	133	138	143	148	153	179	204
5′1″	100	106	111	116	122	127	132	137	143	148	153	158	185	211
5′2″	104	109	115	120	126	131	136	142	147	153	158	164	191	218
5′3″	107	113	118	124	130	135	141	146	152	158	164	169	197	225
5′4″	110	116	122	128	134	140	145	151	157	163	169	174	204	232
5′5″	114	120	126	132	138	144	150	156	162	168	174	180	210	240
5′6″	118	124	130	136	142	148	155	161	167	173	179	186	216	247
5′7″	121	127	134	140	146	153	159	166	172	178	185	191	223	255
5′8″	125	131	138	144	151	158	164	171	177	184	190	197	230	262
5′9″	128	135	142	149	155	162	169	176	182	189	196	203	236	270
5′10″	132	139	146	153	160	167	174	181	188	195	202	209	243	278
5′11″	136	143	150	157	165	172	179	186	193	200	208	215	250	286
6′	140	147	154	162	169	177	184	191	199	206	213	221	258	294

Giampapa Institute of Anti-Aging
Patient Biomarkers Home-Testing Guidebook

Here are six tools to use at home to determine your own aging rate. Chart each result on the Home Biomarker Log and Measurement Graph provided and monitor your results. Remember these are just tools or a gauge to determine your progress. A more accurate determination of your aging rate or profile is available from your anti-aging physician or through the GIAMPAPA INSTITUTE.

1. *Skin Elasticity*:
   Loss of skin starts to be significant around age 45 and is the direct result of the underlying deterioration of the connective tissue, such as collagen and elastin, under the skin surface. This loss of skin tone and turgor contributes to the wrinkling and loose skin around the jowls and neck.
2. *Falling Ruler Reaction*:
   This tool is a measurement of your reaction time, which falls off sharply with aging.
3. *Static Balance*:
   This determines how long you can stand on one leg with your eyes closed before falling over.
4. *Visual Accommodation*:
   With age, the lens of the eye becomes progressively less elastic, resulting in presbyopia, or nearsightedness. While this is not an accurate test of visual accommodation, it will give you some idea of the effect of age on your vision.
5. *Body Composition*:
   The use of this measurement provides you with a fairly good idea of your fat-to-lean mass ratio or percent of body fat.
6. *Body Mass Index*:
   Using the chart provided, you can follow changes in your body composition.

See the following explanations of each tool. Remember to chart your results on the Home Biomarker Log and Measurement Graph and monitor your changes.

To complete your self-evaluation and progress, fill out the three-part subjective questionnaire and repeat at four-month intervals.

1. *Skin Elasticity*:
   Pinch the skin on the back of your hand between the thumb and forefinger for five seconds; then time how long it takes to flatten out completely. Record.

0–1 seconds = 20–30 years of age
2–5 seconds = 40–50 years of age
10–55 seconds = 60–70 years of age

2. *Falling Ruler Reaction*:
   Buy an 18″ wooden ruler. Have someone suspend the ruler by holding it at the top (large number down) between your fingers. The thumb and forefinger of your dominant hand should be 31/2" apart, equidistant from the 18" mark on the ruler. The person lets go of the ruler without warning and you must catch the ruler as quickly as possible between your thumb and forefinger. Do this 3 times and average the results. Record.

   12 inches = 20–30 years of age
   8 inches = 40–50 years of age
   5 inches = 60–70 years of age

3. *Static Balance*:
   Stand on a hard surface (not a rug) with both feet together. You should be barefoot or wearing an ordinary low-heeled shoe. Have a friend close by to catch you if you fall over. Close your eyes and lift your foot (left foot if right-handed, right foot if left-handed) about 6" off the ground, bending your knee about 45 degrees. Stand on your other foot without moving or jiggling it. Have someone time how long you can do this without either opening your eyes or moving your foot to avoid falling over. Do this 3 times and average the results. Record.

   28 seconds = 20–30 years of age
   18 seconds = 40–50 years of age
   4 seconds = 60–70 years of age

4. *Visual Accommodation*:
   Hold the newspaper out in front of you. Slowly bring the newspaper to your eyes until the regular size letters start to blur. Have someone measure the distance between your eyes and the newspaper. Record.

   4.5 inches = 20–30 years of age
   12 inches = 40–50 years of age
   39 inches = 50–60 years of age

5. *Body Composition:*
   Obtain a set of skin-fold calipers from your local health shop or fitness center. Calipers are simple to use once you get the hang of it. You make two skin-fold measurements. Women use the skin fold on the top of the hip bone and on the back of the arm. Men measure the skin fold on the front of the thigh and the back of the arm. Do each measurement 3 times.

(They should not vary more than 1%. If they do, try to be more careful). Try to do the measurements in the same place each time. Record.

6. *Body Mass Index*:

   Measure your height and weight yourself using the same equipment each time. Use the BMI chart provided. Record.

Home Biomarker Log and Measurement Graph

Patient Name__________________

	(Start) Month 1				Month 4				Month 8				Month 12		
Age	20–39	40–59	60–70		20–39	40–59	60–70		20–39	40–59	60–70		20–39	40–59	60–70
Skin 0–1 2–5 sec. 10–55 sec.															
Ruler 12 in. 8 in. 5 in.															
Balance 28 sec. 18 sec. 4 sec.															
Visual 4.5 in. 12 in. 39 in.															
Body Composition Fat %															
Body Mass Index Index number															

Diet and Body Composition Plan

GIAMPAPA INSTITUTE Diet and Body Composition Plan should be a part of every anti-aging diet.

The GIAMPAPA INSTITUTE Diet and Body Composition Plan Is Individualized for You

The meals that the GIAMPAPA INSTITUTE Diet and Body Composition Plan recommends for you take your weight, activity level, sex and body composition into account.

Body Weight: Body weight determines grams of protein and hence grams of carbohydrates, grams of fat and calories per day.

Activity Level: More active individuals will obviously have greater nutritional needs.

Male and Female Body Composition: Males tend to have less body fat and more muscle mass so they burn calories more rapidly. In general, females tend to require less grams of carbohydrates/protein/fat to meet their nutritional needs.

Sample Meal Planner Conversion Chart

Pick a category (A, B, C or D) and follow it in Diet 1, Diet 2 or Diet 3*

WOMEN

ACTIVITY LEVEL (hours per week of exercise)	LOW – MEDIUM (0–4 hours) Use Meal Planner	MEDIUM – HIGH (5–10 hours) Use Meal Planner
Weight Under 140 lbs.	A	B
141–180 lbs.	B	C
181–200+ lbs.	C	D

MEN

ACTIVITY LEVEL (hours per week of exercise)	LOW – MEDIUM (0–4 hours) Use Meal Planner	MEDIUM – HIGH (5–10 hours) Use Meal Planner
Weight Under 140 lbs.	B	C
141–180 lbs.	C	D
181–250+ lbs.	C	D

Note: If you are a competitive athlete or are very active, choose the next higher meal planner category (e.g., C instead of B) or add one meal per day.

We have provided you with plans for delicious and easy-to-prepare meals that contain the correct amounts of carbohydrates, protein and fat you need at each meal.

YOUR MEAL PLANNER IS:________

*Only Diet 1 is included in this sample program.

Sample Diet 1
Hormonal Maintenance

Regular Meals: Day 1

	Amounts for Meal Planners				Recommended Giampapa
BREAKFAST + Optigene Professional Packet (A.M.)*	**A**	**B**	**C**	**D**	______
Fruit Protein Shake					
Protein Powder (1 scoop = 16 g)	1	1	2	$2^{1/4}$	______
Banana, medium, frozen	1/2	1/2	2/3	1	______
Strawberries, frozen	1/2 c.	1/2 c.	1 c.	1 c.	______
Water	3/4 c.	3/4 c.	1 c.	$1^{1/4}$ c.	______
Almonds, raw	2 t.	2 t.	1 T.	2 T.	______
Honey	0	0	1 tsp.	2 tsp.	______

Directions: *Combine all ingredients in a high-speed blender and blend till smooth.*

LUNCH + Optigene Professional Pack (Mid-Day)*	**A**	**B**	**C**	**D**	
Turkey Sandwich					
Wheat Bread, whole grain	1 slice	2 slices	2 slices	3 slices	______
Turkey Breast, sliced	$1^{1/2}$ oz.	3 oz.	3 oz.	4 oz.	______
Lettuce Leaf	1	1	1	1	______
Tomato Slice	1	1	1	2	______
Mayonnaise, reduced fat	1 T.	$1^{1/2}$ T.	$1^{1/2}$ T.	2 T.	______
Fruit	1 plum	1 plum	1 plum	1 apple	______

Directions: *Prepare turkey sandwich and serve with fruit and beverage.*

SNACK	**A**	**B**	**C**	**D**	
Turkey Bagel					
Mini Bagel	1	1	1	1	______
Turkey Breast, sliced	$1^{1/2}$ oz.	$1^{1/2}$ oz.	$1^{1/2}$ oz.	$1^{1/2}$ oz.	______
Cream Cheese, whipped, low-fat	1/2 oz.	1/2 oz.	1/2 oz.	1/2 oz.	______

Directions: *Spread cream cheese on bagel halves and top with turkey.*

or 1 packet Whey Tocotriene*

*Institute products.

DINNER + Optigene Professional Packet (P.M.)*	**A**	**B**	**C**	**D**	
Easy Pizza at Home					
Flour Tortilla, 8–9″	1	1 1/2	2	2	______
Pizza Sauce	3 T.	4 T.	6 T.	6 T.	______
Chicken Strips, precooked	2.5 oz.	3.5 oz.	3.5 oz.	4 oz.	______
Mixed Vegetables (onions, red or green peppers, mushrooms), sliced	1 c.	1 1/4 c.	1 1/2 c.	1 1/2 c.	______
Mozzarella Cheese, low-fat	1 oz.	1 oz.	1 1/2 oz.	1 1/2 oz.	______
Green Salad	2 c.	2 c.	3 c.	3 c.	______
Italian Salad Dressing†	1 T.	1 T.	2 T.	2 T.	______

Directions: *Top tortilla with pizza sauce, chicken, mixed vegetables and cheese. Bake in 400-degree oven until cheese is melted and lightly browned. Serve with a green salad and Italian dressing.*

Beverages: *Always consume an 8 oz. appropriate beverage with each meal.*

*Institute product.

†Italian Dressing Recipe: *Combine 1 T. olive oil, 1 T. red wine vinegar and 1 T. lemon juice in a small jar. Add salt, pepper, fresh garlic or garlic powder to taste. Shake well.*

The photographs on pages 297–299 show an actual patient who has undertaken both cosmetic surgery and an intensive anti-aging age management program. She is shown pre-op (left photo) and two years post-op (right photo).

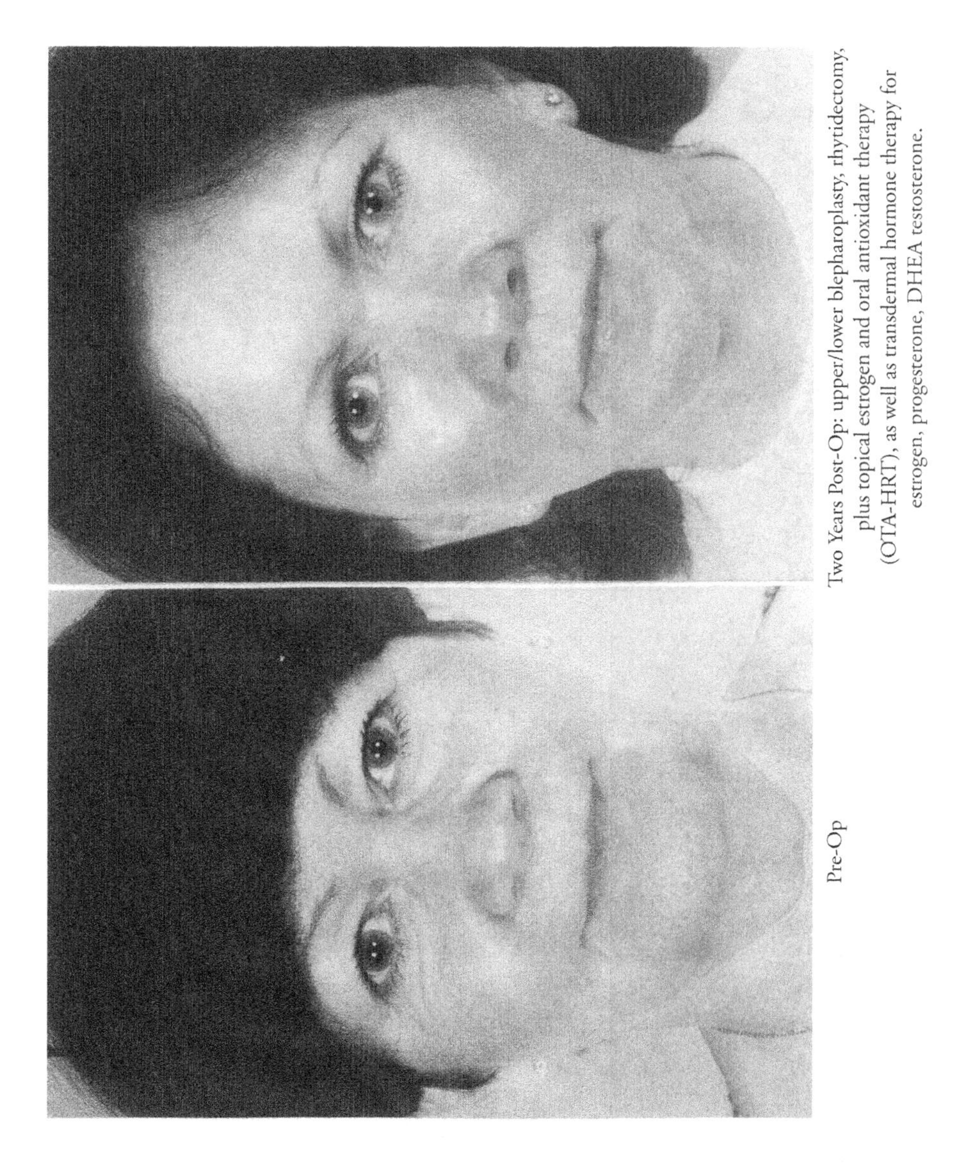

Pre-Op

Two Years Post-Op: upper/lower blepharoplasty, rhytidectomy, plus topical estrogen and oral antioxidant therapy (OTA-HRT), as well as transdermal hormone therapy for estrogen, progesterone, DHEA testosterone.

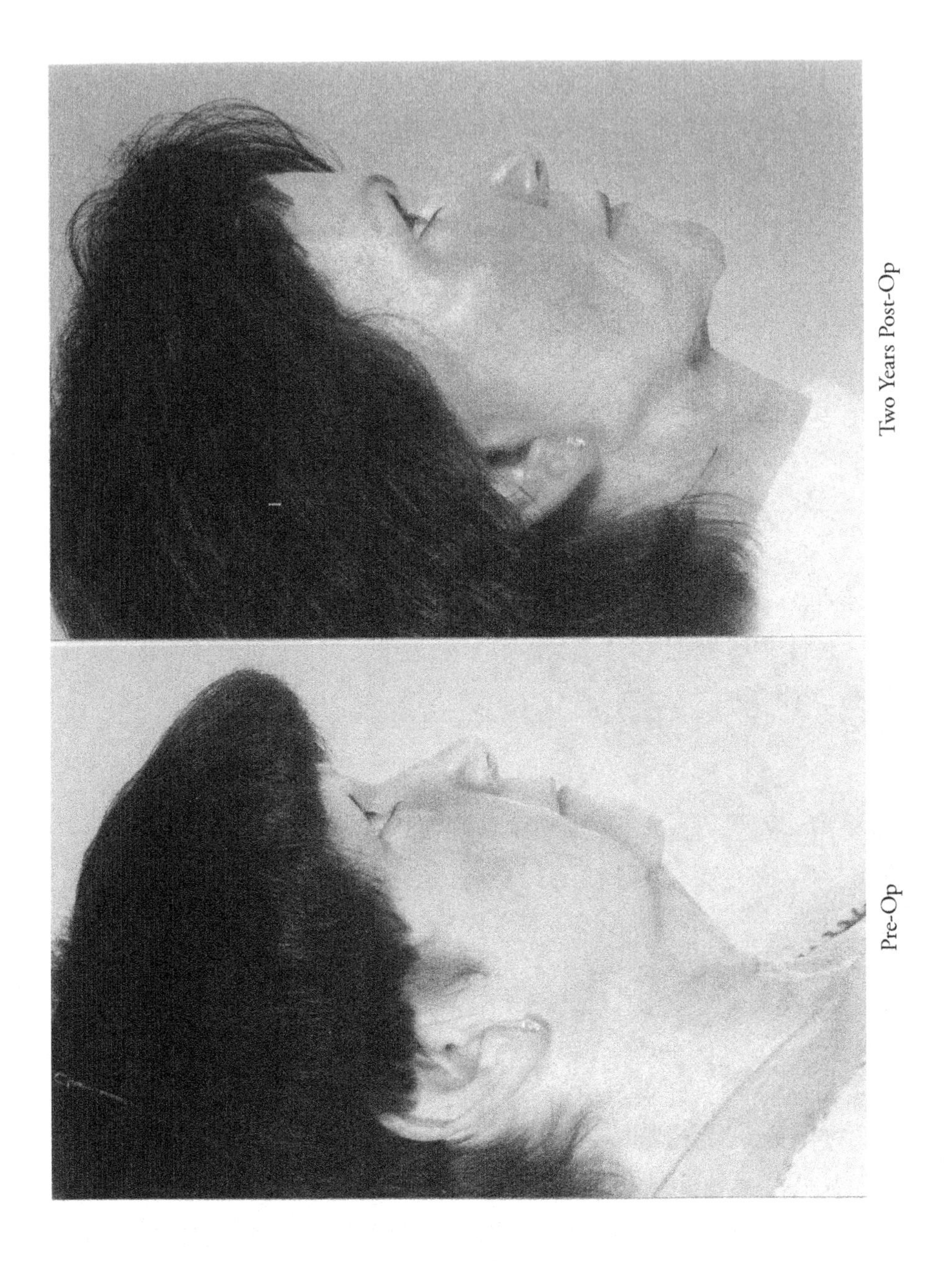

Two Years Post-Op

Pre-Op

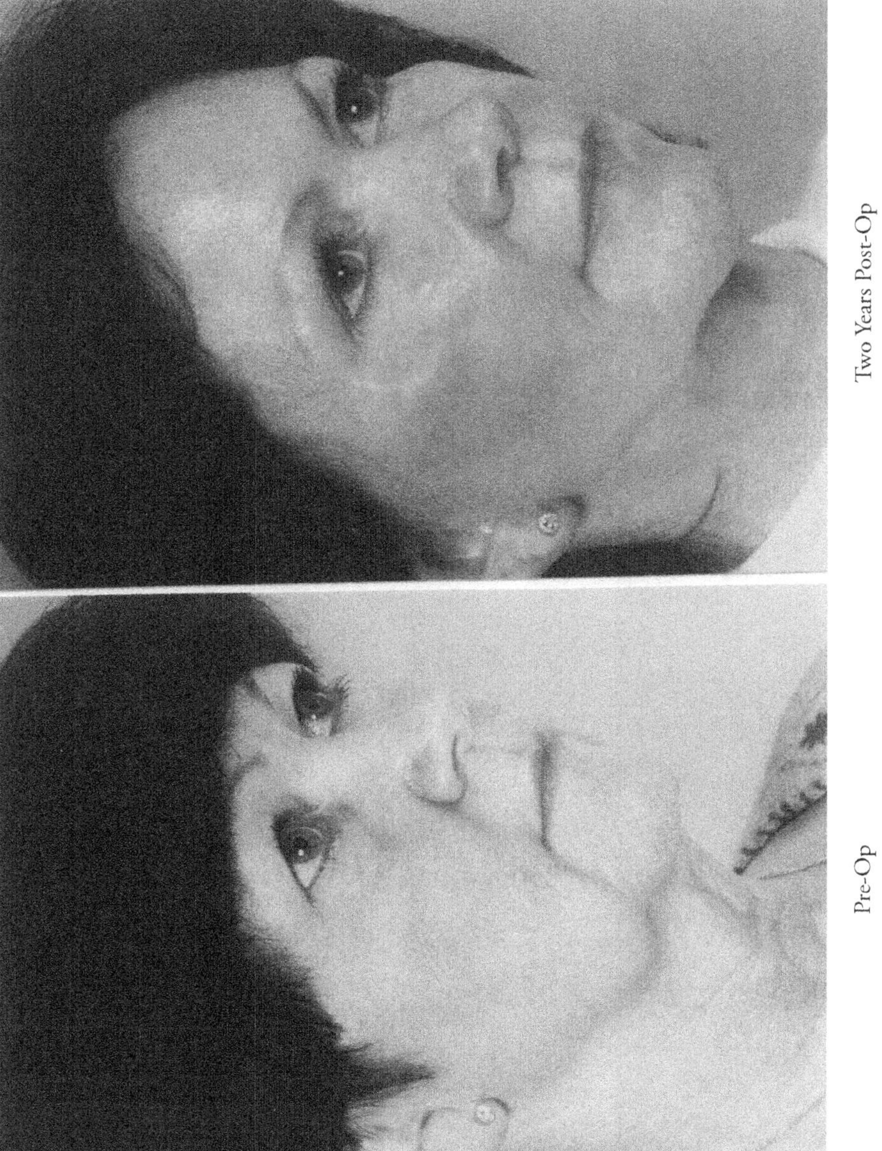

Pre-Op

Two Years Post-Op

Additional Recommended Readings

Arthur JR, Bermano G, Mitchell JH, et al. Regulation of selenoprotein gene expression and thyroid hormone metabolism. *Biochem Soc Trans.* 1996; 24(2): 384–388.

Benecke M. *The Dream of Eternal Life: Biomedicine, Aging, and Immortality.* New York: Columbia University Press; 2002.

Berdanier CD. Diabetes mellitus: a mitochondria genomic error? In: Berdanier CD, Hargrove JL, eds. *Nutrients and Gene Expression.* Boca Raton, FL: CRC Press; 1996: 141–164.

Berdanier CD. Nutrient-gene interactions in lipoprotein metabolism. In: Berdanier CD, Hargrove JL, eds. *Nutrients and Gene Expression.* Boca Raton, FL: CRC Press; 1996: 101–121.

Berdanier CD. Nutrition and genetic diseases. In: Berdanier CD, Hargrove JL, eds. *Nutrients and Gene Expression.* Boca Raton, FL: CRC Press; 1996: 1–20.

Bishop JE, Waldholz M. *Genome.* New York: Simon & Schuster; 1990.

Burzynski SR. Potential of antineoplastons in diseases of old age. *Drugs & Aging.* 1995; 7: 157–167.

Burzynski SR. Topical use of antineoplastons for treatment of hyperpigmentation. 1986; US Patent 4,593,038.

Burzynski SR, Conde AB, Peters A, Saling B, Ellithorpe R, Daugherty JP, Nacht CH. A retrospective study of antineoplastons A10 and AS2-1 in primary brain tumours. *Clin Drug Invest.* 1999; 18: 1–10.

Burzynski SR, Kubove E, Burzynski B. Treatment of hormonally refractory cancer of the prostate with Antineoplaston AS2-1. *Drugs Exp Clin Res.* 1990; 16: 361–369.

Cingolani ML, Re L, Rossini L. The usefulness, in pharmacological classification, of complementary pattern-recognition techniques and structure modeling as afforded by the iterative collation of multiple-trial data banks. *Pharmacol Res Comm.* 1985; 17(1): 1–22.

Deutsch D. *The Fabric of Reality.* New York: Penguin Press; 1997.

Dodd JL. Incorporating genetics into dietary guidance. *Food Technol.* 1997; 51(3): 80–82.

Evans W, Rosenberg IH. *Biomarkers: The 10 Keys to Prolonging Vitality*. New York: Fireside; 1991.

Forcyt JP, Poston WS II. Diet, genetics, and obesity. *Food Technol.* 1997; 5(3): 70–73.

Fraker PA, Teljord W. Regulation of apoptotic events by zinc. In: Berdanier CD, Hargrove JL, eds. *Nutrients and Gene Expression*. Boca Raton, FL: CRC Press; 1996: 189–208.

Frieds J, Crapo LM. *Vitality and Aging*. San Francisco: WH Freeman & Co.; 1981.

Haseltine WA. Discovering genes for new medicines. *Sci Am.* 1997; 276(3): 92–97.

Hesketh JE, Partridge K. Gene cloning: studies of nutritional regulation of gene expression. *Proc Nutri Soc.* 1996; 55: 575–581.

Ho DY, Sapolsky RM. Gene therapy for the nervous system. *Sci Am.* 1997; 276(6): 116–120.

Iritani N. Diet-gene interactions in the regulation of lipogenesis. In: Berdanier CD, Hargrove JL, eds. *Nutrients and Gene Expression*. Boca Raton, FL: CRC Press; 1996: 123–140.

Jones PA, Takai D. The role of DNA methylation in mammalian epigenetics. *Science.* 2001; 293: 1068–1070.

Julius M, Lang CA, Gleberman L, et al. Glutathione and morbidity in a community-based sample of elderly. *J Clin Epidemiol.* 1994; 47(9): 1021–1026.

Juszkiewicz M, Chodkowska A, Burzynski SR, Feldo M, Majewska B, Kleinrol Z. The influence of antineoplaston A5 on the central dopaminergic structures. *Drugs Exp Clin Res.* 1994; 20: 161–167.

Juszkiewicz M, Chodkowska A, Burzynski SR, Mlynarczyk M. The influence of antineoplaston A5 on particular subtypes of central dopaminergic receptors. *Drugs Exp Clin Res.* 1995; 21: 153–156.

Lang CA, Naryshkin S, Schneider DL, et al. Low blood glutathione levels in healthy aging adults. *J Lab Clin Med.* 1992; 120: 720–725.

Lee CM, Weindruch R, Aiken JM. Age-associated alterations of the mitochondrial genome. *Free Radic Biol Med.* 1997; 22(7): 1259–1269.

Lehrman S. Can the clock be slowed? *Harvard Health Lett.* 1995; 20(3): 1–3.

Levin LB. Nutritional modulation of growth and maturation and the development of specific age-related diseases: secondary analysis of NHANES II. Greensboro, NC: University of North Carolina at Greensboro; 1987.

Liau MC, Lee SS, Burzynski SR. Modulation of cancer methylation complex isoenzymes as a decisive factor in the induction of terminal differentiation mediated by antineoplaston A5. *Internat J Tissue Reactions.* 1990; 12(suppl): 27–36.

Liau MC, Luong Y, Liau CP, Burzynski SR. Prevention of drug induced DNA hypermethylation by antineoplaston components. *Internat J Exp Clin Chemother.* 1992; 5: 19–27.

Lipsitz LA, Goldberger AL. Loss of 'complexity' and aging. *JAMA*. 1992; 267(13): 1806–1809.

Mariotti S, Sansoni P, Barbesino G, et al. Thyroid and other organ-specific autoantibodies in healthy centenarians. *Lancet*. 1992; 339: 1506–1508.

McCully KS. *The Homocysteine Revolution*. New Canaan, CT: Keats Publishing, Inc.; 1997.

McFadden SA. Phenotypic variation in xenobiotic metabolism and adverse environmental response: focus on sulfur-dependent detoxification pathways. *Toxicology*. 1996; 111: 43–65.

Moore SD. Gene hunter targets isolated Iceland. *Wall Street*, April 14, 1997, p. 7.

Morley JE, Kaiser J, Raum WJ, et al. Potentially predictive and manipulable blood serum correlates aging in the healthy human male: progressive decreases in bioavailable testosterone, dehydroepiandrosterone sulfate, and the ratio of insulin-like growth factor I to growth hormone. *Proc Natl Acad Sci U S A*. 1997; 94: 7537–7542.

Motulsky AG. Human genetic variation and nutrition. *Am J Clin Nutr*. 1987; 45: 1108–1113.

Motulsky AG. Nutrition and genetic susceptibility to common diseases. *Am J Clin Nutr*. 1992; 5: 1244S–1245S.

Paul HS. Regulation of gene expression of the branched-chain keto acid dehydrogenase complex. In: Berdanier CD, Hargrove JL, eds. *Nutrients and Gene Expression*. Boca Raton, FL, CRC Press; 1996: 21–37.

Pero R, Zimmerman M. *Reverse Aging*. Chico, CA: Nutrition Solution Publications; 2002.

Rastan S. Of men in mice. *Nat Genet*. 1997; 16(2): 113–114.

Richter C. Oxidative damage to mitochondrial DNA and its relationship to aging. *Int J Biochem Cell Biol*. 1995; 27(7): 647–653.

Roush W. Live long and prosper? *Science*. 1996; 273: 42–46.

Rucker R, Tinker D. The role of nutrition in gene expression: a fertile field for the application of molecular biology. *J Nutr*. 1986; 1(16): 177–189.

Satyanarayana J, Shoskes DA. A molecular injury-response model for the understanding of chronic disease. *Mol Med Today*. 1997; 3(8): 331–334.

Schneider EL, Vining EM, Hadley EC, Farnham SA. Recommended Dietary Allowances and the health of the elderly. *N Engl J Med*. 1986; 314(3): 157–160.

Simopoulos AP. Diet and gene interactions. *Food Technol*. 1997; 51(3): 66–69.

Simopoulos AP. Genetic variation and nutrition. *Am J Clin Nutr*. 1995; 30(4): 157–167.

Simopoulos AP. The role of fatty acids in gene expression: health implications. *Ann Nutr Metab*. 1996; 40(6): 303–311.

Steventon GB, Mitchell SC, Waring RH. Human metabolism of paracetamol (acetaminophen) at different dose levels. *Drug Metab Drug Interact.* 1996; 13(2): 111–117.

Stoicheff H, Vital C. Mitochondrial DNA and disease. *N Engl J Med.* 196; 334(4): 270–271.

Sturman SG, Steventon GB, Waring RH, et al. MAO-B and Parkinson's disease. *Mov Disord.* 1990; 5(4): 338–340.

Suarez FL, Savaiano DA. Diet, genetics, and lactose intolerance. *Food Technol.* 1997; 51(3): 74–79.

Sugita Y, Tsuda H, Maruiwa H, Hirohata M, Shigemori M, Hara H. The effect of antineoplaston, a new antitumor agent on malignant brain tumors. *Kurume Med J.* 1995; 42: 133–140.

Symons MC. Electron movement through proteins and DNA. *Free Radic Biol Med.* 1997; 22(7): 1271–1276.

Tomita M, Kitahara K. A model for oscillating chemical reactions. *Biophys Chem.* 1975; 3: 125–141.

Wallace DC. Mitochondrial DNA in aging and disease. *Sci Am.* August 1997; 40–47.

Wei YH, Dao SH. Mitochondrial DNA mutations and lipid peroxidation in human aging. In: Berdanier CD, Hargrove JL, eds. *Nutrients and Gene Expression.* Boca Raton, FL: CRC Press; 1996: 165–188.

Williams A, Steventon G, Sturman S, et al. Xenobiotic enzyme profiles and Parkinson's disease. *Neurology.* 1991; 41(Suppl 2): 29–32.

Williams A, Sturman S, Steventon G, et al. Metabolic biomarkers of Parkinson's disease. *Acta Neurol Scand.* 1991; 84(Suppl 136): 19–23.

Williams AC, Steventon GB, Sturman S, et al. Hereditary variation of liver enzymes involved with detoxification and neurodegenerative disease. *J Inherit Metab Dis.* 1991; 14(4): 431–435.

Wilmut L, Schnieke AE, McWhir J, et al. Viable offspring derived from fetal and adult mammalian cells. *Nature.* 1997; 385: 810–813.

Wishart JM, Horowitz M, Need AG, et al. Relations between calcium intake, calcitriol, polymorphisms of the vitamin D receptor gene, and calcium absorption in premenopausal women. *Am J Clin Nutr.* 1997; 65(3): 798–802.

Wolffe AP. The cancer-chromatin connection. *Sci & Med.* Jul/Aug 1999; 28–37.

Credits

Page 3, Diagram 1: From *Scientific American*, September 2000, p. 12. Reprinted with permission of Terese Winslow.

Page 40, Diagram III-1: From *Scientific American*, September 2000, p. 12. Reprinted with permission of Terese Winslow.

Page 43, Diagram III-6: Copyright 1998 by Richard Hochschild. Reprinted with permission.

Page 52, Diagram III-15: From Maurice E. Shils, James A. Olson, Moshe Shike and A. Catharine Ross, *Modern Nutrition in Health and Disease*, Ninth Edition. Baltimore, MD: Lippincott Williams & Wilkins, 1999, p. 574. Reprinted with permission.

Page 219, Diagram XI-1: Adapted from John Goodman, Histone Tails Wag the DNA Dog, *Helix*, University of Virginia Health System, vol. 17, no. 1, Spring 2000.

Page 251, Diagram XIII-2: From *Scientific American*, September 2000, p. 12. Reprinted with permission of Terese Winslow.

Page 253, Diagram XIII-4: From *Scientific American*, April 1999, p. 65. Artist, Keith Kasnot.

Page 253, Diagram XIII-5: From *Scientific American*, September 2000, p. 58. Artist, Tomo Narashima.

Page 255, Diagram XIII-6a: From *Anti-Aging Medical News*, Spring 2000, p. 4.

Page 256, Diagram XIII-7: From *Advanced Cell Technology* website, 7/31 2002, http://www.advancedcell.com/ht-program.html. Reprinted with permission.

Page 258, Diagram XIII-9: From *Scientific American*, September 2000, p. 60. Artist, Tomo Narashima.

Index

CPSIA information can be obtained at www.ICGtesting.com
Printed in the USA
LVOW03*1407220315

431564LV00013B/92/P